Advances in Information Systems and Business Engineering

Edited by

U. Baumöl, Hagen, Germany
J. vom Brocke, Vaduz, Fürstentum Liechtenstein
R. Jung, St. Gallen, Switzerland

The series presents current research findings based on diverse research methods at the interface between information systems research, computer science, and management science. The publications in this series aim at practical concepts, models, methods, but also theories which address the role of information systems in the innovative design and sustainable development of organizations. Contributions are selected which on the one hand propose innovative approaches how modern information and communication technologies can enable new business models. On the other hand, contributions are eligible which present considerable improvements to existing solutions. The publications are characterized by a holistic approach. They account for the socio-technical nature of work-systems and suggest that the development and implementation of information systems need to consider the multifaceted organizational context driven by people, tasks, and technology.

Tor Tønnessen

Managing Process Innovation through Exploitation and Exploration

A Study on Combining TQM and BPR in the Norwegian Industry

Springer Gabler

Tor Tønnessen
Center for Industrial Production
Aalborg University, Denmark

Dissertation, Center for Industrial Production, Department of Business and Management, Aalborg University, Denmark, 2012

ISBN 978-3-658-04402-2   ISBN 978-3-658-04403-9 (eBook)
DOI 10.1007/978-3-658-04403-9

The Deutsche Nationalbibliothek lists this publication in the Deutsche Nationalbibliografie; detailed bibliographic data are available in the Internet at http://dnb.d-nb.de.

Library of Congress Control Number: 2013955551

Springer Gabler

Printed on acid-free paper

Springer Gabler is a brand of Springer DE.
Springer DE is part of Springer Science+Business Media.
www.springer-gabler.de

Preface and Acknowledgements

Managing processes is important for driving enterprise performance. In order to stay efficient companies need to improve their business processes on a continuous basis. To ensure competitive edge companies need to change radically when new business opportunities occur and when external demands change. Exploitation is about continuous improvement of the existing. Exploration is about radical change. Companies need to do both. However, experience and research show that companies have difficulties managing exploitation and exploration in the same system at the same time. Why is this so difficult, and how can the two development logics be integrated in a mutually supportive manner? This is theme of the book.

To operationalize exploitation in an industrial setting, I use TQM as an approach to continuous process improvement. To operationalize exploration I use BPR as an approach to radical process change. The question of how to integrate the business development approaches that the quality movement and BPR represent is an important enabling factor in the further development and success of other current and actual change concepts such as Business Process Management (BPM). The book contributes to the understanding of the challenges of this integration and ways to accomplish it.

The point of departure for the book is my PhD study at Center for Industrial Production at the University of Aalborg and case study research in Norwegian industry. Many people have given their contributions to this book. First of all I thank my PhD advisor Professor Dr. Harry Boer for all his good comments, discussions, encouraging support, patience and guidance. His approach to Continuous Innovation made me see how many business development puzzles can fit together into an adequate way. I thank Associate Professor Dr. Bjørge Laugen for important contributions in discussions about theoretical and methodical issues, and for having given me quick and good comments along the way.

My colleagues for many years, Professor Dr. Tor Claussen, Dr. Trond Haga, Senior Advisor Ove Gandrud and Senior Advisor Rolf Sjursen have contributed in important ways to develop my skills as a researcher and to widening of my theoretical perspectives. Many years of close cooperation with Pål Lynne Hansen from LO has made me see the importance of employee participation in change.

I thank my case companies, Aker Stord, Serigstad and Titania for opening their doors to my research. I also thank my former employer IRIS for the opportunity to start up and carry out the work with my PhD, and my present employer Statoil for the opportunity to finish it. Dr. Sturle Næss at Statoil has supported me refining and clarifying my theoretical discussions and contributions. I thank Professor Dr. Jan vom Brocke for important contributions to the process of transforming my PhD thesis into a book.

Finally, I thank my wife Sigrunn and my children Espen, Marianne and Ellen for their patience with a husband and father who spent too many hours at the computer the last years.

Stavanger 15.07.13 Tor Tønnessen

Résumé

"Maintaining an appropriate balance between exploration and exploitation is a primary factor in system survival and prosperity" (March 1991:71).

Both exploration for new possibilities and exploitation of the existing are necessary development strategies for organizations to survive and prosper (March 1991). According to March, this balancing is not easy, and the two strategies compete for scarce resources as a trade-off. In other words, they are considered mostly antithetical, where investing more resources in one of these development strategies may result in having less to invest in the other. This way of balancing can be accomplished by separating the two in time, at organization levels or by departments (Volberda 1998).

Lately, some scholars have questioned whether March's theory of balancing the two mostly as a trade-off is always correct, and argue for a "synergistic" approach as a better alternative in many situations (Sutcliffe et al. 2000, Cole 2002, Boer & Gertsen 2003, Osri et al. 2005, Boer et al. 2006, Greve 2007, Isobe et al. 2007, Un 2007). More empirical research and evidence are needed to give insight into and further develop the theory on the balancing of exploitation and exploration (Sutcliffe et al. 2000, Cole 2002, Oshri et al. 2004, Boer et al. 2006, Andriopoulus & Lewis 2009, Fang et al. 2010, Raisch et al. 2010). My research contributes to current knowledge through exploratory empirical research.

In order to make the theoretical constructs exploitation and exploration operational in an industrial setting, and to discuss and further develop March's balancing theory, I have chosen to use two business development concepts to represent forms of exploitation and exploration: Total Quality Management – TQM (Deming 1986, Juran 1989) and Business Process Reengineering – BPR (Hammer & Champy 1993), respectively.

TQM and BPR have a strong influence on the way in which organizational development and the striving for business excellence are being carried out in Norway. In particular was this very visible in the decade before the turn of the century (Røvik 1998, Rolfsen 2000, Neerland 2000, Moltu 2000). TQM and the quality movement is very much associated with continuous incremental improvement. It is important to keep costs and production under control, and to eliminate mistakes and reduce variability and other elements related to the idea of exploitation (March 1991). BPR is associated with radical change. As the name indicates it focuses on processes, like TQM does as well. However, it is not continuous incremental process improvement that is in focus. It is thinking fundamentally new, seeing things in new ways and searching for dramatic changes, new possibilities and other aspects related to March's idea of exploration. Even if processes are in focus, the goal and the outcomes of change processes can also be radically and dramatically directed towards other innovation areas such as technology, markets, raw materials and organization. I use BPR to represent explorative process innovation.

The specific setting that I am referring to is Norwegian industry. Prevailing management traditions and concepts of business excellence are international and have been developed elsewhere within other industrial traditions. Norwegian companies are being exposed to these international concepts and use them to achieve business excellence and innovation. A challenge both from an academic as well as a managerial perspective is to see whether these international management concepts can be combined synergistically. Where do TQM and BPR conflict? What do they have in common? Do we have to choose between them, or can they work in a reciprocally reinforcing manner to increase companies' innovation capability? If adjustments are needed, what kinds of adjustments and what enabling factors seem to be important? My specific contribution is to propose theory on the balancing of these two concepts.

Over the last two decades, these two approaches to process improvement have gradually merged and constitute important elements in Business Process Management (BPM). BPM is a concept for managing business performance by managing end-to-end business processes for successful enterprise performance (Hammer 2010). Consequently, BPM can be considered as one approach to the integration of forms of exploitation and exploration. This book contributes with insight in the challenges of integrating TQM and BPR, provides examples of how to do it and with knowledge of enabling factors. Managing the integration of the kinds of approaches that the quality movement and BPR represent is decisive for successful BPM, which is still in its infancy (Hammer 2010).

My research is part of the two Norwegian national research programs "Business Development 2000" and "Value Creation 2010". These two programs are action research programs, and have been the most comprehensive research efforts related to combining international management concepts and Norwegian industrial traditions over the last decades.

Through my involvement in these two research programs, I have facilitated and participated in development processes in a large number of individual companies and in companies cooperating in networks. The exploratory research reported here is based on three of those projects, one of which is best characterized as a longitudinal case study, while the other two are forms of action research.

In the longitudinal case study I describe and analyze a company's efforts to combine TQM and BPR into one development process. Various methods of data gathering have been used, such as observations, field notes, interviews, reports from meetings and company conferences, analysis of documents from the participating companies and statistics. The results from the case study are discussed in relation to and compared with the results from a national survey on quality management that was carried out together with the University of Stavanger and "Norsk Forening for Kvalitet og Ledelse – NFKL" (Norwegian Association for Quality and Management).

TQM and BPR are analyzed in relation to where and how they conflict with and/or support each other. After having identified conflicting and reciprocally supporting elements, I will present my efforts to combine these two management concepts into one approach in Norwegian industrial context. An important question is whether this kind of combination in the same system is advisable. And if so, what further developments of TQM and BPR practices are necessary in order to promote both incremental and radical innovations, and which enabling factors seem to be important? How can this contribute to the further development of the theory of exploitation and exploration?

Thus, my research objective is to contribute to the further development of the theory of exploitation and exploration through proposing theory on the balancing of TQM and BPR. More specifically, my research question is:

Can TQM and BPR be combined synergistically, – if so, how?

After having discussed and presented my approach, I will give examples of the kinds of change techniques and methodologies that I believe are needed to pursue a new approach to combining characteristics of TQM and BPR synergistically. I will try out some steps towards this new approach and also show the practical use of what I regard as the most important elemens and change techniques.

Two of the new techniques that I developed to pursue my approach will be discussed and analyzed in detail. The "Enterprise Development Conference" is a technique for enabling employee participation in higher-level strategy discussions and innovation. This will be followed by a presentation and discussion of a technique that I have developed to strengthen employee participation in process innovation, the Actor-Process Diagram. These two techniques and my approach to combining TQM and BPR synergistically are tested through action research. The results from the action research projects are presented and discussed in relation to theory.

My main findings are that exploitation capability can support exploration capability, and thereby support exploration, and that TQM and BPR can work reciprocally supportive in the same development process. An industrial democracy (ID) context and a strong management commitment seem to be important enabling factors, and support the synergistic combination of characteristics of exploitation and exploration, thereby promoting both incremental and radical innovations in the same system at the same time. Thus, March's (1991) mostly antithetical theory of balancing exploitation and exploration should be modified.

Finally, important limitations of my research design are summarized, and directions for further research identified.

Contents

1 Introduction

In this introductory chapter I will first briefly present the theoretical discussion that my research contributes to, namely the balancing and combining of exploitation and exploration in business development (March 1991). I will then present my research question and the structure of the book.

1.1 Combining exploitation and exploration

During the past 50 years several theories and approaches to organizational development and change have been developed. Many of these have been presented as a choice between two alternative strategies for organizations to adapting themselves to external circumstances and market changes. For example "Mechanistic" versus "Organic" (Burns & Stalker 1961), "KAIZEN" versus "Innovation" (Imai 1986), "Exploitation versus Exploration" (March 1991), "Continuous Improvement" versus "Reengineering" (Hammer & Champy 1993)," "Control" versus "Exploration" (Sutcliffe 2000), "Operational Effectiveness" versus "Strategic Flexibility" (Boer et al. 2006).

The rationale for and content of these dichotomies are different but they all individually present two alternative approaches to business development that point in similar directions. The first alternative is a gradual, incremental, continuous, efficiency focused strategy for improvement of the existing. The second is a search and change strategy for new and radical alternatives. March's theory of the balancing between exploitation and exploration (March 1991) is one representative of this stream of dichotomies, and has influenced many scholars and studies in the field (Sutcliffe et al. 2000, Cole 2002, Boer & Gertsen 2003, Gibson & Birkinshaw 2004, Petersen et al. 2004, Jacoby 2005, Boer et al. 2006, Greve 2007, Isobe 2007, Andriopoulus & Lewis 2009, Raisch et al. 2009, Fang et al. 2010). Many of these recent studies relate their work to March's theory. I have consequently chosen March's theory as the departure for my discussion, and contribute to the discussion following March's 1991 article. According to March, *"Exploration includes things captured by terms such as search, variation, risk taking, experimentation, play, flexibility, discovery, innovation. Exploitation includes such things as refinement, choice, production, efficiency, selection, implementation, execution"* (March 1991:71).

Organizations are increasingly challenged by intensified customer demands for high reliability products and services. This occurs at the same time as the demand for improved and new products, services and processes is constantly increasing (Sutcliffe et al. 1997). According to March (1991), both the exploration for new possibilities and the exploitation of the existing are necessary development strategies for organizations in order to survive and prosper. The two strategies compete for scarce resources in a trade-off situation (March 1991). In other words, they are considered mostly antithetical, where investing more resources in one of them may result in having less to invest in the other.

The balancing of the two can be done in several ways, where trade-off is one approach. Lately, some scholars have questioned whether March's theory of balancing the two as mostly trade-off is always correct (Sutcliffe et al. 2000, Cole 2002, Boer & Gertsen 2003, Gibson & Birkinshaw 2004, Osri et al. 2005, Boer et al. 2006, Greve 2007, Isobe 2007, Un 2007). My research departs from this question and considers a more synergistic approach to organizational change and business development aimed at applying the two strategies in a more reciprocally reinforcing way. The challenge is to improve these strategies in ways where efforts to improve the capability of one of them simultaneously improves the capability to pursue the other and vice versa.

In order to make the different theoretical approaches operational in an industrial setting and through that discuss and further contribute to the development of March's balancing theory, I have chosen to use two business development concepts to represent exploitation and exploration: Total Quality Management – TQM (Deming 1986, Juran 1989) and Business Process Reengineering – BPR (Hammer & Champy 1993), respectively. Consequently, I do not cover the totality of the idea of exploitation and exploration.

TQM and BPR have influenced the way organizational development and the pursuit of business excellence has been carried out in Norway over the last one and a half decade. TQM is very much associated with continuous process improvement and incremental changes of the existing. I use this concept to represent characteristics of March's idea of exploitation. BPR is associated with radical process innovation. I use this concept to represent characteristics of March's idea of exploration, more specifically, for the process innovation area. Even if the novelty value of TQM and BPR seems to have faded lately, my research and close cooperation with industry show that the basic ideas have survived and are still of utmost importance for business development. Examples in Norway are large companies such as Statoil, Hydro and Aker.

Managing processes is important for enterprise performance (Hammer 2010). Over the last two decades, these two approaches to process improvement have gradually merged and constitute important elements in Business Process Management (BPM). BPM is a concept for managing business performance by managing end-to-end business processes (Hammer 2010). Consequently, BPM can be considered as one approach to the integration of forms of exploitation and exploration. This book contributes with insight in the challenges of integrating TQM and BPR, provides examples of how to do it and with knowledge about enabling factors. Managing the integration of the kinds of approaches that the quality movement and BPR represent is decisive for the further development of BPM as a successful enterprise development concept.

With this as a background, my collaboration with industry has focused on developing and trying out tools, methods and theory that may contribute to building a synergistic approach to combining TQM and BPR. Where do they conflict? What do they have in common? Do we have to choose between them, or can they work in a reciprocal reinforcing manner to increase companies' innovation capability? If adjustments are needed, what kinds of adjustments and which enablers seem to be most important?

1.2 Research question

Based on the introductory discussion above, the overall objective of my research is to contribute to the further development of the theory of exploitation and exploration through proposing theory on the balancing of TQM and BPR. More specifically, my research question is:

Can *TQM and BPR be combined synergistically, – if so, how?*

A brief overview of the structure of the book is given below.

1.3 Structure

Chapter 2 contains a presentation of my theoretical point of departure. I will first present and discuss March's theory of balancing exploitation and exploration. In this discussion valuable contributions to the theory from other scholars will be included and discussed.

I will then in **Chapters 3 and 4** turn to the international management concepts TQM and BPR. They will both be defined, and I will give a brief summary and briefly discuss some important characteristics of the philosophy, change strategy, systems approach and methods used in these management concepts.

Chapter 5 is a theoretical discussion of the question of combining TQM and BPR synergistically. Where do they conflict, and what do they have in common? Can these traditions work reciprocally supportive or do we have to choose between them? Then, after having discussed my theoretical departure and the concepts that I use to operationalize theory into an industrial setting, I present my research model in **Chapter 6**. This chapter also contains definitions and further clarifications of the different elements in my research model.

Chapter 7 presents my exploratory research design and methodology, and the tradition that the research is part of. After some historical lines, the two national action research programs "Enterprise Development 2000" and "Value Creation 2010" are briefly presented. Then I position the role that I have taken in these two programs and my methodological approach. Strengths and weaknesses connected to case study design and action research are discussed and related to other research methods.

My first **case study, Aker Stord** is presented in **Chapter 8**. This study analyses the company's implementation of TQM and the company's effort to introduce BPR into a TQM process. The results from the Aker Stord study are compared with the results from a national survey on quality management and discussed in relation to theory.

After having identified and discussed conflicting and shared characteristics of TQM and BPR both conceptually in the theoretical discussion and also empirically, I present my approach to combining TQM and BPR synergistically into a theoretical framework in **Chapter 9**. A process model showing important enablers and what can happen when TQM and BPR are combined is presented. Among others, Industrial Democracy (ID) is part of this model. The Aker Stord case study shows that ID needs to be considered as an important contextual factor and as part of my framework in a Norwegian setting. The industrial democracy setting in Norway as a contextual and enabling factor is consequently discussed. Then, a practical framework for organizing employee participation in the synergistic combining of TQM and BPR is presented. As part of that some examples will be given of the kinds of change methods and tools that my analysis found appropriate to use and further develop in order to succeed.

Two of the new tools that I found of vital importance to my approach will be presented and discussed more thoroughly by referring to the action research case studies in **Chapter 10.** Section 10.1 presents and discusses a technique for enabling employee participation in business strategy development and innovation, the "Enterprise Development Conference". This is followed by a presentation and discussion in Section 10.2 of a technique that has been developed to ensure employee participation in business process innovation, the Actor-Process Diagram. In addition to testing out the Diagram, this case study is also testing out important elements in my approach to the synergistic combining of TQM and BPR.

Chapter 11 discusses my main findings in relation to my initial research objective and research question, and summarizes my contributions to theory. The chapter also contains a critical reflection of limitations with the research methodology that I used, and relates to the methodological discussion in Chapter 7. Finally, I identify directions and needs for further research.

2 Theoretical departure and approach – balancing exploitation and exploration

I will now present the theoretical background for my study, the theoretical discussion that I take part in and the theory development that my research contributes to. I will do this by discussing different contributions to the exploitation and exploration question.

"Maintaining an appropriate balance between exploration and exploitation is a primary factor in system survival and prosperity" (March 1991:71).

This statement in an article by March (1991), where he discusses the balancing theory, has generated much discussion, efforts to further develop the theory, research and concept development. It is to this discussion that I want to contribute. I will refer to some of the contributions that I find most adequate to my discussion below. But first, what does this theory say? I will start by defining the key elements and then discuss the different associated development strategies and challenges connected to balancing exploitation and exploration. Important is the question of how this balancing should be directed and what kind of consequences the different alternative directions may have.

March (1991) defines exploration and exploitation as follows: *"Exploration includes things captured by terms such as search, variation, risk taking, experimentation, play, flexibility, discovery, innovation. Exploitation includes such things as refinement, choice, production, efficiency, selection, implementation, execution"* (March 1991:71).

In other words, exploration deals with the search for new possibilities while exploitation is about utilizing the existing. Now, to the challenge of balancing the two:

"Systems that engage in exploration to the exclusion of exploitation are likely to find that they suffer the cost of experimenting without gaining many of its benefits" (March 1991:71). *"The distance in time and space between the locus of learning and the locus of the realization of returns is generally greater in the case of exploration than in the case of exploitation, as is the uncertainty"* (March 1991: 85). *"Conversely, systems that engage in the exploitation to the exclusion of exploration are likely to be find themselves trapped in suboptimal stable equilibria"* (March 1991: 71).

In line with this observation Isobe (2000) found that it took on average more than four years for small and medium sized companies in Japan to change their technological competences and over five years to earn profit from them.

It is the appropriate balancing of the exploration of new possibilities and the exploitation of the well-known that, according to March (1991), makes organizations survive and prosper in a constantly changing environment. Organizations need to be efficient and reliable, while they need to be able to be flexible, creative, to adapt to changing environments and demands at the same time. However, is this balancing easy? Research shows that it is not (March 1991, Sutcliffe et al. 2000, Cole 2001, Boer & Gertsen 2003, Boer et al. 2006, Greve 2007, Isobe et al. 2007, Un 2007). According to March (1991), the two strategies compete for scarce resources as a trade-off. In other words, they are considered mostly antithetical, where investing more resources in one of these development strategies may result in having less to invest in the other: *"Both exploration and exploitation are essential for organizations, but they compete for scarce resources. As a result, organizations make explicit and implicit choices between the two. The explicit choices are found in calculated decisions about alternative investments and competitive strategies. The implicit choices are buried in many features of organizational forms and customs, for example, in organizational procedures for accumulating and reducing slack, in search rules and practices, in the ways targets are set and changed, and in incentive systems"* (March 1991:71).

March (1991) finds that refining exploitation more rapidly than exploration is likely to become effective in the short run, but self-destructive in the long run, and that learning processes adapting people's belief to the existing organizational beliefs and codes tend to inhibit exploration (March 1991).

This antithetical view and reciprocal dependency of balancing the two is expressed even more clearly in a later article by March (1995): *"Exploration and exploitation are linked in an enduring symbiosis. Each requires the other in order to contribute effectively to an organization's survival and prosperity. At the same time, however, each interferes with the other"* (March 1995: 433).

March (1995) argues that focus on exploitation discourages the experimentation and variation that are essential to long term survival. Exploitation as a development strategy results in sticking to one organizational capability to such an extent that there is little exploration of others. It is also the other way round: *"In a similar fashion, exploration undermines exploitation. Efforts to promote experimentation encourage impatience with new ideas, technologies and strategies. They are likely to be abandoned before enough time has been devoted to developing the competence that would make them useful. The impatience of exploration results in unrealized dreams and unelaborated discoveries. As a result of the ways in which exploration and exploitation tend to extinguish each other, organizations persistently fail to maintain an effective balance between the two"* (March 1995:433).

As mentioned in the introduction, we see this antithetical view not only in March (1991), but also in several theories and approaches to organizational development and change. Among the many academics that have contributed to further developing this balancing theory or have questioned it (for example Sutcliffe et al. 2000, Cole 2002, Boer & Gertsen 2003, Petersen et al. 2004, Jacoby 2005, Oshri et al. 2005, Boer et al. 2006, Greve 2007, Isobe 2007, Un 2007), Sutcliffe et al. (2000) find that this trade-off and antithetical approach have neglected the need to deal with the relationship between them. The common practice has been to deal with the two strategies individually and in a disintegrated manner.

Volberda (1998) distinguishes between three main ways of arranging the balancing between the two. First, exploration and exploitation can be separated horizontally or vertically in the firm. The separation of exploitation and exploration in a production department and a research department, respectively, is an example of a horizontal separation. They can also be separated vertically by delegation to parts of the organization inside one department. Second, they can be separated in different firms where one firm specializes in exploration while others specialize in exploitation. The division of labor between biotechnical firms that explore new products and larger companies that specialize in efficient production, marketing and distribution is an example of that. Third, the balancing between the two can be separated in time. A firm can explore during a certain period of time and then exploit during another period of time (March 1991, Volberda 1998).

Sutcliffe et al. (2000) question whether balancing by separating the two is always the best way of thinking about the two approaches and whether there might be conditions under which this is not the best solution at all. In line with later contributions (for example Cole 2002, Oshri et al. 2004, Boer et al. 2006, Greve 2007, Isobe 2007, Un (2007), Sutcliffe et al. (2000) find that little empirical research and evidence have been produced to support, give insight into and further develop the theory of balancing exploitation and exploration. To systematically examine the balancing theory, Sutcliffe et al. (2000) propose a conceptual framework for future research. Using the terms "Control" and "Exploration", their terminology differs slightly from March (1991).

"Control emphasizes systematically clarifying that which is shared or convergent" (Sutcliffe et al. 2000:318). Or, in other words continuous improvement, efficiency and highly reliable processes for delivering processes and services. Schedules and criteria for performance are specific, understood and shared. Purpose, roles and tasks are clear. Tightly coupled action plans and close coordination between operational units and individuals are stressed. Systems and procedures are designed to reduce needless variation. Conformance to standards and specification are stressed, and also the tightening of these specifications as a result of experience.

"Exploration emphasizes systematically discovering that which is unforeseen and divergent" (Sutcliffe et al. 2000:318) and requires flexible, boundariless, learning-oriented processes and capability to adapt to changing conditions and explore new, breakthrough technologies or systems. This stream emphasizes distinct and independently defined goals and performance criteria carried out by independent operating units. Processes are more independent and there is a focus on the development of innovative ideas. Greater variance is seen as a way to discovery and capacity building. Accepted standards and specifications are challenged, and new opportunities are looked for rather than avoiding and minimizing risk.

These two streams require and pursue different kinds of processes. The first one pursues reliability – and control-oriented processes. The other pursues learning – and exploration-oriented processes (Sutcliffe et al. 2000). Most organizational research in the past has focused on structure and function. Little research has focused on processes, which is where I want to contribute. Sutcliffe et al. (2000) argue that one reason for that is that processes tend to be harder to perceive than structures such as departments, functions and tasks.

To understand, consider and develop different approaches or alternatives to March's trade-off theory, research designs that enable and focus on the analysis of processes seem valuable. As I will return to below, both the two management concepts that I use to represent forms of exploitation and exploration have a process focus. How this is reflected in the research design will be discussed in Chapter 7 and demonstrated in the case studies.

An important part of the framework that Sutcliffe et al. (2000) have developed is a division of alternative process models and the logics underlying these models. I will use these models in the discussion of others' and my own theoretical contributions.

2.1 Process models

Sutcliffe et al. (2000) distinguishes three models for dealing with the relationship between control-oriented and exploration-oriented processes; the "Singular", the "Binary" and the "Dual" Process Model.

The Singular Process Model emphasizes the achievement of control by stressing the importance of variance reduction. Much of the literature on quality management and other process improvement methods assume a singular model. It is important to achieve high levels of control and high reliability through standard routines and procedures. Carrying out unified organizational goals through disciplined implementation and by means of motivational procedures and comprehensive measures is emphasized.

The appropriateness of quality management methods and process management techniques for improving routine operations is well documented (McDonald 1992, Hackman & Wageman 1995, Sutcliffe et al. 2000). But, there is little documentation showing that these kinds of methods are well suited for non-routine, unpredictable situations where situational conditions are changing. The latter requires a more flexible, experimental and improvisational approach (Sutcliffe et al. 2000). The singular model assumes that there is one approach that is universally applicable and independent of situational factors.

The Binary Process Model assumes that the choice of approach to business development depends on situational factors. A control-oriented approach is well suited for handling routine operations where conditions are stable. Non-routine operations, where adapting to changing situational requirements is important, needs an exploration-oriented approach. Thus, these two kinds of situations require different approaches. It is either or. The binary model does not examine the relationship between the two types of processes, the control-oriented and the exploration-oriented.

Sutcliffe et al. (2000) argues that the development of an appropriate approach to process management that is tailored to the latter, i.e. uncertain situations, can enhance performance. In other words, traditional quality management and process management techniques are considered to be well suited for dealing with routine situations. The development of new process management methods for dealing with non-routine situations is believed to be beneficial to enhance performance in these situations. It is argued that such methods can help people to handle unclear and changing environments because such practices help to more rapidly build intuition, flexibility and provide a structured way of handling uncertainties.

Thus, this led Sutcliffe et al. (2000) to propose that "highly uncertain situations would require an alternative to standard process management practices (i.e., standard quality control practices)" and also that "matching process management techniques to situational requirements would enhance their effectiveness" (Sutcliff et al. 2000:320). I will return to this discussion in Chapter 4 on BPR, and illustrate different aspects and challenges related to this position in my case studies. I developed the Actor-Process Diagram as a contribution, among others, to meet this kind of needs.

The Dual Process Model reflects that *"organizations cannot typically focus on one type of goal (as simple contingency) but that they must do both simultaneously to remain adaptable under different conditions"* (Sutcliffe et al. 2000:321). At this point the dual model is in line with March's theory of balancing as a trade-off between the two: *"Thus the recognition and awareness of both types of processes is not sufficient; rather, this approach stresses how the decision of how to balance the two approaches necessary involves potential trade-offs in resource and attention allocation"* (Sutcliffe et al. 2000: 321).

March's trade-off theory is primarily an "antithetical" approach to the balancing between the two. The two tend to compete for scarce resources. Antithetical conditions emerge in situations where control and exploration processes do not need to be intertwined and where they draw upon the same limited resource pool. An important management task in these kinds of situations is first of all to acknowledge that an antithetical situation is present, and then to determine the most appropriate distribution of resources between the two.

Another approach to the relationship between exploitation and exploration is the "orthogonal" way. This reflects the notion that there are situations where the two types of processes can be pursued independently of each other due to slack resources. The two kinds of processes can be pursued without interfering each other. The outcome from one of them does not affect the outcome of the other. Use of resources to pursue one of them does not affect the amount of resources made available for pursuing the other. It is not a trade-off situation.

According to Sutcliffe et al. (2000) orthogonal situations can be handled by structurally partitioning the two. An example of that is the above-mentioned separation of a research department from production units. It can also be handled by developing the capability to carry out both kinds of processes in the same organization/department and using switching rules to handle the transitions between them. This way they can be separated in time due to changes in situational conditions. For example switching from high reliability "shelf technology" production to a situation where the organization develops new designs beyond customer requirements.

A third approach to the relationship between control-oriented and exploration-oriented processes is the "synergistic" way. This approach relates to a dual process model where the focus is on the interrelationship between exploration and exploitation. The synergistic approach refers to conditions where the effective pursuit of the two is mutually supportive. In other words, doing one better simultaneously improves the ability to do the other. The antithetical and the orthogonal way focus on differences between control-oriented and exploration-oriented processes. The synergistic view stresses their complementarities and interdependencies (Sutcliffe et al. 2000). This way of arguing challenges March's antithetical trade-off position and is important for my research. I will therefore refer to this in some detail.

Reliability is important for short-term performance and considered to be an effective strategy under conditions of stability. According to Sutcliffe et al. (2000) resilience is more important than reliability for organizations under conditions of change. Effective coping and adaptability depends on resilience, and reliability and resilience are mutually reinforcing. Referring to Kobasa (1979) and Sternberg & Kolligan (1990), Sutcliffe et al. (2000) argue that resilience in individuals is enhanced by experience that allow for the exercise of judgment, discretion, imagination, by the ability to make and recover from mistakes and by observing role models that demonstrate these behaviors. Thus: *"As individuals gain control over key task behaviors and exercise discretion in performing those behaviors, they develop a sense of competence. As a sense of competence increases, individuals are able to respond effectively in unfamiliar and or changing situations, and effective action subsequently reinforces a sense of competence. Resilience is an outcome of the self-reinforcing nature of this cycle"* (Sutcliffe et al. 200:327).

The authors argue that similar processes play a role at group and organization level. Collective beliefs can have a positive effect on performance (Wood & Bandura 1989). As organizational units better understand their capabilities, competencies and identity, collective beliefs about a unit's capability for action can be important for facilitating resilience and for achieving synergies between control and exploratory processes. Sutcliffe et al. (2000) argue that better trained and multifunctional groups develop better sensing and coping capabilities: *"As the capabilities for action increase, work groups that perceive many possibilities for action may be better able to grasp variations in their environment. The more an entity sees in a situation, the greater the likelihood that it will see specific changes that need to be made"* (Sutcliffe et al. 2000:327).

In other words, individuals and work groups that are given the possibility to carry out and build competence in both control processes and exploratory processes will increase their ability to see where they can intervene to improve performance, they will be more motivated to do so and they will also have better capabilities for adequate action.

2.2 Exploitation and exploration capabilities. Trade-off or mutually reinforcing?

Sutcliffe et al.'s (2000) complementary and mutually supportive competence and capability building challenges March's argument that resources put into, and attention paid to, the one kind of development processes comes at the expense of the other. March holds that refining exploitation more rapidly than exploration is effective in the short run but self-destructive in the long run (March 1991). This seems logical if exploitation comes at the cost of exploration. However, if the two development strategies work mutually supportive in a synergistic way, this may not be the case.

March (1991) further argues that exploration of new alternatives reduces the speed with which skills related to existing practices are improved, and that improvements in competencies at existing procedures make experimentation with others less attractive. Will this always be the case? According to the synergistic approach, it will not.

March (1991) argues that when people adapt to organizational codes and requirements, this improves exploitation more rapidly than exploration. With reference to other research (Argyris & Schön 1978, David 1985), March argues that the reason for this is that each increase of competence in an activity increases the likelihood of rewards for engaging in that activity and thereby for further competence building in the activity. Accordingly, he finds that the tendency of adaptive processes to increase exploitation and reduce exploration make adaptive processes potentially self-destructive (March 1991).

March finds the understanding of the mutual learning process between organizations and their members important for managing the trade-off between exploitation and exploration. This mutual learning process can be summarized as follows: *"Organizations store knowledge in their procedures, norms, rules and forms. They accumulate such knowledge over time, learning from their members. At the same time individuals in an organization are socialized to organizational beliefs"* (March 1991:74).

According to March (1991), learning the organizational code and imitation will inhibit experimentation. The faster members of an organization are socialized to the existing organization, the less the beliefs of its members will contribute to adjustments in the code. And oppositely, the slower the members are socialized, the more the organization will learn from its members' beliefs. The code can learn only from individuals who deviate from the code. Related to this, he introduces the term knowledge equilibrium. This is a state where all individuals in the organization and the organizational code share the same belief with respect to each dimension. Then the knowledge equilibrium is stable.

The faster the members learn the code, the faster equilibrium is reached. Accordingly, slower learning leads to higher knowledge equilibrium because individual beliefs will have greater influence on the code. It is important that individuals do not adjust so fast to the code that the code learns little from them. In a closed system the highest knowledge equilibrium will be reached in a situation where the organization learns fast from its members' individual beliefs and the members learn the code slowly. Slower individual learning maintains diversity longer thereby providing the exploration that allows the knowledge found in the code to improve.

March (1991) develops his theory through simulation. He finds that a mix of fast and slow learners in the organization is the ideal situation. Slow learners are a significant contribution to organizational learning. Rapid socialization tends to reduce exploration. On the other hand, rapid socialization and knowledge about the code makes performance more reliable. As techniques are learned and work is standardized, both the time required to accomplish the job and also the outcome of the job in terms of quality is likely to improve. Again, March argues for a balance between exploitation and exploration and the choice between the two. The mixture of fast and slow learners contributes to this balancing. The slow learners represent variation, creativity, choice and innovation. The fast learners increase reliability, quality, efficiency and refinement of the existing. Furthermore, he finds that people who have been in the organization for a while are less likely to contribute to new knowledge to the code. Newcomers are more likely to contribute.

However, what if the organizational beliefs focus on exploration and develop routines that favor creativity and innovation? What if the organizational beliefs and routines are in line with a synergistic approach to the balancing of exploitation and exploration? Will adaptation still favor exploitation or can both be improved in complementary and reciprocally supportive ways? What if fast learning of the code motivates for more and new learning? What if creative efforts and learning from exploitative activities like production optimization and variance reduction increases motivation for explorative activities and innovations like new production methods, new technology, new raw materials, new markets, new processes, new organizational solutions, etc.? What if beliefs, routines and practice favor exploration? Will "fast learners" still threaten exploration?

In a study of Japanese shipbuilding in the period 1971-2000, Greve (2007) found that the rate of launching exploitation innovations increased when the duration since the last exploitation innovation decreased. Conversely, the exploitation innovation rate decreased when the time period since the last exploitation innovation increased. In other words, practicing and learning the code of exploitation generates more exploitation. This finding supports March's theory that learning the code promotes further learning of the code and exploitation. The same pattern was found for exploration innovation (Greve 2007). The rate of exploration innovation decreased when the duration since the last exploration innovation increased. The correlation in this latter case was not significant and the findings are therefore not conclusive. But the reported pattern may indicate that fast learning of an organizational code where beliefs, routines and activities support exploration innovation, can promote more exploration innovation.

March's concept of "slow learners" can be contrasted to Sutcliffe et al.'s (2000) synergetic approach. March argues that the slower employees learn the "codes" of the organization, the more likely is it that they will contribute to exploration, while the faster people learn the organizational codes the less likely is it that they will contribute to exploratory activities. Sutcliffe et al. (2000) argue otherwise when discussing the synergistic approach. The more people learn about the organization and its activities, the more motivated and capable they will be to participate both in exploitative and explorative activities. In other words, they will not only be better in doing what they already do, they will also transfer their knowledge, creativity and motivation towards new areas and exploratory activities. Process and product optimization activities that belong to Sutcliff et al.'s (2000) concept of "Control" and to

March's concept of "exploitation" can, according to Sutcliffe et al. (2000) also increase exploration capabilities.

March's trade-off balancing theory may be challenged by some other findings reported by Greve (2007). Among others, Greve did not find that exploitation reduces exploration. On the contrary, his findings suggest that exploration innovation and exploitation innovation were generated by similar processes. Furthermore, he found that situations of performance below aspirations in the organizations significantly increased both the rate of making exploration innovations and exploitation innovations at the same time. Managers not only turned to exploitation as a solution, they tried exploration at the same time. However, Greve is careful when concluding and clearly expresses the limitations of his study and the problem of generalization. The study builds on a limited set of data from one branch, and he calls for further research on the subject (Greve 2007).

The arguments from Sutcliffe et al. (2000) and Greve (2007) are supported by the results from a survey of 302 small and medium-sized manufacturing companies, all members of the Osaka Industrial Association in Japan, reported by Isobe et al. (2007). In line with March (1991), the authors (p.3) state that *"the capability to refine proprietary assets (refinement capability) and the capability to reconfigure the asset structure (reconfiguration capability) are the critical sources of a firm's competitive advantage"*. They find that the understanding of how these two capabilities co-evolve and interact is poor. This is where they want to contribute.

Based on the results from the survey, they conclude that refinement capability facilitates exploitation, whereas reconfiguration capability facilitates exploration. In other words, the capability to refine proprietary assets and portfolio tends to improve operational efficiency. The capability to reconfigure the asset structure and integrate new assets improves strategic performance. These results are not surprising. One should expect that knowledge, skills and capacity to refine the existing facilitate operational performance. Likewise, that knowledge, skills and capacity to look for alternatives and innovate, facilitates strategic performance.

However, Isobe et al. (2007) also found that firms with superior refinement capability tend to possess superior reconfiguration capability as well. They further found that a firm's refinement capability significantly enhances its reconfiguration capability, which in turn enhances its strategic performance. This latter observation does not support March's idea of balancing the two mostly as a trade-off between exploration and exploration. On the contrary, these results suggest that refinement capability improves reconfiguration capability, which in turn facilitates strategic performance. Isobe et al. (2007) did not however find the same effect for reconfiguration capacity on refinement capability. They concluded: "*A possible interpretation of this finding is that refinement capability and reconfiguration capability have different implications for a firm's asset configuration; whereas refinement capability may leverage resources for reconfiguration capability, reconfiguration capability may limit the availability of resources for refinement capability*" (Isobe et al. 2007:15).

This reasoning is in line with March regarding the possible negative effect of exploration on exploitation: *"Exploration undermines exploitation"* (March 1995:443). The first part of the argument from Isobe et al. (2007) is, however in line with Sutcliffe et al. (2000), who argue that when people become better in doing what they already do, they will also transfer their knowledge, creativity and motivation towards new areas and exploratory activities. "Fast learners" can enhance reconfiguration capability and thereby strategic performance, not the contrary as March (1991) suggests. Based on their observations Isobe et al. (2007) finally concluded that refinement capability and reconfiguration capability work dynamically and complementary in developing a competitive advantage for a firm.

Another study of 111 US importing companies concluded along some of the same lines. Yalcinkaya et al. (2007) found that a company's exploitation capability is an important foundation for the development of the company's exploration capability. Exploitation capability is in this case defined as *"the ... ability to improve continuously ... existing resources and processes"*. Exploration capability is defined as *"the ... ability to adopt new processes, products, and services that are unique from those used in the past"* (Yalcinkaya et al. 2007:66). However, Yalcinkaya et al. (2007) also found exploitation capability to be negatively correlated to product innovation. This finding does not challenge March's trade-off position. But, the study indicates that exploitation capability is important to product development in an indirect way. Exploitation capability enhances the development of exploration capability, which in turn is important to product innovation. The authors did not analyze and explain this finding thoroughly, but call for further research to understand the processes behind these complicated relationships. They concluded that the relationship between exploration and exploitation is an intricate one. Through my case study research I want to contribute to the building of more knowledge about this intricate relationship.

In a study of R&D activities in a sample of 57 pharmaceutical firms Cardinal (2001) found that organization-wide control enhanced both radical and incremental innovation. Cardinal defined control as *"any process by which managers direct attention, motivate and encourage organizational members to act in desired ways to meet the firm's objectives"* (Cardinal 2001:27). Her definition of control is wider and not quite identical to Sutcliffe et al. (2000). She differs between input control, behavior control and output control. Among others she found that both input control, behavior control and output control enhanced radical innovation, in this setting defined as new drugs. The interesting point here is that while some have concluded that control may inhibit the generation of new ideas and the creativity stages of innovation (Duncan 1976, Zaltman et al. 1973), this study concludes otherwise and more in line with Sutcliffe et al. (2000), Greve (2007) and Isobe et al. (2007): *"The pervasiveness of the view that control is inherently negative has limited our ability to further explore how control may help solve the unique challenges of managing and challenging R&D activities. In closing, the results provide evidence that technological innovation in fact can be managed with some forms of organizational control previously thought to be detrimental to the innovation process"* (Cardinal 2001:31).

Cardinal argues that within the pharmaceutical industry formal control mechanisms coexist with strong informal professional norms that influence values, activities and outcomes. Control may enable and make sure that scientists effectively carry out their work in a way that is aligned with these professional norms.

Furthermore she suggests: *"While it is commonly accepted that incremental and radical innovation should be managed differently, the results of this study suggest otherwise. In this instance the management of R&D activities may be considered more similar than previously thought"* (Cardinal 2001:19).

Thus, in the case of pharmaceutical R&D Cardinal found that some of the mechanisms that enhance radical and incremental innovation can be the same. For this specific branch, this point in the same direction as the findings from Greve (2007), who suggests that "exploration innovation" and "exploitation innovation" were generated by similar processes.

In a study of 22 innovations in 13 companies Tushman et al. (2006) found that the ability to innovate both incrementally and radically depends on organizational design and also on top management focus. "Ambidextrous" designs where highly differentiated units exploit and explore but are tightly linked by an executive team that manages the organizational separation, may permit a business unit to evolve through both incremental and radical change. The authors found that ambidextrous designs had positive effects on both incremental and radical innovations (Tushman et al. 2006). The study did not focus on the effect of exploration on exploitation or vice versa, but none of the firms experienced performance losses when they attempted to initiate multiple product innovations at the same time. So exploration activity did not reduce efficiency and quality.

Ambidextrous designs are found to have a positive association with firm performance in general (Raisch et al 2009). However, several research issues related to ambidexterity as a concept remain unexplored, ambiguous or conceptually vague. Ambidexterity can be pursued in a variety of ways, other than by differentiating between units (Gibson & Birkinshaw 2004, Raisch et al 2009). Gibson & Birkenshaw (2004) found contextual ambidexterity (behavioral capacity to simultaneously demonstrate alignment and adaptability across an entire business unit) to be positively related to business results.

According to Stacey (1992), handling both control and learning at the same time results in organizational tension, paradoxes and contradictions. However, this provokes conflict and learning and is, consequently also a source of creativity: *"Successful organizations – that is, continually innovative organizations – cannot choose between tight, formal control systems and structures on the one hand and loose, informal processes that provoke learning on the other. Whether they are large or small they must do both at the same time. This is because they must all simultaneously handle both the knowable, closed changes involved in the day-to-day running of the business and the unknowable, open ended changes involved in the innovative development of the business"* (Stacey 1992:19).

In spite of these kinds of good logical reasoning and later research results, the literature does not provide many good examples of organizations with the capability to succeed with both development strategies in the same system simultaneously (Sutcliffe et al. 2000). Research shows that this balancing is not easy (March 1994, Boer & Gertsen 2003, Petersen et al. 2004, Boer et al. 2006, Greve 2007, Andriopoulus & Lewis 2009, Raisch et al. 2009, Fang et al. 2010). I find that there is a need for solid theoretical and methodological approaches to support efforts to balance exploitation and exploration synergistically.

Thus, based on the discussion above, I conclude that recent research supports March's (1991) idea of the necessity of both exploitation and exploration for prosperity and survival in a constantly changing environment. However, March's idea of balancing exploration and exploitation as a trade-off is being questioned. The capability to refine may enhance the capability to reconfigure, but not necessarily the other way around. Some suggest that exploration has negative effects on exploitation. Some suggest that it does not. So it seems to be wise to look for conditions that decide whether, when and how the one affects the other, and to investigate the possibilities for creating businesses where exploitation and exploration can work synergistically. This is an area where there is a need for more knowledge and understanding (Tushman et al. 2006).

In their research, Sutcliffe et al. (2000) found examples of all three ways of dealing with the relation between control-oriented and the exploration-oriented processes; the antithetical, the orthogonal and the synergistic approaches. They also found that little research had been done to examine the synergistic approach. My research has the ambition to contribute to filling that gap.

2.3 Summary

Maintaining an appropriate balance between Exploitation and Exploration is a primary factor in system survival and prosperity (March 1991). However, views and approaches to this balancing differ. March (1991) represents a mostly antithetical approach and see the two development strategies conflicting and competing for scarce resources. This view has been challenged by, among others, Sutcliffe et al (2000) and Boer et al (2006). These scholars question whether this is always correct, and advocate a more synergistic approach to the balancing of the two. The two development strategies may work mutually reinforcing. However, more research and empirical evidence is needed to leverage our knowledge about this balancing. This is the departure for my research.

I will now proceed by clarifying what I mean by TQM and discuss the appropriateness of using this concept to represent forms of exploitation. Then BPR will be clarified and the appropriateness of using this concept to represent forms of process exploration will be discussed.

3 Total Quality Management

3.1 Background

The attainment of quality has always been one of the goals of human effort. But it was not until the last decades that it really started to occupy the attention of management (Juran 1995). Quality then became an increasingly important means of competing on the world market (Bergman & Klefsjø 1994). The emergence of Japan as an economic superpower was largely the result of its quality revolution (Juran 1995), and one important factor in the increased attention to quality as a means of competition (Bergman & Klefsjø 1994).

According to Juran (1995) the 20th century has been the century of production and the 21st century will be the century of quality. In the 20th century there was an explosive growth in science and technology. Huge new industries emerged to translate new technology into benefits for people. The new technologies required complex designs and precise executions. The empirical methods of earlier centuries were unable to provide appropriate product and process design, so process yields were low and field failures high. Companies tried to deal with low yields by adding inspection to separate the good from the bad. They tried to deal with field failures through warranties and customer service. These solutions were costly and created dissatisfaction among customers. The need emerged to prevent defects and failures (Juran 1995).

The major forces that demanded the quality revolution in the late 20th century have been summed up by Juran (1995:630):

- Greater complexity and precision of products.
- Threats to human safety and health, and to the environment.
- Government regulation of quality.
- The rise of the consumerism movement.
- Intensified international competition on quality.

By the 1980s it became clear that quality leadership could not be achieved only by bringing in different tools for different purposes. Instead it became clear that it was necessary to apply the entire array of quality know-how throughout the entire organization, to all functions and at all levels and to do so in a coordinated way. The shorthand expression for this view became "Company Wide Quality Control" in Japan and "Total Quality Management" in the USA (Juran 1995).

Total Quality Management has been developed by a considerable number of people and institutions in a number of countries. Most important are Japan and the USA. Much of the literature has been written in the USA or in English for an American audience. The current American version grew out of the Japanese quality movement as it developed in the 1960s and evolved through the 1980s. The Japanese tradition combined ideas from American authors like Deming and Juran with large-scale worker participation and training in continuous improvement initiatives (Bergman & Klefsjø 1994).

Different academics such as Juran (1989), Deming (1986), Feigenbaum (1991), Ishikawa 1985) and Crosby (1979) have influenced developments in different directions. However, these different directions have a lot in common. The "Excellence" model of the European Foundation for Quality Management (EFQM 2003) has been used as a basis for the TQM efforts in some of the large industrial companies that I have cooperated with during my research, like Aker. This model, including both the "Large Company Operational and Business Unit version" (EFQM 2006) and the "Small and Medium Enterprise version" (EFQM 2006), has been used as a basis for the TQM efforts in the different companies that have participated in the action research projects. This model will shortly be referred to below, and can be considered as an expression of the current status of TQM thinking in Europe.

3.2 Definitions

I will now define quality, total quality and Total Quality Management. The number of definitions of these terms is numerous (Bergman & Klefsjø 1994). Below are some frequently cited examples. Based on these, I will then present the definitions that I found appropriate for my action research projects in a Norwegian industrial setting.

3.2.1 Quality defined

Below are some definitions put forward by some of the most influential academics.

Juran (1989:15): *"Fitness for use"*, which has a dual meaning: Product features that meet customer needs, and freedom from deficiencies

- Crosby (1979:17): "*Conformance to requirements*"
- Feigenbaum (1991:7): "*The total composite product and service characteristics of marketing, engineering, manufacture, and maintenance through which the product and service in use will meet the expectations of customers*"
- Ishikawa (1985:44): To manufacture products that can: "*Satisfy the requirements of consumers*"

These definitions lead into the same direction and can be summed up by translating the ISO (International Standardization Organization) standards into "plain" English. Here "quality" has been defined as follows:

"A quality product or service is one that meets the needs and expectations of customers" (Praxiom Research Group 2005).

This definition is in accordance with the definitions of the above-mentioned academics. Besides, most of the companies that I have cooperated with in ED 2000 and VC 2010 use the ISO standards. Accordingly, this has been the most appropriate definition for me to use.

3.2.2 Total Quality defined

In ED 2000, I chose to consider Total Quality as an ideal state/goal for an organization to strive for. What this goal should look like in practice will vary for different organizations. In general, and with reference to the ISO standards (International Standardization Organization 2000), Total Quality should facilitate the achievement of an organization's purpose:

- *To identify and meet the needs and expectations of its customers and other interested parties (people in the organization, suppliers, owners, society), to achieve competitive advantage, and to do this in an effective and efficient manner, and*
- *To achieve, maintain, and improve overall organizational performance and capabilities* (International Standardization Organization 2000).

Accordingly, Total Quality does not only concern an organization's product or services, but also the way these are produced, including the cooperation with other organizations and the surrounding society. Furthermore, Total Quality is not a static condition. Customer demands and societal standards are changing continuously, which in turn demands continuous improvement of products, services and the way they are produced. Total Quality is a moving target. The term "the way they are produced" refers to the totality of an organization's processes and structure.

3.2.3 Total Quality Management defined

I see Total Quality Management as a management concept and a means to continuously move the organization towards a state of total quality. Like "quality" and "total quality", this term has also been given many definitions, and it is necessary to make some clarifications before I proceed.

Walter A. Shewart is known to be the first to have used statistical methods for controlling and improving production processes. This happened in the United States in the post World War 1 period. Important goals were to improve and control work processes, reduce variation and increase customer satisfaction (Bergman & Klefsjö 1994).

Both the earlier mentioned academics W. Edwards Deming and Joseph M. Juran, who have been of great importance for the quality revolution in Japan in the 1950s, were influenced by Shewart's thoughts. Deming, in particular, heavily stressed the importance of quality control by the use of statistical methods. They also added other important elements to what should become Total Quality Management such as process and customer focus, and the importance of management commitment to quality.

In Japan, the involvement of employees through quality circles and training at all levels became very important (Juran 1989) and, unlike the West, the Japanese did not differentiate between quality development and organizational development. In practice this meant that the development of quality tools, approaches and methods was done in the same operation as the organization that was to use these methods was being developed (Neerland 2000). This is an approach that I have adopted in my action research.

Thus, the following eight central features from the development of TQM are the ones that are most important for my understanding of TQM:

Process focus

- Customer focus
- Management commitment
- Employee involvement
- Training at all levels
- The integration of quality development and organizational development
- Statistical means to improve and control production
- Continuous improvement

Based on these features, the definitions above, and on my understanding of the meaning of the terms Quality and Total Quality, I define Total Quality Management as: "*A way of managing an organization's efforts to achieve customer satisfaction through the involvement of all employees in a process for continuous improvement of the totality of an organization's products, services and methods of production, including cooperation with customers, suppliers and the surrounding society*" (definition I used in my action research setting translated into English).

Accordingly, Total Quality Management is used as the term for the management concept that shall manage a process for continuous improvement needed to approach the ideal state of Total Quality. The definition does not explicitly mention strategic, tactical and operational management of day-to-day processes, nor does it mention quality tools like quality assurance and inspection. In order to have a short definition, the term "methods of production" in the definition was used very broadly and in a way that also included these aspects.

3.3 TQM as a complete management concept

For the introduction of TQM in organizations in an action research intervention setting I found it important to have a complete management concept. According to e.g. Juran (1989), Deming (1986), Feigenbaum (1991), Ishikawa (1985), Crosby (1979) and following TQM concepts presented by international consultants, e.g. PA Consulting Group, TQM encompasses:

- A management philosophy
- A strategy for change in line with the philosophy
- A systems approach to the structuring of an organization's resources in accordance with the philosophy and strategy
- Adequate methods, means and tools to support a change process in accordance with philosophy and strategy.

Philosophy. TQM as a management philosophy and its most important elements have been commented on during the discussion of the definitions above; e.g. the concept's customers focus.

Strategy. An example of a frequently cited strategy for implementing TQM philosophy is condensed in Deming's (1986) 14 points management list for TQM:

- Create constancy of purpose for improving products and services.
- Adopt the new philosophy.
- Cease dependence on inspection to achieve quality.
- End the practice of awarding business on price alone; instead, minimize total cost by working with a single supplier.
- Improve constantly and forever every process for planning, production and service.
- Institute training on the job.
- Adopt and institute leadership.
- Drive out fear.
- Break down barriers between staff areas.
- Eliminate slogans, exhortations and targets for the workforce.
- Eliminate numerical quotas for the workforce and numerical goals for management.
- Remove barriers that rob people of pride of workmanship, and eliminate the annual rating or merit system.
- Institute a vigorous program of education and self-improvement for everyone.
- Put everybody in the company to work accomplishing the transformation.

A systems approach to the organization of resources. An example of how to structure and organize a TQM process is provided in detail by Juran (1989). To summaries very briefly: The first step is to establish a quality council, drawn from the ranks of senior managers. This council should be assisted by the quality department. The latter may serve as secretary for the quality council. Then, a number of facilitators and/or quality improvement managers should be appointed and trained. Improvement projects should be nominated and started. For each project a project team with a project manager should be appointed. Work force personnel may be involved in the project teams, and participate in arrangements like quality circles and suggestion systems.

Tools and techniques. Academics and consultants have listed a considerable number of quality techniques to be used for different purposes and at different stages in an improvement process. Important tools are "the seven QC tools" (Ishikawa 1982) and "the seven management tools" (Bergman & Klefsjø 1994).

These four elements constitute the TQM concept. As mentioned earlier, many variants exist. In a European setting, the EFQM "Excellence" model may represent the current TQM thinking. TQM is not a trendy term any more, and the TQM way of thinking is being further developed under the term "Excellence". But, as I will show below, the basic ideas are very much the same and are being further developed under this heading.

3.4 The EFQM Excellence model

The European Foundation for Quality Management (EFQM) is a membership-based not-for-profit organization created in 1988, *"with a Mission to be the driving force for sustainable excellence in Europe and a Vision of a world in which European organizations excel"* (EFQM 2003: 2).

EFQM was created to promote world-class approaches to the management of European organizations that will lead to sustainable excellence. The EFQM model was introduced as the primary framework for assessing and improving organizations, in order to achieve a sustainable advantage (EFQM 2003). The model is based on the fundamental concepts of excellence presented in Figure 1.

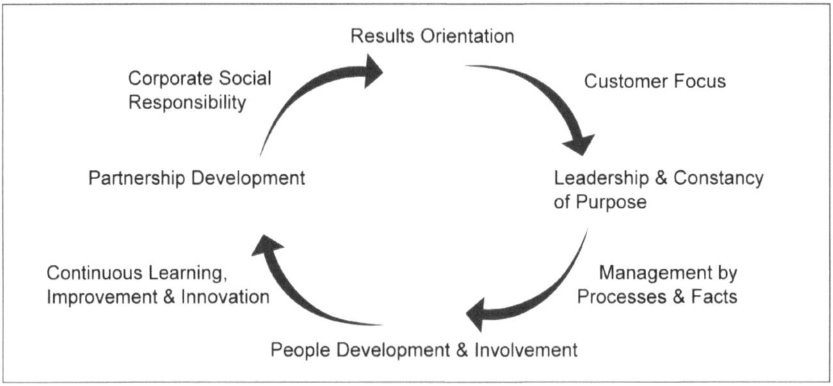

Figure 1 The EFQM fundamental concepts of excellence

Results orientation. Excellence is achieving results that delight all the organization's stakeholders. That includes all those individuals or groups that have an impact on the organization such as customers, employees, partners, suppliers, the society in which the organization operates, and those with a financial stake in the organization.

Customer focus. Excellence is creating sustainable customer value.

Leadership and constancy of purpose. Excellence is visionary and inspirational leadership, coupled with constancy of purpose.

Management by processes and facts. Excellence is managing the organization through a set of interdependent and interrelated systems and facts.

People development and involvement. Excellence is maximizing the contribution of employees through their development and involvement.

Continuous learning, innovation and improvement. Excellence is challenging the status quo and effecting change by using learning to create innovation and improvement opportunities.

Partnership development. Excellence is developing and maintaining value-adding partnerships.

Corporate responsibility. Excellence is exceeding the minimum regulatory framework in which the organization operates and striving to understand and respond to the expectations of their stakeholders in society.

It is on the basis of these concepts of excellence that the EFQM Excellence model has been developed (EFQM 2003). The model is presented in Figure 2.

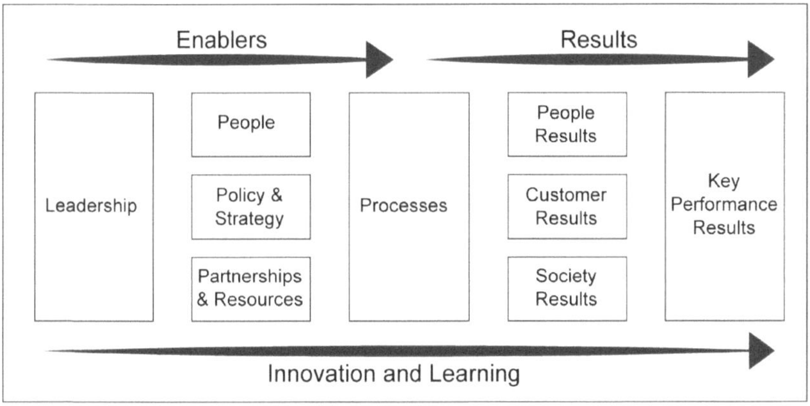

Figure 2 The EFQM Excellence model (EFQM 2006)

The model can be used for different purposes in an organization:

- As a tool for self-assessment by measuring where the organization is on the path to excellence.
- As a basis for a common vocabulary and a way of thinking about the organization that is shared across all functions.
- As a framework for positioning existing initiatives, removing duplications and identifying gaps.
- As a structure for the organization's management system (EFQM 2003).

The model is also used as the basis for assessment for the EFQM Excellence Award (EFQM 2003) and for national and regional quality awards.

As illustrated in Figure 2, the model is based on nine criteria for assessing an organization's progress towards excellence: five enablers and four result criteria. These enablers and success criteria are further divided into sub criteria and a number of measures and questions related to each sub criterion. The five enablers are Leadership, People, Policy & Strategy, Partnership & Resources and Processes. The four result criteria are People Results, Customer results, Society Results and Key Performance Results.

It is this understanding of TQM that I use in my further development of a synergistic approach to combining TQM and BPR in a Norwegian industrial democracy context.

3.5 Some challenges

One challenge when introducing TQM to a Norwegian industrial setting is that the concept builds on a consensus way of thinking. It does not deal with the fact that many organizations have an opposing part that often is represented by one or several trade unions. Especially, is this the case in Norwegian industry.

Another important challenge is the "cascading" way of thinking about TQM introduction and training. It starts with top management, who then takes it to the level below, and so on (Juran 1989). It can take some time until the message and activities reach the shop floor. Different levels of the company are not hit by the new concept and related activities at the same time.

Furthermore, research has shown that some of the arrangements that were developed in a Japanese setting have not been as successful in other settings. An example of this is quality circles (Mak 1999).

Yet another challenge is the practice of establishing specific quality councils, steering committees and arrangements (Juran 1989), which work alongside, and addition to the already existing formal organization and arrangements for cooperation in the organizations.

Development takes place partially in a stepwise fashion through a number of minor improvements and partly through radical innovations, i.e. more dramatic changes in technology, processes, organization, market or raw materials. The TQM concept has primarily focused on establishing an ongoing process of continuous improvement at all levels and in all parts of the business, and in practice it is the many small improvements that have been focused on. This is partly due to the strong influence on the concept of the Japanese way of thinking – "Kaizen" (Imai 1986).

3.6 TQM as one approach to the building of exploitation capability

In the TQM description and definition above I have used different terminology compared to the definition and description of "Exploitation" and "Control" in Chapter 2. None of these words have identical content, but point in the same business development direction. To Sutcliffe et al. (2000) “Control” represents a direction that emphasizes the need for continuous improvement, efficiency, highly reliable processes, little needless variance and conformance to standards.

March defines exploitation through terms such as refinement, choice, production, efficiency, selection, implementation and execution (March 1991). His definition is not very precise, but he points to a development approach that focuses on making the best use of current practices while continuously trying to improve them at the same time. March uses the word “practices” in a wide sense when he describes the idea of exploitation. Among the areas that are mentioned are technology, organization, procedures, forms, routines, practices, strategies, costs, markets, competencies and paradigms (March 1991, 1995). He does not seem to exclude any improvements areas as irrelevant for exploitation. On the contrary exploitation seems to be relevant to the totality of improvement options in an organization. *“The essence of exploitation is the refinement and extension of existing competencies, technologies and paradigms*” (March 1991:85).

Regarding TQM, it is the KAIZEN version (Imai 1986) – the gradualist approach to development – that I refer to as TQM practice. The EFQM model (2006) also includes radical innovations. However, it is continuous incremental improvement that has become the prevailing TQM practice (Imai 1986, Hammer & Champy 1993, Willoch 1994). Like exploitation, TQM covers all possible improvement areas for an organization, and excludes none. It covers both the results of an organization’s effort and the way these results are produced. TQM (often referred to more widely as Quality Management, Total Quality, Total Quality Control, Continuous Improvement or the Quality Movement) is not identical to exploitation. However, considering TQM as one approach to the building of exploitation capability is a common and accepted practice among many scholars in the field, including March (Davenport 1993, March 1995, Sutcliffe et al 2000).

3.7 Summary

TQM is a variety of practices and have been referred to by different expressions. Very often these expressions are not defined, differently defined or poorly defined when they are used. Consequently I have found it necessary to give some clear definitions of the different quality expressions that I use in my research. When I refer to TQM, I refer to the definition I have given in this chapter and the KAIZEN practice described by Imai (1986) – the gradualist approach.

In my efforts to challenge March’s theory of balancing exploitation and exploration, I use TQM to operationalize the idea to a practical industrial setting and to represent forms of exploitation. They are not identical, but both exploitation and TQM focus continuous improvement, control, refinement, efficiency and incremental changes of to-days’ practice.

In addition to the small, stepwise improvements, the TQM concept also prescribes radical changes in the enterprise's value creating processes, e.g. in the European "Excellence Model" (EFQM 2006). "Continuous improvement" must accommodate both small, stepwise improvements and more radical and dramatic changes. In practice TQM has focused little on radical change and has been criticized for that (Hammer & Champy 1993, Willoch 1994). It is on the latter category that Business Process Reengineering (BPR) focuses. As mentioned previously, BPR is often presented as an alternative to TQM (Hammer & Champy 1993), and as an approach to achieving major and radical changes. However, a growing number of scholars (Cole 1994, Kelada 1996) and enterprises, such as Aker and Statoil are gradually becoming more concerned with combining these two concepts and utilizing the best of each of them in one and the same development process. In other words, they develop in the direction of a more synergistic approach (Sutcliffe et al. 2000) towards balancing forms of exploitation and exploration. I will revert to this topic after a presentation and discussion of BPR.

4 Business Process Reengineering

Business Process Reengineering has been defined as:

"The fundamental rethinking and radical redesign of business processes to achieve dramatic improvements in critical, contemporary measures of performance, such as cost, quality, service and speed" (Hammer & Champy 1993: 32).

This definition consists of four central and powerful key words: *"fundamental"," radical"," dramatic"* and *"processes"*. According to the authors, these key words should be understood as follows.

Fundamental

"The first key word is 'fundamental'. In doing reengineering, business people must ask the most basic questions about their companies and how they operate: Why do we do what we do? And why do we do it the way we do? Asking these fundamental questions forces people to look at the tacit rules and assumptions that underlie the way they conduct their businesses. Often, these rules turn out to be obsolete, erroneous or inappropriate" (Hammer & Champy 1993:32).

Thus: "*Reengineering begins with no assumptions and no givens*" (Hammer & Champy 1993:32). And further down the same page: "*Reengineering first determines what a company must do, then how to do it. Reengineering takes nothing for granted. It ignores what is, and concentrates on what should be*" (Hammer & Champy 1993: 32-33).

Radical

"The second key word is 'radical', which is derived from the Latin word 'radix', meaning root. Radical redesign means getting to the root of things: not making superficial changes or fiddling with what is already in place, but throwing away the old. In reengineering, radical redesign means disregarding all existing structures and procedures and inventing completely new ways of accomplishing work. Reengineering is about business reinvention – not business improvement, business enhancement or business modification" (Hammer & Champy 1993: 33).

Dramatic

"The third key word is 'dramatic'. Reengineering isn't about making marginal or incremental improvements but about achieving quantum leaps in performance" (Hammer & Champy 1993: 33-34). And then further down the same page: "Reengineering should be brought in only when a need exists for heavy blasting. Marginal improvements require fine tuning; dramatic improvements demand blowing up the old and replacing it with something new" (Hammer & Champy 1993: 33-34).

Processes

"The fourth key word in our definition is 'processes'. Although this word is the most important in our definition, it is also the one that gives most corporate managers the greatest difficulty. Most business people are not 'process oriented'; they are focused on tasks, on jobs, on people, on structures, but not on processes" (Hammer & Champy 1993: 35).

The authors consider a business process as: *"A collection of activities that takes one or more kinds of input and creates an output that is of value to the customer"* (Hammer & Champy 1983: 35). This focus on a process-centered way of designing business activities is demonstrated in Figure 3.

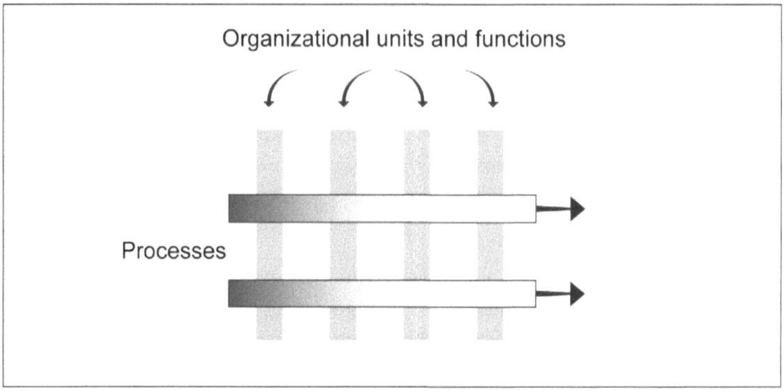

Figure 3 Business processes crossing hierarchical organizational units and functions

Figure 3 is based on a similar figure from Ernst & Young & Willoch (1996: 17) and illustrates how business processes cross an organization's units and functions. Each unit usually has an appointed manager and resources. People are organized in units and belong to units. It is the business processes that create value, and people are allocated to the processes for shorter or longer periods of time. Each process should have an appointed process owner (Hammer & Champy 1993). Business Process Reengineering highlights business processes and process centered organizations. This means that all employees recognize and focus on processes (Hammer 1996).

The fundamental, radical and dramatic character of Business Process Reengineering and its focus on processes rather than units and hierarchy is well demonstrated by one of Hammer & Champy's classic redesign examples, the IBM credit issuance process (Hammer & Champy 1993: 36-39). The IBM Credit Corporation is in the business of financing customers' purchases of the computers, software and services that the IBM Corporation sells. Initially and before reengineering the entire credit issuance process consumed six days on average. The performance improvement achieved by the redesign was extraordinary. IBM Credit slashed its six days turnaround to four hours. It did so without an increase in head count. In fact it achieved a small head count reduction. At the same time the number of deals that it handled increased a hundredfold (Hammer & Champy 1993).

This case is a good example of "fundamental rethinking" and "radical redesign" of a "business process" that gave "dramatic" improvement in important performance measures. The company achieved a dramatic 90% reduction in cycle time and a hundredfold improvement in productivity (Hammer & Champy 1993: 39). In other words, the total design of the whole process at stake was questioned and not only improvements in parts in the process.

4.1 The process hierarchy

According to BPR practice a company's activities may be broken down into a hierarchy of processes. On top you have the main processes. Each main process can be divided into a set of sub-processes, which can be further divided into work processes, which in turn can be divided into actions with a subject, a verb and some descriptors. Willoch (1994) uses several cases to show what a process oriented way of arranging a business's main value creating processes might look like. A brief version of Willoch's (1994:48) BP case is given in Figure 4.

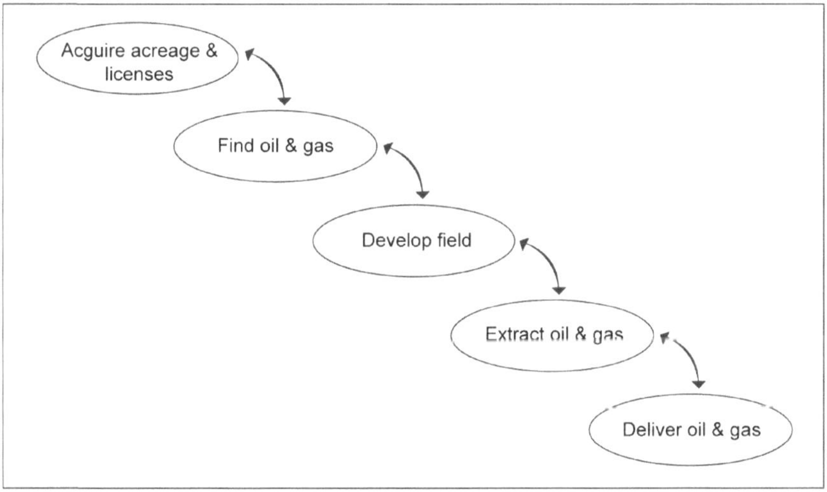

Figure 4 The mapping of the main processes in an oil company – BP

In this case BP has five main processes: 1) Acquire acreage & licenses, 2) Find oil & gas, 3) Develop field, 4) Extract oil & gas, 5) Deliver oil & gas. Each of these main processes can in turn be divided into a set of sub-processes, work processes, activities and actions. For example, one of the sub-processes in main process 3 will be "Develop the drilling program", a work process at the next level may be "Hiring a drilling contractor" etc. To illustrate this principle for the companies I cooperated with in ED 2000, I developed a general sketch of a process hierarchy. This sketch is shown in Figure 5.

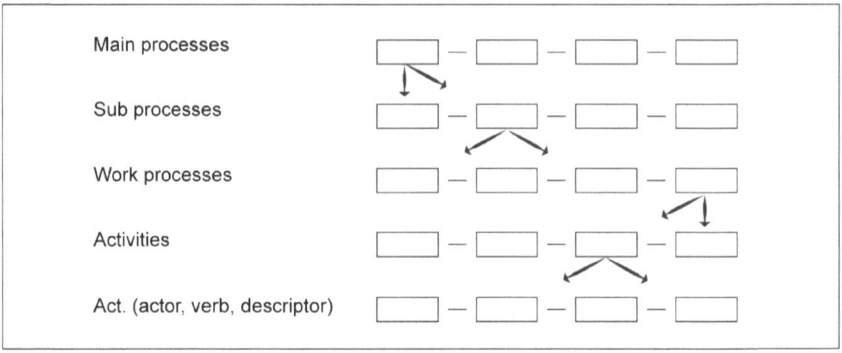

Figure 5 The process hierarchy

4.2 Some challenges, contradictions and vagueness in the BPR concept

The BPR concept encompasses some fundamental contradictions and is not very clear (Rolfsen 2000, Moltu 2000). For example, one of the fundamental contradictions is that the time needed to make radical improvements is described very differently. In Hammer & Champy (1993) the radical change should take place very quickly. Davenport (1993) expresses a very different view:

"Process improvement can begin soon after changes in a process are identified, and incremental benefits can be achieved within months. Because of the magnitude of organizational change involved, process innovation often takes a much longer time" (Davenport 1993).

Another contradiction is the role of information technology (ICT). Hammer & Champy claim: *"A company that cannot change the way it thinks about information technology cannot reengineer"* (Hammer & Champy 1993: 83).

This statement is supported by the process redesign cases used in the book (Hammer & Champy 1993). The IBM credit issuance process redesign case referred to above is one of them. However, many reengineering efforts are performed without any relation to ICT, but are called reengineering projects nevertheless (Moltu 2000). In *"The Reengineering Revolution. A Handbook"* (Hammer & Stanton 1995) the role of ICT is not mentioned at all.

However, these apparent contradictions may have one advantage. The concept suits different situations and needs. The vagueness of the BPR concept is best illustrated by the earlier cited definition of process: "*We define a business process as a collection of activities that takes one or more kinds of input and creates an output that is of value to the customer*" (Hammer & Champy 1993).

Process is defined in different ways by different BPR authors (Harrison 1995, Moltu 2000). Davenport (1993:5) states: "*A process is simply a structured, measured set of activities designed to produce a specified output for a particular customer or market. It implies a strong emphasis on how work is done within an organization, in contrast to a product focus' emphasis on what. A process is thus a specified ordering of work activities across time and place, with a beginning, an end, and clearly identified inputs and outputs: a structure for action*".

The focus here is on how work is done and on the structuring of activities. The human aspects are absent. This is different from the process thinking that we find in the social sciences and in the organizational development tradition (Schein 1987, 1992, 2002, Moltu 2000). To Schein, for example, a process is not the "what" but the "how". To Schein *"it is not the formal structure of an organization, but the actual behavior that occurs within that structure"* (Schein 2002: 1).

4.3 Common stages in BPR processes

Even if the concept is vague in certain respects, even if contradictions may be found between different BPR authors and there is no theory for change, there are some similarities in BPR practice (Butler 1994). A British study of 48 consultancy firms found that the number of stages in BPR practice varied from three to eleven, with four as the most popular (Archer 1996).

The EU project "Cobra" found the six stages in Figure 6 to be the most common stages in BPR processes (Coulsen-Thomas 1995). The Figure shows how these stages interrelate.

According to the Cobra methodology (Coulsen-Thomas 1995) five of these stages are compulsory for a full BPR approach. Some BPR practitioners believe that reengineering ought to be based on a "green field" approach and that stage 3, process analysis, is an unnecessary distraction from creativity and innovation. Stage 3 is found in many BPR methodologies, but is left out in others (Coulsen-Thomas 1995). The Cobra project also identified a large set of commonly used techniques. A few of these techniques are shared with the quality management tradition, such as brainstorming, SWOT analysis, cause & effect analysis and benchmarking (Coulsen Thomas 1995).

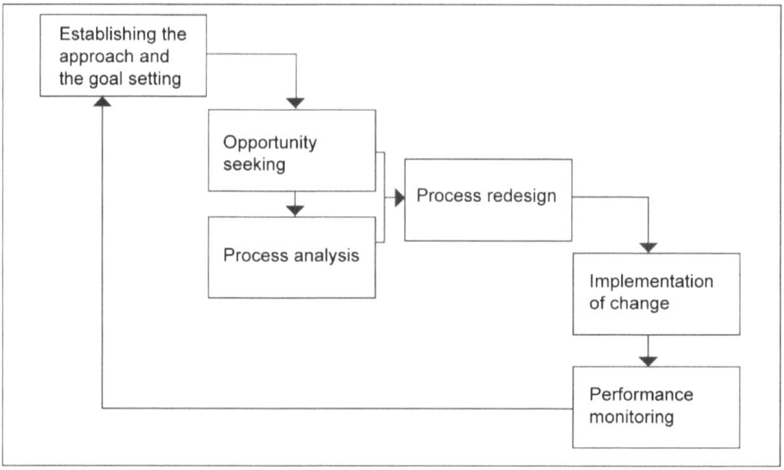

Figure 6. Cobra methodology. Stages in a BPR project

4.4 BPR as one approach to the building of radical process exploration capability

BPR neither represents the totality of the idea of exploration nor the totality of areas that can be explored. As we see from the definition above it focuses on process innovation, but also fundamental rethinking of what the business does and how it does it. This may include and result in radical changes in other innovation areas as well, such as raw materials, technology, market, organization and products. Different words and phrases are used in the definitions of BPR and exploration. March uses *"search, variation, risk taking, play, flexibility, discovery, innovation"* for describing exploration (March 199171). Sutcliffe et al. (2000) refer to a stream that focuses on innovation and explores breakthrough new technologies, systems and opportunities. Thus, the wording is different between the authors, but they point to a similar development process.

Like exploitation, exploration is not clearly defined by March (1991) either. As for exploitation, the application area for the idea of exploration is wide. Among others, March mentions new technology, products, procedures, forms, routines, practices, new ideas, markets, relations, variation and knowledge. *"The essence of exploration is experimenting with new alternatives"* (March 1991:85). He does not exclude any innovation or improvement area from the exploration sphere.

I have chosen to use Business Process Re-engineering as one approach to the building of process exploration capability. Innovation is important in March's idea of exploration (1991), and processes are one of the innovation areas that exploration may be directed towards.

Like with TQM, there are many BPR variants and the variants are also presented under different headings such as Business Process Redesign (Davenport & Short 1990), Re-engineering (Davenport 1990), Process Innovation (Davenport 1993), Business Process Re-engineering (Willoch 1994), Process Reengineering (Roberts 1994), and Process Redesign (Tenner & DeToro 1997).

Common to all these variants is that reengineering facilitates the transformation of businesses from being primarily functional organized and oriented to process oriented. If the company succeeds, this is a shift to a fundamentally new way of thinking about management. It is an innovation. Rather than focusing on functions such as production, product development, marketing, etc., and considering the tasks that each function performs, complete processes are considered across functions.

Thus, BPR focuses on redesigning the process as a whole and supports organizations in fundamentally rethinking how they do their work in order to dramatically improve performance. Optimizing the performance of sub-processes can result in some benefits, but cannot give dramatic improvements if the process itself is fundamentally inefficient and outmoded.

Many authors have expressed that reengineering is different from continuous incremental change and consider BPR as an approach to radical process innovation (Davenport 1990, 1993, Hammer & Champy 1993, Roberts 1994, Kelada 1996, Tenner & DeToro 1997). Davenport (1993) uses the term Business Process Innovation and contrasts innovation with improvement. Improvement has existing processes as the starting point, and the level of change is incremental. Process Innovation is radical and the starting point is *"clean slate"* (Davenport 1993:11). Hammer & Champy's approach starts by asking fundamental questions about what the company does and why it is done the way it is, and use expressions such as *"fundamental rethinking", "disregarding all existing structures and processes"* and *"inventing completely new ways of accomplishing work", "business reinvention", "radical", "dramatic", "blow up the old and replacing it with something new"* (Hammer & Champy 1993:329).

4.5 Summary

TQM and BPR share a few of the same techniques, and both concepts focus on process identification, process analysis and process change. However, the goals and approach for bringing about change are fundamentally different. Whereas TQM promotes incremental improvement on a continuous basis, BPR promotes fundamental rethinking and radical redesign. Both TQM and BPR efforts may be directed towards various innovation areas such as technology, markets, raw materials, products and organization.

I will go into further detail on TQM and BPR in the following chapters where the potentials for combining TQM and BPR are analyzed. I will also summarize and contrast characteristics of TQM and BPR.

5 Is a synergistic approach to the combining of TQM and BPR appropriate?

Hammer & Champy's classic "*Reengineering the Corporation*" (1993) was introduced as an answer to the shortcomings of TQM and as an alternative way to change management: *"Nor is reengineering the same as quality improvement, total quality management (TQM) or any other manifestation of the quality movement. To be sure, quality programs and reengineering share a lot of common themes. They both start with the needs of the process customers and work backwards from there. However, the two programs also differ fundamentally. Quality programs work within the framework of a company's existing processes and seek to enhance them by what the Japanese call kaizen, or continuous incremental improvement to process performance. The aim is to do what we already do, only to do it better. Quality improvement seeks steady incremental improvement to process performance. Reengineering, as we have seen, seeks breakthroughs, not by enhancing existing processes, but by discarding them and replacing them with totally new ones. Reengineering involves, as well, a different approach to change management from that needed by quality programs"* (Hammer & Champy 1993:49).

Against this background, many consultants introduced BPR as a counterpart and a better alternative to business development than TQM (Willoch 1994). This is a view that is compatible with the 1993 Hammer & Champy classic, but not in line with views expressed by Hammer (1996) in his writings a few years later when he argues against those who see these two approaches as incompatible: *"... TQM and Reengineering fit together over time in the life of a process. First the process is enhanced until its useful lifetime is over, at which point it is reengineered. Then, enhancement is resumed and the entire cycle starts again"* (Hammer 1996:82).

In other words, a process is improved by TQM and then, at a later stage, reengineered. Accordingly, an organization should have the ability to go for excellence in both process exploration and exploitation (March 1991) in the same system, but not necessarily at the same time and on an ongoing basis.

Hammer's understanding of the TQM concept is rather limited compared to e.g. the "Excellence model" developed by the European Foundation for Quality Management – EFQM (2006). When comparing TQM to Reengineering Hammer claims: *"TQM assumes that the design of the process is sound and that all it needs is some minor adjustments"* (1996:82). Further down the same page: *"But if the world has changed dramatically since the process was first (or most recently) designed, the current design may be fundamentally flawed and incapable of delivering the required performance. Reengineering is then called for. Reengineering does not merely enhance the individual steps of the process but entirely reconsiders how they are put together"* (Hammer 1996:82).

It is the "KAIZEN" (Imai 1986) version of TQM Hammer (1996) is referring to. The EFQM model has a more comprehensive approach in which both challenging the status quo and (radical) innovation are included: *"Excellence is challenging the status quo and effecting change by using learning to create innovation and improvement opportunities"* (EFQM 2006:14).

This approach overlaps with Reengineering goals, but practice and the road to success and methodology differs very much. For example, EFQM focuses more on thorough evaluations and revisions of existing processes as a means to further development than Reengineering. Reengineering may even skip the evaluation phase when processes are reengineered (Coulsen-Thomas 1995).

5.1 The interdependence between radical innovations and continuous improvement

Sutcliffe et al. (2000) argue for a synergistic, balanced and combined approach where processes for "greater control" (continuous improvement/exploitation) and "disjunctive change" (radical innovation/exploration) are mutually reinforcing and facilitate and contribute to the effectiveness of the other. Cole finds that a series of small wins often precede and follow large changes. Small wins pave the way for large changes by providing momentum and basic learning, and eliminate the impediments to optimizing new products or processes. Small wins make large-scale changes possible. Also, when many revolutionary changes are analyzed, they are often found to be based on a series of small wins (Cole 2001).

Many see the patient accumulation of small improvements accounting for the bulk of technological progress (Tushman et al. 1997). Many creative solutions are associated with continuous improvement. Vice versa, Cole (2001:10) argues that *"for large scale disContinuous Innovation to be successful, there has to be a great deal of continuous improvement surrounding it – before, during, after"*.

One of my ambitions is to explore the possibilities of building on the strengths of both TQM and BPR ways of thinking and at the same time see if some of the weaknesses associated with each concept can be neutralized. For example, in TQM it is important to build a thorough method for organizing a process for continuous improvement (Deming 1986, Juran 1989, Imai 1986). On the other hand, the TQM concept has been criticized for having a focus mainly on incremental changes (Hammer & Champy 1993) and not on radical change, where BPR is meant to have its strengths.

BPR has been criticized for stopping at the implementation phase. Most of the methodologies stop here (Vakola 2000). Even if Reengineering was introduced as a better, more potent and competing alternative to TQM (Hammer & Champy 1993, Davenport 1993), many support the view that a company's ability to business excellence depends on the ability to handle both incremental and radical innovation in the same system. This view is in accordance with research and experience in several countries (Davenport 1993, Kelada 1994, Cole 1994, Cole 2001, Zairi & Sinclair 1995, Weston 2001, Boer & Gertsen 2003, Petersen et al. 2004, Boer et al. 2006). To recap, the Cobra methodology (Coulsen-Thomas 1995) presented above is an example of a methodology that recommends a continuous improvement phase after implementation. Experience with the use of the eight-phase CONDOR BPR methodology at three large European construction companies showed that this continuous improvement phase is a critical success factor (Koala 2000).

Davenport (1993: 14) claims: *"However different in character, continuous process improvement and process innovation present similar challenges. Both require a strong cultural commitment and a high degree of organizational discipline, a process approach, a measurement orientation, and a willingness to change. A company that is unsuccessful at one will probably not succeed at the other."* This challenges March's (1991) antithetical approach to balancing exploitation and exploration in the process innovation area.

TQM and BPR may be combined in the same system and development process. An organization may improve existing practices from 80-90%, then further on from 90-99.9%. But then it stops, and a radical innovation is necessary in order to continue improving (Kelada 1994). On the other hand, systems and procedures for radical change alone are not considered sufficient. A radical change has to be followed up by continuous improvement in routines, maintenance, equipment, training, organization etc. in order to obtain the potentials of the investment.

Furthermore, the fact that both concepts share some of the same language, the same focus on customer satisfaction, and some of the improvement and change methodology, can make it easier to combine the two concepts. The probably most important common characteristic is that both concepts are process oriented: *"Reengineering and TQM are merely different pews in the church of process improvement. The two share an orientation towards process, a dedication to improvement, and a dogma that one begins with the customer. After that they diverge"* (Hammer 1986:81-82).

A similar argument is expressed by Davenport (1993:14): *"But although continuous quality improvement may be good practice, it is not a prerequisite for success at process innovation. The skills and enablers are different. Success at quality initiatives is a qualification for success in process innovation, ...",* and further down on the same page: *"A company will attempt to stabilize a process and begin continuous improvement, then strive for process innovation. Lest it slides down the slippery slope of process degradation, a firm should then pursue a program of continuous improvement for the post-innovation process."*

The temporal interdependence between continuous improvement and radical change for an organization's innovative ability may be demonstrated in Figure 7. The figure demonstrates that in order to improve business performance you need periods of radical change that are followed up by continuous improvement. It also demonstrates that radical change without continuous improvement may actually result in a performance decline (Imai 1986). The meaning of the term "continuous improvement" has in practice been very close to the Japanese concept "KAIZEN", which primarily includes elimination of defects and incremental improvements (Imai 1986). However, the inclusion of radical innovations in the same process for continuous improvement is in line with TQM philosophy (Deming 1986, Juran

1992, Finster 2001, Cole 2001, European Foundation for Quality Management 2005), even if practice has focused mostly on the elimination of defects, incremental improvements and problem solving (Hammer & Champy 1983, Hammer 1986, Cole 2001).

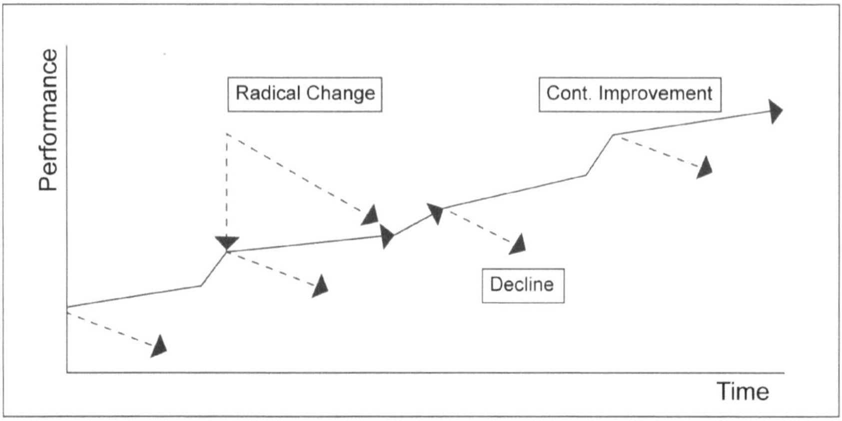

Figure 7 Radical change and continuous improvement in the same system and in the same business development process (inspired by Imai 1986: 27).

5.2 Slow or fast growing economies – different needs at different times, or?

The 1980s were considered a decade of Japanese supremacy, and continuous improvement was often held to be superior to radical innovation (Imai 1986, Gomory 1989, Florida & Kennedy 1990). However, the 1990s saw a decline in the Japanese economy while the American economy grew at the same time. A renewed emphasis was put on radical innovation in the US industry (Brown & Eisenhardt 1998). It was argued that continuous improvement was un-American and inconsistent with the American cultural emphasis on improvisation and innovation (Hammond & Morrison 1996). Imai (1986) argued that continuous improvement worked best in slow-growth economies and that radical innovation was more suited to fast-growth economies. Cole (2001) argues that a more convincing explanation is that conventional continuous improvement works best when firms are playing catch up. Therefore they pretty much know the direction they need to go by observing those ahead of them. Thus continuous improvement fitted large Japanese companies that focused on catching up after World War 2. However, when companies are operating on the frontiers of technological knowledge, more "disContinuous Innovation" is required. During the late 1990s the Japanese fell behind in many key areas (Cole 2001).

In periods of rapid technological change and rapidly changing high tech markets, coming late to market reduces profits significantly (Vesey 1991). Managers seek to compress development, production and delivery time and integrate these operations into as seamless a process as possible. A common element is speed (Cole 2001). As mentioned, continuous improvement was developed in slow speed economies and the systematic problem solving methodlogy, e.g. the PDSA cycle: Plan-Do-Study-Act (Deming 1986), was considered to take some calendar time. However, even in slow moving industries, companies very often need speed in their problem solving efforts. Car producers like Volkswagen and Ford have developed their own tools for rapid problem solving ("quick hitters"). Selected teams meet to identify their concerns, generate ideas for action and recommend solutions in a few hours or a day (Cole 2001).

Many fast moving industries in competitive technology-intensive global markets, as referred to above, see more discontinuous forms of innovation as advantageous compared to continuous improvement. However, Cole (2001) raises the question whether Continuous Innovation, which is one approach to the synergistic combination of exploitation and exploration, has a contribution to make towards promoting disContinuous Innovation. Learning from mistakes is very important to TQM and continuous improvement (Deming 1986, Imai 1986, Juran 1992). Learning is also a very important element in Continuous Innovation, defined by Boer et al. (2006:10) as *"the effective, ongoing interaction between operations, incremental improvement, learning and radical innovation aimed at combining operational effectiveness and strategic flexibility through excellence in exploitation and exploration"*. The traditional quality way would be to eliminate chronic waste in product development processes. Cole (2001) argues for a way to build quality improvement, through Continuous Innovation, into the development processes also for radical innovations. He sees the "probe and learn" approach as important here, and argues that it has the potential to serve as a sustained energizing force in such processes. This will be discussed below.

5.3 Learning as a driving force to a synergistic combining of continuous improvement and radical innovation

Organizations are being exposed to an accelerating speed of technological change (Fine 1998). The speed at which firms develop and come up with new products has become an increasingly critical competitive issue. Shorter product life cycles mean that firms have less time to make profit out of their investments and to come up with the right products with the right quality (Cole 2001). For example, the product life cycle in the PC industry was about a year in the mid 1980s. By 1997 it was reduced to approximately three months (Currey & Kenney 1998).

In order to build quality improvement through Continuous Innovation into the development process in a highly competitive and rapidly changing business environment, Cole (2001: 12) argues that *"the first step in such thinking is to understand that product development in turbulent sectors, like high tech, is an emergent process in which the premium is on learning and rapid incorporation of that learning into subsequent as well as previous development processes. This severely limits the contribution of conventional planning, so much the hallmark of the traditional approach to incorporating quality into the product development process. It also paradoxically encourages the successive generation of error, early and often, as part of the learning process".*

Cole (2001) further argues that product development in a turbulent environment and under high levels of uncertainty requires a nonlinear process and that problems and errors are inevitable. The traditional quality movement stresses prevention, both upstream through planning quality into design and downstream production. Errors have been the enemy. In the concept of "probe and learn" errors are considered not only inevitable, but also desirable if they contribute to learning (Cole 2001).

I find this desire for errors very controversial in an industrial setting. Cole uses examples from the IT industry that releases products very early to have them tested out by customers. Customers respond by reporting errors or shortcomings, the products are revised and new versions are released in a more or less continuous manner. These revisions are inexpensive and accepted by customers. So, in this kind of industries, probe and learn seems an ideal strategy both for the supplier and for customers who get the new products very early and can influence the further development.

In other industrial settings like nuclear plants, the space industry, oil & gas, etc., errors are not desirable, and may cause a lot of harm. In that kind of environment one cannot afford "probe and learn". Furthermore, the contradiction between learning and reliability/prevention of failure may be a bit exaggerated. The quality tradition gives learning a high priority (Imai 1986, Deming 1986, Juran 1989).

Referring to Argyris (1996), Cole argues that errors that cause employees to question the underlying values and policies of an organization and thereby lead to more efficient behavior are especially desirable. Argyris (1996) differs between two forms of learning. "Single loop learning" focuses on solving the problem at hand. "Double loop learning" focuses on the recognition that the way you go about defining and solving problems may be a problem in itself. Solving one problem is similar to single loop learning. Questioning the way the business is done in general, organizing for continuous improvement and "probe and learn", stimulates double loop learning.

Argyris (1996) defines "single loop learning" *as "instrumental learning that changes strategies of action or assumptions underlying strategies in ways that leave the value of the theory of action unchanged"* (Argyris 1996:20*). "Double loop learning" means "learning that results in a change in the values of the theory-in-use, as well as in its strategies and assumptions"* (Argyris 1996:21).

An important aspect of Continuous Innovation (Cole 2001, Boer & Gertsen 2003, Petersen 2003, Boer et al. 2006) is to train employees to continuously question the way things are done as well as underlying theories-in-use, policies and values. The ambition is to encourage and enable people to participate in testing out new products, processes, materials and markets, to learn from tests and changes, to reflect on the learning processes, to learn how to learn and to continuously look for incremental and radical innovation possibilities.

Much of the organization literature in this area has suggested that reliability and learning are antithetical and that the pursuit of one is presumed to preclude the pursuit of the other (Sutcliffe et al. 2000). Furthermore that organizations face a choice; either to compete by being highly reliable in exploiting what they already know and are known for, or to compete by being a leader in exploring new, breakthrough technologies or systems (Sutcliffe et al. 2000). There seems to be a growing recognition among academics and practitioners that we need to move beyond framing this choice in either/or terms.

Sutcliffe et al. (2000) have challenged the task of integrating greater reliability and control (and thereby preventing mistakes) with greater learning and exploration. However, there is very little research examining the competences, investments and performance trade-offs of trying to balance the pursuit of both types of processes. The task is challenging, and further research is needed (Sutcliffe et al. 2001).

5.4 Business Process Management (BPM)

BPM is an example of a recent and actual business management concept that builds on and integrates different management concepts and traditions, among others, key elements from TQM and BPR. The key building blocks of BPM are the IT tradition, the quality movement and the business process movement (Brocke & Rosemann 2010, Hammer 2010, Harmon 2010). BPM has been defined as "a structured, coherent and consistent way of understanding, documenting, modeling, analyzing, stimulating, executing and continuously changing end-to-end business processes and all involved resources in light of their contribution to business success" (The Australian Community of Practice 2013).

BPM is a holistic management approach (Harmon 2010), however, still in its infancy (Hammer 2010). To further develop and mature the concept, important areas for further research and improvements have been identified (Brocke et. al 2011, Hammer 2010, Harmon 2010). To find answers, case studies and action research are called for (Brocke et. al. 2011).

Many of the identified improvement areas for BPM are dealt with in this book both theoretically and methodologically. My case studies provide insight in and examples of how some of these issues have been successfully dealt with in a Norwegian industrial context. Among others; process redesign and organizational change, the role and involvement of people in process evaluation and design, process change methodology, process orientation and cultural change/motivation, how to coordinate several redesign initiatives, process management and innovation, and how to handle both radical and incremental innovations in the same redesign process. In general the most important contribution from my research to BPM is my discussion of the challenges of integrating TQM and BPR approaches in a mutually reinforcing way and ways of accomplishing it. This is the main theme of the book, and a necessity for successful BPM. Is a synergistic, dual approach to the integration of TQM and BPR a good approach? In that case, when should this approach be chosen, how should it be done and what are the enabling factors? When are other alternatives, such as separating TQM and BPR approaches in space and time, a better solution?

5.5 Summary

In theory it seems reasonable to combine the TQM and BPR ways of thinking into the same development process in the same system, at the same time and on a continuous basis. The theoretical discussion above and some of the empirical studies referred suggest that exploitation capability can have a positive effect on companies' exploration capability. In my case this means that continuous improvement capability can constitute a good foundation for the building of process innovation capability. However, is it necessary or even smart to look for synergy, considering the conflicting elements in the two concepts? Would it be more reasonable to handle the two separately in the same organization, as "multi-standard organizations" have the ability to do so well (Røvik 1998) or keep them separate in time in line with Imai (1986) and Hammer (1996)?

Important in recent and actual management concepts such as BPM is the integration of the kinds of approaches that TQM and BPR represent. The road to synergies between the approaches needs to be discussed. My research can contribute to this discussion and to the search for good theoretical and methodological solutions.

On the background of this theoretical discussion, I will now present my research model and further clarify how I use some terms that are important in the book.

6 Research model and definitions

In the introduction my research question was delimited to: "Can TQM and BPR be combined synergistically, if so, how? My research model is presented in Figure 8.

6.1 Research Model

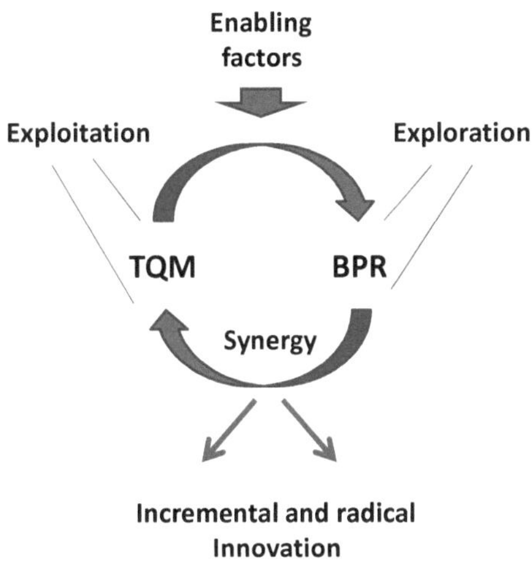

Figure 8. Research model. The synergistic combination of TQM and BPR

My overall research objective is to contribute to the further development of the theory of exploitation and exploration (March 1991) by studying the possibly synergistic interaction between exploitation and exploration characteristics in Norwegian industry. I use TQM and BPR to represent forms of exploitation and exploration, respectively. Operationalized this means that I investigate the interaction and the potential for synergy between TQM and BPR to promote both incremental and radical innovation in the same system and at the same time. In that case, how can this be accomplished and what seems to be important enabling factors?

6.2 Clarifications and definitions

With reference to my theoretical discussions above, I see that different authors' use the terms that are important in my research differently and some of the definitions are not very precise. Before I enter my empirical research I will therefore further clarify how I use the terms "innovation", "exploitation" and "exploration". On the basis of that, I will clarify how I use the terms "incremental" and "radical innovation". TQM and BPR have been described and defined in Chapters 3 and 4. In this section I will summarize and contrast the most important characteristics of the two.

6.2.1 Innovation

Many researchers have contrasted "continuous improvement" with (radical) "innovation" (Imai 1986, Davenport 1993, Hammer & Champy 1993, Cole 1994, 2001, Zairi & Sinclair 1995). Table 1, based on Imai (1986), compares the Japanese way of running continuous improvement, called "KAIZEN", with (radical) innovation.

	KAIZEN	Innovations
Effect	Long-term and long-lasting but undramatic	Short-term but dramatic
Pace	Small step	Big step
Timeframe	Continuous and incremental	Intermittent and non-incremental
Change	Gradual and constant	Abrupt and volatile
Involvement	Everybody	Select few "champions"
Approach	Collectivism, group efforts, systems approach	Rugged individualism, individual ideas and efforts
Mode	Maintenance and improvements	Scrap and rebuild
Spark	Conventional know-how and state of the art	Technological break-troughs, new inventions, new theories
Practical requirements	Requires little investment but great effort to maintain it	Requires large investment but little effort to maintain it
Effort orientation	People	Technology
Evaluation criteria	Process and efforts for better results	Results for profits
Advantage	Works well in slow growth-economy	Better suited to fast-growth economy

Table 1 Characteristics of KAIZEN and (radical) innovation (Imai 1986:24)

Imai calls these two roads to progress the "gradualist" and the "great leap forward" approaches (Imai 1986:23). I find these expressions appropriate for summing up the features of "KAIZEN" and (radical) "innovation". In line with many of the later theoretical contributions I have referred to, Imai is well aware of the interdependence and connection between these two approaches to progress for an organization's success. However, how useful is this division between continuous improvement and radical innovation and what about overlapping and shared features between the two?

Cole (2001) has questioned the usefulness of the common categorization of "continuous improvement" and "innovation": *"The common assumption is that continuous improvement is small scale and that innovation is discontinuous and large scale. Yet, there is no logical reason to associate the term innovation with large-scale discontinuous change. Consistent with a dictionary definition, innovation is best associated with creative solutions and these can occur at a small as well as a large scale, and can be more or less discontinuous. Put more bluntly, there are plenty of innovations that occur in the course of Continuous Improvement"* (Cole 2001: 9).

Kanter (1988:170) has defined innovation as

***"The creation and exploitation of new ideas"*.**

This wide definition covers both continuous improvement and radical change. It covers exploration and exploitation. It covers both sustaining innovations and disruptive innovations (Christensen 2003, Anthony 2004). It also covers different innovation areas like new products, new methods of production, new markets, new organization and new sources of supply/raw materials (Schumpeter 1934). It is Kanter's understanding and definition of the term innovation that I use in my empirical research.

However, both the preceding and also the following discussion show that different authors use the word differently. Some authors use the word innovation to express radical innovation, e.g. Imai (1986). When I discuss different authors, I will refer to their definitions and ways of using the expression.

6.2.2 Exploitation and exploration

As with the terms continuous improvement and innovation, I find the terms exploitation and exploration, and the borderline between the two overlapping and vaguely defined in the literature. These pairs of terms are not synonymous, but represent similar pairs of development directions. However, in some cases an action can be considered to belong to the exploitation paradigm or to the continuous improvement paradigm from one viewpoint and to the exploration paradigm or the radical innovation paradigm from another viewpoint. For example, the change from one raw material to another can be considered explorative through the eyes of a small production department in a large concern, while the board of directors of that concern may consider the change exploitative. In my research I use Imai's distinction between KAIZEN and (radical) Innovation to clarify what I mean with forms of exploitation and exploration respectively. There will still be some challenges with the overlapping and borderline between the two terms, but I find Imai's key features useful to describe the overall direction related to each of them, the "gradualist" and the "great leap forward" approach. It should be mentioned, however, that Imai (1986) tends to focus on technology when he refers to innovation, I will follow March's wider approach when I refer to exploration.

In line with March (2001), Levinthal & March (1993) and Yalcinkaya et al. (2007) I define **exploitation** as:

> ***Improving existing resources and processes.***

Exploration is defined as:

> ***Adopting new processes, products, and services that are unique relative to those used in the past.***

6.2.3 Incremental and radical innovation

I will use the word **incremental innovation** for:

> ***Changes that belong to the exploitation paradigm***

Radical innovation will be used for:

> ***Changes that belong to the exploration paradigm***

Being aware of the fact that these two different forms of innovation overlap and can be considered differently depending on the perspectives and the eyes of the actors affected, I will take the affected company or department perspective when I categorize in my empirical studies and analyses.

6.2.4 Contrasting characteristics of TQM and BPR

When referring to TQM, I use my definition from Chapter 3. When referring to BPR I use Hammer & Champy's (1993) definition that was presented and explained in Chapter 4. The detailing and contrasting of TQM and BPR in Table 2 is based on Deming (1986), Imai (1986), Juran (1989), Hammer & Champy (1993) and Hammer 1996). I use table 1 (Imai 1986) as a point of departure, add on and edit on the background of my preceding theoretical discussion. Using Table 1 as a point of departure also enables the visualizing of BPR as one variant of (radical) innovation.

Table 2 starts out with the same 12 variables as in Table 1 for comparison reasons, but I have added 12 more. All these have been commented on in my theory discussions. They will therefore not be further commented on here. A general comment is that the characteristics of (radical) innovation as they have been listed by Imai (1986) can be used for BPR as well. We also see that BPR shares some characteristics with TQM, while others are very different.

It is these TQM and BPR characteristics that I use to represent actual characteristics of exploitation and exploration. In order to avoid repeating all these characteristics in my case study analyses, the "gradualist" approach (Imai 1986) represented by TQM can be summed up as follows:

Change is typically done in small steps, it comes gradually, details are important and everybody is involved. Management leads the change processes, but the change process and activities are very much bottom up. Groups and collectivism are important, conventional know-how and state of the art solutions are prevailing, a lot of effort is put into keeping the process going – but no large investments. Information is open and shared, and everybody takes part in a learning process.

The "great leap forward" approach (Imai 1986) represented by BPR can be summed up as follows:

Changes typically come in big steps and can be abrupt and volatile. The change process is top down, few employees are involved, specialist-competence is needed,

	TQM	**BPR**
Effect	Long-term and continuous improvement	Short-term and radical
Pace	Primarily small steps	Big steps
Timeframe	Continuous	Intermittent, speed, short term
Change	Gradual and constant	Abrupt, volatile and radical
Involvement	Everybody	A few "champions"
Approach	Collectivism, group efforts, systems approach	Rugged individualism, individual ideas and efforts
Mode	Maintenance and improvements	Rethink, scrap and rebuild
Spark	Conventional know-how, state of the art and status analyses	Innovation break-troughs, new theories and new solutions
Practical requirements	Requires little investment but great effort to maintain it	Requires large investment but little effort to maintain it
Effort orientation	People	Processes, technology and business operation
Evaluation criteria	Improvements of many operating variables	Dramatic improvements in a few operating variables, quantum leaps in performance and results for profits
Advantage	Works well in slow growth-economies	Better suited to fast-growth economies
Leadership style	Strong commitment, visionary, visible, role model, coach, democratic	Strong commitment, visionary, visible, role model, coach, authoritarian/consultative
Customers	Customer focus	Customer focus

	TQM	BPR
Processes	Process focus	Process focus
Training	Improvement techniques	Process understanding and process design
Control	Statistical process control, revisions and quality assurance	Business results, costs and risk analysis
Suppliers	Supply chain management and close customer-supplier cooperation	New concepts and costs
Organization	Functional and cross-functional	Mostly cross-functional
Stakeholders	All the organizations stakeholders	Shareholder focus
Management	TQM management system	Process management
Information	Shared	To a few
Implementation	Many small improvement projects and activities	One or a few large projects
Culture	Shared values, goals and common commitment to continuous improvement and customer value	Process oriented professionals understanding and designing business and work processes for business performance and customer value

Table 2. Contrasting characteristics of TQM and BPR

technology – especially information and communication technology, plays an important role, the change often requires large investments, the state of the art knowledge in the company and existing in house practice is often of little importance, and information sharing is scarce.

6.3 Summary

In order to study the balancing of Exploitation and exploration in an industrial setting empirically, I use TQM to represent forms of Exploitation and BPR to represent forms of Exploitation. I investigate the interaction and possible synergies between them. If these two concepts should work mutually reinforcing in the same system at the same time, how can this be accomplished and what seems to be important enabling factors?

Before I turn to my empirical research, I will position my research in the setting where it belongs, and give a presentation and discussion of my exploratory methodological approach.

7 Research design

"A research design is the logical sequence that connects the empirical data to a study's initial research questions and, ultimately to its conclusions" (Yin 2003: 20). My research departs from a fifty year old Norwegian action research tradition on business development (Gustavsen 1992, Greenwood & Levin 1998). My research design is a case study approach (Campbell 1975, Eisenhardt 1992, Ragin 1992, Wievorka 1992, Geertz 1995, Stake 1995, Stake 2000, Yin 2003, Flyvbjerg 2006).

I use two variants of case study design. For my initial study of combining TQM and BPR in a Norwegian industrial context (Aker Stord) I use a single case study design in line with Yin (2003). In this case the researcher takes the role as an outside observer. For the development of and testing of my contributions to a more synergistic approach to the combining of TQM and BPR, I use action research (Greenwood & Levin 1998) in the two companies Serigstad and Titania. In action research the researcher is still an outsider, but shares responsibility for the actions being studied with the insiders. I consider action research to be a variant of case study design. This will be discussed below.

My discussion of research methodology will cover the following topics:

- Background and historical lines. The Norwegian Action Research tradition and the national research programs that my research is part of
- Methodological choice; case study design
- Single case study design
- Comparing my methodological choice to other case study approaches and research methods
- Data collection and analysis. A description and discussion of my role as a researcher
- Verifying my research methodology
- Additional and specific considerations related to action research
- A critical discussion, including a discussion of validity

7.1 Background and historical lines

– The Norwegian Action Research tradition and the national research programs this research is part of

As mentioned in the introduction, my research has been part of two comprehensive national research programs, the five-year research program "Enterprise Development 2000" (ED 2000) and the ten year follow-up program, "Value Creation 2010" (VC 2010). These two programs are joint national efforts in which a considerable number of enterprises, the Confederation of Norwegian Businesses and Industry (NHO), the Norwegian Federation of Trade Unions (LO), the Research Council in Norway (NFR), Innovation Norway, several

Norwegian counties and research institutes from different regions in Norway cooperate. The two research programs are part of a historical line and tradition that will be summarized below before presenting the two programs.

7.1.1 The heritage from Kurt Lewin

The historical lines and tradition that my research departs from can be summarized in the following quotation from Gustavsen (1992:12): *"How to create a better working life through processes of change to which research makes a significant contribution?"*

Kurt Lewin posed this kind of question in the 1930s, and was the first to introduce action research as a research methodology to the social sciences (Gustavsen 1992). Kurt Lewin was a social psychologist. His special interest was social change and how to promote it. He envisaged a process whereby one could construct a social experiment with the aim of achieving a certain goal (Greenwood & Levin 1998). E.g. he carried out a social experiment during World War II where the goal was to substitute beef with tripe as part of the regular diet of American families. In that period a lot of beef was needed for the troops and there was a shortage in the supply of beef. Housewives were trained in the art of cooking tripe for dinner, and the research question concerned the extent to which housewives could be encouraged to use tripe for dinner instead of beef (Greenwood & Levin 1998).

Another example from his successors (Coch & French 1953) illustrates how his approach was applied to work life. The abbreviated version is taken from Gustavsen (1992: 12-13): In the factory Harwood a technical change was underway. Researchers persuaded the local parties to do an experiment. The workers were divided into three groups. The first group was treated as usual. The other group was given the right to elect special representatives who could go into discussions with the management about the change process. The third group was given direct participation. Various variables were measured such as job satisfaction and productivity. The results of the experiment showed that the group with representative participation recovered quicker from the change than the group that was treated as usual. The group with direct participation recovered even faster, and reached a higher level of productivity.

It was in connection with this type of projects that the notion of action research was originally introduced (Lewin 1946) with basic elements such as:

- Involvement from research
- Research played an active role
- The intervention took the form of an experiment
- The project was designed on a basis of theory and hypotheses about participation
- The results of the intervention were measured on several variables
- The results were presented as a verification of a hypothesis
- The project contributed to the long-term theory building in social research (Gustavsen 1992)

Lewin's thinking about experimentation became the main strategy for the Norwegian Industrial Democracy Program (Greenwood & Levin 1998). The ideas behind this program were developed in London at the Tavistock Institute, with which Kurt Lewin was associated. During the rebuilding of British industry after World War II the Tavistock researchers were called on by the British government to do research on the introduction of new technology in coalmines. The background was that an expected increase in productivity due to the introduction of some new technology had not appeared. In short, the researchers found that the lack of improved performance could be explained by the incompatibility between demands created by the new technology and what was beneficial for the workers as a group of interacting human beings (Trist & Bamforth 1951).

These insights led to a break with the traditional approach to work in the coal mines, where research focused on the technically most efficient way to organize workers into separate responsible groups dealing with a clearly identifiable and bounded element in the production cycle. These insights became an important background for the shaping of the later industrial democracy movement in Norway (Greenwood & Levin 1998).

Doing experiments in real life situations is difficult, and the relationship between employers and unions in the post war period in Great Britain did not allow experimentation with the organization of industrial work, which was therefore not continued in Great Britain. The industrial setting and relationships between the labor market parties were quite different in Norway at that time. It was more peaceful, there were fewer conflicts and the cooperation was better organized (Gustavsen 1992).

The cooperation between the labor market parties in Norway goes back to 1935 when the first main agreement – referred to as the "Basic Agreement" – was signed. In addition, during the rebuilding period after the Second World War works councils were introduced in Norwegian working life (Gustavsen et al. 2001). Einar Thorsrud from Norway made links to the Tavistock Institute, and together with Eric Trist and Fred Emery the Industrial Democracy Program in Norway was sketched out very much in line with Kurt Lewin's approach (Gustavsen 1992). The industrial setting in Norway made larger scale experimentation in real working life situations possible.

The Industrial Democracy Program has had an important influence on later research, development, legislation, agreements between the labor market parties and reforms in Norwegian working life. It gave rise to stronger and more continuous links between the labor market parties. In order to oversee the experiments in the program, the labor market parties set up a joint committee. Over time this committee developed into a more complex institutional form of cooperation with the aim of supporting work place projects with a participative design (Gustavsen et al. 2001). In 1990 this cooperation was formed into "Hovedorganisasjonenes Fellestiltak – Bedriftsutvikling, HF-B" (The Confederation of Norwegian Businesses and Industry and the Norwegian Federation of Trade Unions' joint action program – Business Development).

This background is important for the understanding of my own action research and the role of the labor market parties. The historical line and the reasoning can be followed up in time to the two national research programs that my research is part of. I will now briefly describe these two programs and how they relate to this 50 year action research tradition on industrial democracy, job design and productivity.

7.1.2 Enterprise Development 2000

Enterprise Development 2000 (ED 2000) was set up in 1994 as a joint national effort between the labor market parties and the Norwegian Research Council (NFR). The initiative to the program came from HF-B (Gustavsen et al. 2001). The subtitle of the program was "Conceptually Managed Productivity Development and Organizational Renewal in Working Life". The primary objective was expressed as follows: *"The program shall contribute to the formation of value and help ensure employment by developing knowledge about strategies, methods, ways of work and infrastructure which is necessary to create organizational and inter-organizational development processes that will enable an increasing number of Norwegian businesses to participate in the front lines of international competition"* (The Research Council of Norway 1995:4).

The program itself represented an innovation in several respects. NFR usually distinguishes between research driven programs and user driven programs. ED 2000 should overcome this split. The research institutes became contract partners and were given the funds. The program expected the researchers to recruit enterprises, and the content of the action research should depart from needs and priorities expressed by the participating companies. The program integrated researcher driven research and user driven research.

Furthermore, there was a focus on value creation. Jointly pushing this aspect was not obvious for most of the labor market partners, especially not at regional and local level. This was the case even if the parties had a part B in their General Agreement promoting and regulating business development cooperation between the two main parties (NHO & LO 2006).

In ED 2000 there was also a focus on knowledge creation of relevance to approaches for sustaining and initiating organizational and inter-organizational improvement processes. The focus was not primarily on implementing given structures such as teamwork and democratic leadership, but on means of relevance to the development of such structures.

In previous research programs the labor market parties had often participated, for example as members of program boards, but now the cooperation went beyond this to encompass joint responsibility for funding as well as for the operational side of the program (Gustavsen et al. 2001). This operational cooperation took place not only at the national level, but also at regional, local network and company level. The action research was organized in seven modules spread throughout the country where more than a hundred companies participated in local partnerships.

7.1.2.1 The RF module and the cooperation with the industry network in Sunnhordland

The research institute RF (currently IRIS), was given responsibility for the program in the two counties Hordaland and Rogaland. The researchers organized a regional steering committee where the regional labor market parties were represented together with research, higher education and county administration. At the company level all the activities were planned and carried out in cooperation with the management and the local union(s). The point of departure was local challenges and needs. To start with theory in the participating companies was neither possible nor a goal. However, it was possible for researchers to have influence on the direction, but only as one of the actors in partnerships of actors with equal influence and power.

The researchers in the RF module cooperated with three industrial networks. These three network co-operations were organized as projects. I had the role of project manager for one of them, the Industry Network in Sunnhordland (IfS). This network had 13 member companies. The companies were in the offshore, shipbuilding and metal-melting industry. Their size varied from 13 to 1800 employees. The local activities were managed by a local board where both research, companies, network administration, the network board and unions were represented.

The overall objective of the RF module was related to the question of how international management concepts can be combined with Norwegian work life traditions. This objective departed from the researchers' cooperation with industry, and especially with the Industry Network in Sunnhordland (IfS), and was in line with the ED 2000 goals and ambitions. The leading company in the network, Aker Stord, was in the process of implementing TQM at the time when ED 2000 started up. They needed some assistance and wanted TQM to be a network activity in the region. They wanted to support both their suppliers and their neighbor companies through a joint and coordinated TQM effort. Aker Stord had long lasting traditions for union-company cooperation and employee participation, and found the ED 2000 program suitable and in accordance with the company strategy. It was also a company decision at that time to increase the cooperation with research. This point of departure set the scene and decided the goal and the content of important parts of my research for the following 16 years.

Two main goals, each with several sub goals were derived from the overall objective:

1) To initiate development processes in the participating companies based on union-management cooperation implying among others:
    - Establishing new arenas for employee – employer cooperation.
    - Initiating projects where employees were involved directly or indirectly (through elected representatives).
    - Creating a culture for continuous improvement.

2) To build a co-operative structure between the enterprises and the labor market parties in the network.

During the 5 year project period the IfS project developed its own local approach to Continuous Improvement. This concept combined important elements from TQM, HSE management and Norwegian work life traditions. New means, ways of work and methods were implemented in all the participating companies. The tools that are presented in my case studies later were all developed in this setting, and further improved in the next follow up national research program Value Creation 2010 (VC 2010).

7.1.3 Value Creation 2010

Like ED 2000, VC 2010 focuses on organizational development. However, value creation and innovation became more important issues in VC 2010. Like ED 2000, VC 2010 is a co-operation between the Confederation of Norwegian Businesses and Industry, the Norwegian Confederation of Trade Unions and the Research Council of Norway. In addition Innovation Norway joined the program both on the national and the regional level. The program was initiated in 2001, and was intended to continue until 2010. However, in 2007 the VC 2010 program was merged into a new program called "*Virkemidler for Regional Innovasjon – VRI*" (Means to Regional Innovation) which is planned to continue until 2016.

The main objective of VC 2010 is: *"To encourage and contribute to organizational development and innovation, both within individual enterprises and in learning networks between enterprises, based on new forms of cooperation between the industrial (social) partners and other actors of significance in the value creation processes. This is done by active use of researchers as development partners. The program is supporting the development of regional innovation strategies in regional partnerships"* (The Research Council of Norway 2006:1).

Like in ED 2000, the enterprises and networks are encouraged to actively use social science as a resource in their development, and the researchers in VC 2010 should produce scientific material and publications to increase the general knowledge in the field of work-life research.

While ED 2000 had seven research modules in Norway, VC 2010 had eleven modules organized as "Main Research Projects". Like in ED 2000, each main project is composed of a group of researchers working with enterprises, networks, local governmental authorities and higher education in regional development coalitions. The action researchers who worked in the program acted as development agents in the participating companies, they initiated the building of new industry networks, took active roles in the development of the networks and in the further development of the regional innovation systems.

7.1.3.1 The VC 2010 main project in the Development Coalition Hordaland & Rogaland

The regional partnership where my empirical data comes from, and where I, as an action researcher, had the role as main project manager, is presented in Appendix 1. The VC 2010 main project and the regional partnership Hordaland & Rogaland cooperated in 2007 with 10 networks representing about 150 companies. My empirical data and the case studies come mainly from the industry network in Sunnhordland, the industry network in Dalane and from one of the individual companies that participated in the program. Our action research was not with isolated companies, but in communities where ideas and good practice were generated, discussed and spread among the companies. All the individual companies that participated in the program had close customer relations to one or more of the network companies.

These local community network settings and the regional partnership have been important for the development of the tools and the action research results that will be presented later. The regional partnership and all the networks present a unique opportunity for diffusion of research results, experiences and ideas. Diffusion has been reported to be a challenge to this kind of action research in industry (Gustavsen 1992). The goal for IRIS' (the successor of RF) main research project in VC 2010 was to further develop the cooperation between the labor market parties, local authorities, research and higher education – "triple helix" (Leydesdorff & Etzkowitz 1998) in order to increase regional innovation at company, local network and regional level, where the two counties Hordaland and Rogaland were seen as one industrial region. Sub-goals were:

- The integration of continuous incremental improvement, radical change and industrial democracy in business development
- Better integration of governmental funds and tools for innovation and value creation in industry
- The development of new local and regional triple helix networks for increased innovation and learning in industry
- Participation in the further development of the networks that we co-operate with
- International cooperation and exchange of experiences and learning

7.1.4 Overall research objective for the action research at IRIS

A main objective of our action research at IRIS in the two above mentioned national research programs has been to develop knowledge and strategies, methods, means, ways of work and infrastructure that may make the integration between international management concepts like Total Quality Management (TQM) and Business Process Reengineering (BPR) and Norwegian industrial traditions more effective (Grønhaug & Hansen 2001).

In the research program ED 2000 our research group had a special focus on Continuous Improvement, a main strategic element in TQM, which was the prevailing management concept in that period. Towards the end of the ED 2000 program, the focus gradually changed towards an approach, where balancing TQM and BPR synergistically became the focus. The reason for this change was very much the gradually increasing popularity of the management concept BPR and of radical process innovation towards the turn of the century and also national and international innovation agendas.

7.2 Methodological choice; case study design

Combining the management concepts TQM and BPR in a Norwegian industrial context synergistically is not an easy task. The researchers, including myself, were familiar with the national and international literature on management concepts, innovation theory, organization theory, learning theory, theories of industrial democracy, the Norwegian work environment legislation, the Norwegian work life traditions and action research. We also had some ideas on how to run these action research projects, but a lot was at stake both for the companies and for the researchers. We were uncertain and needed more knowledge and experience before we started a comprehensive development activity in a network of companies.

At the time when I started my research in ED 2000, Aker Stord was the largest company in the Industry Network in Sunnhordland (IfS), and was at that time considered to be the most important company for the industrial development in the local industrial community. It was part of a concern, but operated as an individual company. Aker Stord took the role as the locomotive company in the network cooperation on TQM. The company had a capable staff in organizational matters, capable and experienced shop stewards, and wished to cooperate with researchers in the further development of their own and the local industrial community's Continuous Improvement capability. In meetings with representatives from the Industry Network in Sunnhordland and Aker Stord, we agreed that an action research approach was the ideal way to cooperate for all parties. This solution gave the researchers valuable access to the companies over time and willingness in the companies to spend valuable company time with the researchers. The companies were given facilitator support over time both individually and as part of a network activity.

Before starting our action research in 1995, we (the researchers) did a comprehensive mapping of a large number of variables like turnover, strategy, work environment etc. in the participating network companies in Sunnhordland. This gave the researchers insight into the participating companies, including Aker Stord. We also met a lot of employees through comprehensive interviewing.

As Aker Stord was in the process of introducing TQM when we started the cooperation with the company and the industry network, we discussed how to start the cooperation to the mutual benefit of all parties. The company decided to do an audit on their TQM journey with a special focus on employee participation in the process in 1996, two years after they had started. The researchers were invited to take part in that audit. The researchers needed more knowledge about the practical use of TQM tools and methods, and I participated in designing the audit to make sure that it could also be used for research purposes and for data gathering, in accordance with my initial research objective. In addition the company decided to run an

EFQM self-evaluation. I was invited to participate to get additional information about the company and its TQM practice. The data collection and the audit will be described in more detail in the Aker Stord case narrative. The initial mapping of the company, the EFQM self-evaluation, the 1996 audit, two equivalent audits in the following years and further data gathered through the ED 2000 period form the basis for the Aker Stord case study. BPR was not an issue at Aker Stord at the time we started. Radical change was considered to be included in the TQM concept in accordance with the EFQM model. However, BPR very soon became an important part of the development process and the case study.

So, the Aker Stord case study had two main goals. 1) To prepare for action research projects in the participating companies. 2) To study the interaction between TQM and BPR. The role of the researchers in the initial Aker Stord study was primarily the outside observer. In the next sections I will discuss the advantages of choosing a case study design compared to alternative research designs and how I solved specific methodological challenges by choosing the specific case study strategy I used for the Aker Stord single case study and the action research projects at Serigstad and Titania.

7.2.1 Advantages of choosing a case study design in my research setting compared to alternative methods

Yin (2003) distinguishes five main research design strategies: 1) experiment, 2) survey, 3) archival analysis, 4) history and 5) case study. The different strategies have different advantages, different shortcomings and suit different purposes. I chose to select the case study approach, for reasons outlined below.

"Case studies are the preferred strategy when "how" or "why" questions are being posed, when the investigator has little control over events, and when the focus is on a contemporary phenomena within some real life context" (Yin 2003:1).

When planning the Aker Stord study, I considered several options. In short:

- Experiments require control of behavioral events (Yin 2003). This kind of control was neither possible nor suitable for my purpose. It was not possible because the TQM process was the result of an Aker Concern decision, it had started before the researchers entered the scene, and would continue with or without research assistance. In addition, the alternative of setting up an experimental design in this kind of situation would be too time consuming, and it would probably take years before the results of the study could be analyzed.

- A survey was no feasible alternative. The advantage of a survey/questionnaire and large samples is breadth and the main problem is depth (Mintzberg 1979, Eisenhardt 1989, Yin 2003, Flyvbjerg 2006). I needed detailed in-depth information about processes, practices and methods, and be able to follow up information from respondents by follow-up questions. A survey would not give me in-depth information about the TQM implementation processes nor the possibility to follow up information in-depth from respondents there and then. However, qualitative data and qualitative analysis can supplement quantitative in-depth data in case studies, and increase validity (Eisenhardt 1989, Yin 2003).

- Archival analysis was used as one of several data collection methods when the research group did the initial status mapping, the EFQM self-evaluation and the audits on employee participation in TQM. However archival analysis alone would not give me the necessary information about the use of practical tools and techniques, respondent's feelings, opinions, experiences and information that could lead to good practice, the reasons for successes, failures, resistance etc.

- History was not relevant in the initial phase of the project. The process was contemporary and had just started. We needed in-depth knowledge in a few weeks time, and could not wait for history to take place.

- Case study is an appropriate method for studying complex social phenomena (Yin 2003). I found the method appropriate for obtaining an in-depth and holistic understanding initially of how a TQM process in a Norwegian industrial company was carried out. I needed to know what worked well, what did not work quite that well, and why. I needed information about methodology and tools. I needed to know how different tools fit together as a whole and was used at different company levels and in various parts of the company. I needed to know how the development process was managed and coordinated, and how different actors and parties cooperated. I had no influence or control over the process that was ongoing, and I needed to study the process there and then in order to prepare for the preceding action research projects. It was also important to use a method that allowed me to study the phenomena of interest in its real life content. A case design made it possible to study the process in a Norwegian industrial context and in an external setting similar to the succeeding action research projects.

Consequently, I found a case-based design optimal for preparing for the next phase of my research, where I use the Aker Stord case experiences as a basis for developing and analyzing approaches to a more synergistic combining of TQM and BPR together with industry. For that purpose I chose an action research approach. The breadth and the "how" and "why" questions are still in focus, but in action research the investigator shares control of the events with the company representatives. The additional challenges of using action research as a case study approach will be discussed in Section 7.2.5.

7.2.2 Research challenges and case study strategy

To structure the discussion of my case study research strategy and discuss how I solved important methodological challenges through this strategy, I use Yin's (2003) recipe for case study designs as a point of departure. Contributions from scholars like Ragin (1992), Stake

(2000), Flyvbjerg (2006) and others will be discussed in relation to this and also used to complement Yin's approach.

According to Yin (2003) the case study as a research strategy comprises an all-encompassing method – covering the logic of design, data collection techniques and specific approaches to data analysis. Yin's definition of case study design is as follows:

"A case study is an empirical inquiry that

- *investigates a contemporary phenomenon within its real life-life context, especially when*
- *the boundaries between phenomenon and context are not clearly evident"* (Yin 2003:13).

Furthermore:

"The case study inquiry,

- *copes with the technically distinctive situation in which there will be many more variables of interest than data points, and as one result*
- *relies on multiple sources of evidence, with data needing to converge in a triangulation fashion as another result*
- *benefits from the prior development of theoretical propositions to guide data collection and analysis"* (Yin 2003:13-14).

In the Aker Stord case the boundaries between phenomena I was going to study and context were not clearly evident. For example, initially I did not know much about the company's position in the concern it was part of, its position in the region, in the networks, cooperation with NHO and LO, with consulting companies, with customers, former business development processes and other situational factors of importance for understanding the specific kind of business development process I was about to study. I needed in-depth and inside information.

The variables that I needed to study were many. Some were evident on the basis of my research objective. Others became evident during the research process. I did not have a full picture of all the variables when I started up. I needed more detail knowledge. In many cases data on different variables did not exist or were not available. Some variables, for example employee participation in idea generation at Aker Stord, needed data from more than one source. Interviews gave me insight into the process of idea generation and also information about reporting procedures and involvement. Statistics gave me the number of ideas that was generated from the different departments.

A case study should not be seen only as an exploratory phase of another research strategy (Yin 2003). E.g. even if I initially use the Aker Stord case as a strategy for data gathering, for competence building, to develop propositions and for planning in the initial phase of a more comprehensive action research program, I consider the Aker Stord study to be an independent inquiry in itself. The same holds for the two action research projects. Action research is the research strategy that I use to test out new approaches to the synergistic combining of TQM and BPR, to generate new propositions and to theory support. I will return to the different elements in more detail further down in the discussion.

7.2.2.1 Intrinsic or instrumental?

Stake (2000) distinguishes between "intrinsic" and "instrumental" case studies. A case study is intrinsic *"if the study is undertaken because, first and last, one wants better understanding of this particular case"* (Stake 2006: 445). A case study is instrumental *"if a particular case is examined mainly to provide insight into an issue or to redraw a generalization. The case is of secondary interest, it plays a supportive role, and facilitates our understanding of something else"* (Stake 2000:445).

The Aker Stord study was primarily instrumental. The research group was preparing itself for action research in seven companies and needed ideas on how to do it. The action research studies at Serigstad and Titania are also instrumental. I needed to test out approaches and tools developed that could promote the synergistic combining of TQM and BPR in a Norwegian industrial context. The distinction between intrinsic and instrumental case studies is especially important for the discussion of external validity which I will return to later.

7.2.2.2 The choice of a single compared to a multiple case study design?

Yin (2003: 39) distinguishes four kinds of case studies: 1) Single case holistic design, 2) Single case embedded design, 3) Multiple case holistic design and 3) Multiple case embedded design. In the initial phase of the ED 2000 project, I needed one or a few instrumental cases (Stake 2000). Ideally, some representative Norwegian cases might have been perfect for my purpose. However, the question of representativeness is challenging when doing case studies and not necessarily optimal. First of all, even in larger collection of cases, the sample size will be too small to warrant random selection (Stake 2000), as e.g. when doing surveys. Secondly, representativeness may not be ideal at all. Stake's (2000) recommendation is to choose a case where we feel we can learn the most: *"That may mean taking the most accessible or the one we can spend most time with. Potential for learning is different and sometimes superior criterion to representativeness"* (Stake 2000:451).

Flyvbjerg (2007:229) argues likewise: *" When the objective is to achieve the greatest possible information on a given problem or phenomenon, a representative case or a random sample may not be the most appropriate strategy. This is because the typical or average case is not the richest in information. A typical or extreme case often reveals more information because they activate more actors and more basic mechanisms in the situation studied."*

In the initial phase of ED 2000 Aker Stord was our cooperation partner, was available at the time, had decided to increase their cooperation with research, represented a company in a large concern with long lasting industrial traditions in Norway, and represented a longitudinal case for the researchers. In short, it suited our purpose, and we were invited to spend time in the company. However, it was not considered to be either an extreme or an atypical case. In a similar way, the two action research cases are also single case designs. The companies were available, motivated to cooperate with researchers and to try out new approaches to business development.

A single case study is holistic when it is the totality of the case that is the unit of analysis. In an embedded study (Stake 2000, Yin 2003) attention is given to subunits. Aker Stord was a large company. In that case I found that an in-depth analysis of the TQM process could best be done through sub units, which differed both in size, TQM maturity, attitudes, composition and many other variables. I could learn from different departments, with different objectives, organization, competencies, managers etc. This way of running an embedded design can improve validity and be as advantageous as a multiple case design (Stake 2000). The action research case companies were smaller. At Titania, all employees were invited to the Enterprise Development Conference, which formed the basis for a common development plan. At Serigstad, the main focus was on business and work processes across the whole plant, and department barriers were broken down. The action research cases were approached holistically.

7.2.2.3 Exploratory, descriptive and explanatory purposes

Yin (2003) divides case studies into three categories: 1) exploratory, 2) descriptive and 3) explanatory. My case studies carry elements of all three.

The main purpose of the Aker Stord case is primarily exploratory in the sense that the research group initially wanted to increase its knowledge about how an industrial company practiced TQM. The Aker Stord study is also descriptive in the way that I wanted to describe the company's way of running the TQM processes and good practice. Finally, it is explanatory: I wanted to understand why something worked well, or not.

The two action research cases are also primarily exploratory. I had some ideas about how to make my framework for the combining of TQM and BPR more complete that I wanted to try out and learn more about. I had started experimenting with two new tools in some other companies. I wanted to further develop these tools together with industry and to explore the potentials of the tools. The two action research cases can also be considered descriptive because I describe how the tools work in practice. Finally, they have explanatory elements as well. I analyze the results that are obtained by using the tools to find out what works and why.

7.2.3 The components of my case study designs

Yin (2003) finds the following five components of case study to be especially important:

1) A study's questions.

2) Its propositions, if any.

3) Its unit(s) of analysis.

4) The logic linking the data to the propositions.

5) The criteria for interpreting the findings (Yin 2003:21).

My presentation and discussion of the case studies will be related to these five components. I will briefly explain the five components and how these were formed to meet my research challenges. I will then discuss challenges regarding research quality when doing case studies. But first, I will review the most common components of a case study design in relation to my case studies.

7.2.3.1 The study's questions

In line with my overriding research objective, research question and the company's need for evaluating employee participation in TQM, the initial research questions specific for the **Aker Stord** study were how and why questions:

- How does a Norwegian industrial enterprise in practice carry out a Total Quality Management process?
- How are employees involved directly and indirectly?
- What seems to work well and why?
- Where are the challenges and why?
- How can important challenges be met?

The Aker Stord case study was a learning process, a mapping of status quo and also a preparation for long-term action research projects. I needed in-depth understanding. A" descriptive design" alone would be inadequate – I had to include "why", that is, explanatory questions. The question of integrating TQM and BPR synergistically was not ready matured at this stage.

Specific research questions were also developed for the two action research studies. Based on learning from the Aker Stord case and in line with my overriding research objective and research question, the specific questions for the **Titania** study were:

- How can direct employee participation in exploitation at shop floor level be expanded to also comprise forms of exploration and higher-level decisions in the enterprise (company vision, values, objectives, strategy and overall planning)?
- If this expansion is possible, what can be the consequences for a company's innovation capability, innovation activity and learning?
- Can the ED Conference contribute to the completeness of an approach that promotes the synergistic combining of TQM and BPR? If so, how?

The specific research questions for the **Serigstad** case were:

- Does the Actor-Process Diagram promote the synergistic combining of TQM and BPR?

- Will a participatory approach to redesign improve the company's capability to come up with ideas for incremental and radical innovation synergistically?
- How far and-wide should this participation be taken?
- Who should participate: the employees directly involved, representatives from other departments, shop stewards or all?
- Can TQM and BPR be combined synergistically?

7.2.3.2 The study's propositions

A proposition directs the attention to something that should be studied within the scope of a study. It directs our attention to where we should begin to look for some kind of evidence that reflects our theoretical issues (Yin 2003). Based on the initial status mapping of Aker Stord, a strong belief in employee participation in TQM expressed by representatives from the management and from the trade unions, and in line with the specific research questions for this case, the propositions were formulated as follows:

- Employee participation in TQM at Aker Stord supports the TQM process
- The TQM process can be anchored in arenas for cooperation described in the main agreement between the company and the unions
- A reciprocally supportive co-ordination of direct and indirect employee participation in business development can support TQM implementation

The development of the first audit was based on these propositions. The EFQM self-evaluation study followed the EFQM recipe. However, some additional information about employee participation in the TQM process was gathered here as well. Remember, the question of integrating TQM and BPR synergistically was not ready matured at this stage.

Based on experiences from 10 ED Conferences preceding the Titania ED Conference, and with reference to my research question, the following propositions were developed for the **Titania** study in line with the specific research questions for this case:

- The ED Conference, including follow-up, generates a considerable number of ideas for incremental and radical innovation initiatives synergistically.
- The ED Conference, including follow-up improves employees' direct and indirect participation in innovation synergistically.
- The ED Conference contributes to the completeness of an approach that promotes the synergistic combining of TQM and BPR.

Based on the Aker Stord experiences and with reference to my research objective and research question, the following propositions were developed for the **Serigstad** study in line with the specific research questions for this case:

- The Actor-Process Diagram generates a considerable number of incremental and radical innovation initiatives synergistically
- The Actor-Process Diagram improves employees' direct and indirect participation in innovation synergistically
- The Actor-Process Diagram promotes the synergistic combining of TQM and BPR
- TQM and BPR can be combined synergistically

All these propositions are discussed in the case narratives and concluded on in the final discussion and summary.

7.2.3.3 Units of analysis

The unit of analysis for a case study must be in accordance with the research questions, and can be an individual, a process, a group, an organization etc. (Yin 2003). In my research, the unit of analysis is the three case companies. However, the case companies were not studied in isolation. The companies were situated in the western part of Norway. Aker Stord and Titania were corner stone companies in the local communities. All three companies were part of a Norwegian industrial tradition. The business development cooperation between the labor market parties was regulated by the Main Agreement between the Association of Norwegian Businesses and Trade (NHO) and the Confederation of Trade Unions (LO), and the three plants were part of concerns.

As to Aker Stord, the TQM process was a concern initiative. Some material, like videos and handbooks were common for the different companies in the concern, some benchmarking processes were carried out between different units inside the concern, and there were planning and follow-up meetings at concern level. Apart from this, the TQM process was carried out as a company effort at Aker Stord.

Titania cooperated with and received valuable training support from managers and shop stewards from their most important customer, Tinfos Titan & Iron (TTI), on how to involve employees in innovation. TTI had been an active participant in ED 2000 and VC 2010 for several years when Titania started up.

The iron foundry at Serigstad was part of the Serigstad Concern and also part of the industrial network TESA. TESA had participated in ED 2000 and VC 2010 from the start, and Serigstad did some important benchmarking with TTI and Aker Stord. All these and other kinds of contextual factors were important to take into consideration for understanding the specific cases. I will now describe how the preparations before data collection were done, and how evidence was collected.

Preparations before data collecting

After having chosen the unit of analysis, some preparations have to be made before data collection can be started:

1. The desired skills of the researchers have to be assured through adequate actions.
2. A protocol for the investigation has to be made

Case study skills

As part of the researchers' preparations for ED 2000, we arranged a two-day action research and case study conference together with experts on research methodology from other universities. All the researchers who participated in ED 2000 attended this conference. In addition, a comprehensive reading list came with the conference. Reading and discussing that literature was an important part of our training and of the conference.

In short, a good investigator should be able to ask the right questions, be a good listener, be adaptive and flexible, should have a firm grasp of the issues being studied, be responsive to contradictory evidence and be unbiased (Yin 2003). The conference contained all these issues. As to the latter issue – the bias challenge, this is one of the main criticisms of case study designs compared to other research designs (Flyvbjerg 2006). This issue became an important part of our preparation and our discussion during data collection and analysis. Bias will be discussed further below and in the final discussion section of the book. Furthermore, the action research case study variant requires something in addition of the role of the researcher compared to research designs where the researcher has a more distant role. This will be discussed in Section 7.2.5.

Protocol

A good protocol increases the reliability of a study and guides the researcher through the data collection, analysis and presentation of results (Yin 2003). All three case studies were organized in line with Yin (2003):

- **An overview of the case study project.** The three cases shared the research objective, the theoretical framework and the research question. The purpose and content of the three cases differed, and project plans for each individual project were developed in the form of agreements between the research institute RF (eventually IRIS) and the companies. In the Aker Stord case, agreements were made for each of the three audits in the form of project proposals. For the action research projects project proposals were made for the cooperation with the company as a whole and attached as appendixes to the contracts with the companies. All these documents are found in the case protocols.
- **Field procedures.** At Aker Stord this included a listing of the departments to be involved, a listing of informants to be interviewed and by whom, a timetable, reference to written material to be included, information meetings and procedures. In the action research projects it included activities, milestones, timetables, descriptions of the planned interventions, data gathering procedures, roles and responsibilities. Details are given in the case narratives.

- **Case study questions.** At Aker Stord, flexible interview guides with open questions were used. The interview guides were developed in three versions tailored to the different categories of informants. At Titania and Serigstad the case study questions were related to the appropriateness of respectively the ED Conference and the Actor-process Diagram to generate incremental and radical innovation initiatives synergistically, and to the appropriateness of these two tools to contribute to the building of a more complete framework for employee participation in the synergistic combination of TQM and BPR. For details, see the case narratives.
- **A guide for the case study report.** At Aker Stord this was partly developed on the basis of earlier versions of audit reports for the three audits in order to meet company needs, and partly for research purposes and in accordance with propositions and questions raised for this specific study. Plans for presentation and review of results were included in the "field procedures". For the action research projects, guides for the case study reports were based on agreements with the case companies in the above mentioned project proposals. These were attached to the contracts with the companies, and based on the specific propositions and questions that were raised for each of the cases.

Collecting the evidence

According to Yin (2003) evidence from case studies may come from six kinds of sources: documents, archival records, interviews, direct observation, participant observation and physical artifacts. At Aker Stord all these were used, except direct observation and participant observation. In the action research projects all of these were used. See the case narratives for details.

7.2.3.4 Linking data to propositions

"Linking data to propositions" are the least developed components in case study design. (Yin 2003). Yin describes five techniques for analyzing case studies; 1) "Pattern matching", 2) "Explanation building", 3) "Time series analysis", 4) "Logic models" and 5) "Cross case synthesis". 1 – 4 will be commented on below. The last one is not applicable in single case study designs.

Yin (2003) holds pattern matching (1) to be one of the most desirable techniques. Pattern matching logic is used to compare an empirically based pattern with a predicted one. If the patterns coincide, the results can strengthen the study's internal validity (Yin 2003). This technique is used in all three cases. The predicted pattern as it was presented in the form of propositions is compared to the empirically based one. Based on the initial status mapping and initial interviews at Aker Stord, I e.g. predicted that employee participation would support the TQM process in Aker Stord, and compare this prediction to the empirical based pattern that I find in the case study. I further predicted that the TQM process could be adequately managed through the use of existing arenas for two-party cooperation between the company and the employees, and compare this with the pattern and the results that I find when I analyze the empirical data. I also predicted that a reciprocally supportive coordination of direct and indirect employee participation would support TQM implementation and compare this with the pattern and the results that I find in the case study analysis. This is described in detail in Chapter 8.

In the Titania case I predicted, in the form of propositions, that the ED Conference would generate a considerable number of incremental and radical innovation initiatives synergistically. I also predicted that the ED Conference would improve employees' direct and indirect participation in innovation synergistically. These predictions were then compared with what I found empirically. See Section 10.1 for details.

For the Serigstad case, I in the same way predicted that the Actor-Process Diagram would generate a considerable number of incremental and radical innovation initiatives synergistically, and that the Actor-Process Diagram would improve employees' direct and indirect participation in innovation synergistically. These predictions were then compared with what I found empirically. See Section 10.2 for details.

Explanation building (2) is a special type of pattern matching. Here the case is analyzed by building an explanation about the case. That is, to stipulate a presumed set of causal links about it. Explanation building is covered in my pattern-matching alternative (1). In my three case studies the predicted pattern is related to the connection between the independent and the dependent variables of the studies. The predicted pattern is compared to the results that I found when I analyzed the empirical based pattern, thereby explaining causation. To Yin (2003) explanation building is less desirable than pattern matching. The causal links in the predicted pattern that relates the independent variables to the dependent variables, and the causal links in the empirical based pattern are explained and compared. See Chapters 8 and 10 for details.

Time series (3) analyses means tracing changes over time in the same case (Yin 2003). The three Aker Stord audits on employee participation were carried out over three years, and gave me the possibility to do time series analyses. This strengthened the quality of this case study. Time series was not relevant for the two action research projects, mostly because of the time frame agreed on with the companies.

Developing logic models (4) generates a complex chain of events over time (Yin 2003). The events are arranged in repeated cause – effect – cause – effect patterns over time, whereby an independent variable at one stage becomes a dependent variable at the next stage. This technique was not used. I did not design my case studies to analyze dependent and independent variables changing place over time.

7.2.3.5 The interaction between research phases

Yin's recipe for case studies seems very structured and orderly. In many studies there are no clear stages. According to Stake: *"Issue development continues to the end of the study, and writing begins with preliminary observations"* (Stake 2000:453).

This was particularly the case with the Aker Stord study. The initial issues or research questions were developed, revised and widened throughout the different stages of the study. All three audits generated new questions. A new management concept, Business Process Reengineering – BPR (Hammer & Champy 1993) was introduced to the company between our first and second audit. This raised new and quite different research questions. The case study itself has now come to an end. However, the learning from the case study and the insights it generated is still raising new research questions in the continuing industry–research cooperation in the VS 2010 follow up program VRI.

The two action research case studies followed a more structured order. Before the Titania case I had arranged 10 similar conferences. Hence, it was easier to plan the Titania research phases in a structured order. The project period was also very short compared to the Aker Stord study.

Regarding the Serigstad case, the most important change after the project had started was a widening of the scope from the production process to the whole plant. This change resulted in some issues modifications. However, also this case study followed a structured research process. Especially the Aker Stord experiences made it easier for me to plan and to see what to expect in the Serigstad case.

7.2.4 Validity and reliability challenges

The most common criteria used to evaluate the quality of empirical research are:

- Construct validity by evaluating whether correct operational measures for the concepts being studied have been established.
- Internal validity (for explanatory or causal studies only and not for descriptive and exploratory studies) by evaluating whether a causal relationship can be established, whereby certain conditions are shown to lead to other conditions, as distinguished from spurious relationships.
- External validity by evaluating whether a domain has been established to which a study's findings can be generalized.
- Reliability by evaluating whether the operations of a study – such as the data collection procedures – can be repeated with the same results (Yin 2003:34).

I will briefly present these four criteria below and give some comments on each of them in relation to the case studies. However, I will give some additional critical comments to the questions of validity and reliability in my methodological reflections in Chapter 11 after the case study analyses. This will be based on my own experiences with the case studies and related to relevant literature.

7.2.4.1 Construct validity

To test whether the correct operational measures for the concepts and relationships being studied have been established can be challenging in social research where abstract concepts are not always directly observable (Remenyi et al. 1998, Yin 2003). This is the case for many of the variables included in my case studies, such as attitudes and meanings. To test construct validity two steps must be covered (Yin 2003):

1) The types of changes that are going to be studied have to be selected and related to the original objectives of the study.
2) It has to be demonstrated that the selected measures reflect the specific changes that are studied.

I will comment on both steps below:

1) Learning more about TQM practice was the initial objective of the Aker Stord case study. The research theme of balancing exploitation and exploration was not expressed explicitly in the first meetings with the company. However, both development logics were implicit in the way the TQM concept was defined and intended to be carried out in Aker Stord's TQM approach TKA (Total Kvalitet i Aker) and also in the EFQM model (EFQM 2003a). TKA did not differ between incremental and radical change, but included both and their integration. Balancing the two development logics evolved as a research theme in the initial phase of our cooperation with the company, and especially when BPR was introduced. So what I study is directly related to my research objective.

The changes that I study in the Titania action research case are meant to see whether the ED Conference promotes a kind of employee participation that support the synergistic combination of incremental and radical idea generation related to my main research question about the synergistic combining of TQM and BPR characteristics. Consequently, what I study in this case is directly related to my research objective. Likewise with the Serigstad case. The specific objective of this case is to further develop and test the Actor-Process Diagram as a means to radical process redesign and the combining of TQM and BPR characteristics synergistically.

2) In order to demonstrate and justify that the selected measures of these changes reflect the specific changes that I have selected to study, three techniques are available to increase construct validity: a) The use of multiple source of evidence, b) To establish a chain of evidence and c) Have the draft case study report reviewed by key informants and people who have detailed knowledge about the ideas, concepts and relationships being studied (Remenyi et al. 1998, Yin 2003). I will comment on all three below.

Multiple sources of evidence

In order to study employees' participation in improvement activities in TKA at Aker Stord several sources that gave information about the same activities were used. This is described in detail in the Aker Stord case narrative. The Aker Stord case is a longitudinal study where the data gathered through three audits in the years 1996, 1997 and 1998/99 play a central part. In order to develop the interview guides for the audits, a list of basic conditions and criteria for employee participation in TKA was developed. This list was mainly based on previous experiences of Aker Stord personnel and the RF researchers. The list was developed by the audit team in a seminar that was facilitated and prepared by the researchers, and further developed as we gained more experience and knowledge about employee participation in TKA through the audits and the analysis of the Aker Stord case. The list as a whole is not scientifically tested. This can be considered a weakness. I could not find any equivalent study in the literature. However, the different elements are tested out through the Aker Stord case and supported by theory (Howell & Dipboye 1986, Deming 1986, Juran 1989, Hammer & Champy 1993, Hammer 1996). The final version of the list is presented in Chapter 9, and used as a guideline and background for my proposal for the organization of employee participation in a synergistic approach to the combining of TQM and BPR, for the choice of innovation methodology and also for the development of new innovation tools.

My time series study is primarily based on interviews with managers, shop stewards and employees representing different parts of the organization. Managers, shop stewards and employees are asked to give information about the same improvement activities, thereby representing multiple sources of evidence.

In addition to asking different actors from different parts and levels in the organization the same questions, I collected a considerable amount of documents, statistics and reports and compared the data thus collected to the results from the interviews. This includes the EFQM evaluation, reports from Departmental Committee meetings, TKA presentations, monthly reports and statistics on improvement proposals and change projects reported in the TKA system, statistics from the company's Suggestion Box system, statistics showing the number that participated in TKA training in the different departments and statistics showing how many internal customer–supplier meetings that were carried out in the different departments. All this is described in detail in the case narrative. I made sure to use multiple sources of evidence wherever possible. E.g. I both asked employees about their department's engagement in improvement activities, and compared their answers with documentation from the improvement proposals database for the actual department for the same time period.

Whereas the Aker Stord case depends very much on interviews, I use participant observation together with registration of ideas from employees, meeting reports, company statistics and a few interviews to get facts about follow up processes in the two action research cases. Consequently, I also in these cases use different sources of evidence. I observe the idea generation and follow-up processes in both companies by being present and by facilitating the development processes. I am present in important activities and meetings, and I also gather company information and statistics about the changes that I observe. The interviews are mostly used to get in-depth information about the statistics, company documentation and activities in the companies' task forces that follow up prioritized innovation activities.

In addition I use data sources from outside the company. E.g. statistics from "Brønnøysund Registeret" in Norway are used to see whether returns and investments at Titania changed after the ED Conference. These kinds of data can indicate whether or not the innovation initiatives that the IRIS researchers documented as a result of the ED Conference move the company in the expected direction. An example of the use of data sources from outside the company from the Serigstad case is data from an external company that conducted a work environment survey. This company interviewed a number of employees at the same time as I did my participant observation and in-depth interviewing. For details, see the case narratives in Chapter 10.

Establish a chain of evidence

A "chain of evidence", from the initial research questions to the case study conclusions, is established for the case studies in the way the case narratives chapters are built up.

On the basis of the specific research questions and propositions for The Aker Stord case referred to above, the case starts with a definition of TKA, and the most important "essentials" and "elements" are described. On this background a theoretical definition of employee participation in TKA is developed. This definition is then translated into an operational definition covering three stages of employee participation in TKA, two basic ways of participating and five organizational levels for participating. Based on this framework all TKA activities are then sorted into one matrix for direct employee participation and one matrix for indirect employee participation. It is this framework that the data gathering initially was based on, and the interview guides that were used in the audits initially were built on. Remember, the research questions were matured and new questions emerged over time, and especially when BPR was introduced to the company. The link between the specific arrangements being studied and the data being gathered is described. After that follows a discussion and a summary of the case experiences in relation to the case specific questions and propositions that are being raised. The discussion following the data-analyses and the reports from the case study refers to the same framework, anchors the framework and the discussion in my main research question and highlights the findings that I find of importance to my research question and to my overall research objective.

TKA at Aker Stord includes a considerable number of tools and arrangements, and the content of TKA also changed over time. The action research cases focus more on one specific tool each. In these cases the chain of evidence is less complex, but follows a similar pattern. At Titania the most important question is: Can the ED Conference contribute to an approach that promotes the synergistic combing of TQM and BPR? If so, how? The objective of the ED Conference is to engage employees in developing a considerable number of incremental and radical innovation ideas. The ideas generated are collected, described and analyzed. The conference also includes some follow up procedures, and the further data gathering is planned and done in accordance with these. The data gathered are then analyzed in relation to the specific research questions for that case, my main research question and my overall research objective.

The chain of evidence in the Serigstad case is built up in a similar way. On the background of the case specific research questions described above, the Actor-Process Diagram as a tool for promoting employee participation in process innovation and the synergistic combining of TQM and BPR characteristics is tried out and analyzed. The data gathered measures results by documenting the incremental and radical ideas that the tool generates and how these are followed up in the company. In a similar way as the Titania case, the data gathered and analyzed are then related to the specific research questions for that case, my main research question and the objectives of my research.

Building a chain of evidence not only strengthens the validity of a case study. It also strengthens reliability. It contributes to making it easier for later researchers to follow the same case study procedures (Yin 2003).

Review of the draft case report by key informants and people with detailed knowledge

At Aker Stord the draft reports from all three audits, which are central in my Aker Stord case study and case analysis, were presented to "key informants" and others who had detailed knowledge about Aker Stord, such as staff members, managers and shop stewards, in half day seminars. These presentations also included an overview of the informants to show that the different parts and levels of the organization were represented. The results were discussed, and the adjustments to the reports that were agreed on in the seminars were incorporated in a second draft version. Two staff members who had detailed knowledge about the TKA process also read and commented on the second version before final reporting. The same two members of staff also read thoroughly through and commented on the whole case study chapter, including the discussions before I finalized it.

At Titania the draft report from the conference was sent to the CEO, the head shop steward and Titania's project leader. All comments were discussed, agreed on and included in the final version of the conference report (Bjørnevik et al 2005). The case narrative builds on and refers to this report. However, the case narrative in Section 10.1 was written three years after the action research was finalized, and has not been verified by the company. This is a weakness. This was also the case for the Serigstad project. The case narrative in Section 10.2 was written six years after the cooperation with the company came to an end. However, all company documentation that the case narratives build on and refer to can be found in my case study protocols.

7.2.4.2 Internal validity

The most relevant threat to internal validity is the occurrence of spurious effects (Yin 2003). In all three cases the causal and explanatory parts are very straightforward. E.g. in the Aker Stord case I want to evaluate the practical effects that different TKA methods has for employee participation. One of these is the effects of brainstorming and yellow post-its for idea generation. The activity this method creates is directly related to the method and managerial follow up.

In the Titania and the Serigstad case, I measured the innovation activity resulting from the ED Conference and the Actor-Process Diagram, respectively. The connection between the use of the tools and the activities they generated is very straightforward.

7.2.4.3 External validity

External validity is a challenge when doing instrumental cases studies (Stake 2000). The question is whether the results from these three single case studies can be generalized. Compared to surveys that rely on statistical generalization in theory building, case studies rely on analytical generalization (Yin 2003). However, theory must be tested by replicating the findings in other cases. Things might turn out quite differently in other companies in different contexts.

All the three case companies value employee participation. Efforts to use indirect employee participation in similar processes in companies where employee participation is less valued, can give different results. The companies are situated in a part of Norway where equality among people is highly valued, where the local communities are small and where people meet each other in many different settings independent of professional position. Cooperation between management and union members may encounter more difficulties in larger cities where the relations between the different actors outside work can be more distant. Situational factors like these make generalizations from case studies difficult.

To increase external validity I took some actions. In the Aker Stord case the first audit was replicated the two following years. This made my analyses less dependent on specific situational factors connected to a specific year, but it is still only one company in one specific setting. I also compared the Aker Stord results with the results from the national quality survey that we carried out (Hansen 1997). However, this survey is not representative for Norwegian companies. All companies were members of the Norwegian Association for Quality and Management. It is reasonable to believe that these companies have more knowledge about and interest for quality management than other companies. The Titania case builds on 10 preceding conferences arranged in other companies (Tønnessen 2000), and supports the experiences from these conferences. All these companies value employee participation, they participated in ED 2000 and VC 2010, and are situated in the southwestern part of Norway. The Serigstad case builds on the positive experiences from similar efforts to engage employees in process redesign at Aker Stord. However, the methodology that was used was not quite identical, the context was different and the scope of work very different.

Thus, even if some of the above mentioned factors can contribute to increase external validity, it is difficult to generalize. The question of external validity and generalization will be further critically discussed in Chapter 11 where I reflect on my own experiences with these case studies.

7.2.4.4 Reliability

The question here is whether later researchers who follow the same case study procedures for the same case will arrive at the same findings and conclusions (Yin 2003). I developed case study protocols for all three cases where all procedures and data gathering are well documented. For Aker Stord, all the interviews were both recorded on tape and notes inserted in the interview guides. For Titania all ideas that the ED Conference gave as a result were documented both on paper and electronically. The same was done at Serigstad for all incremental and radical initiatives that the use of the Actor-Process Diagram, including associated techniques and follow up gave as a result. For all three companies these data, along with all other company information that were gathered, were filed in the case study database. This includes information such as organization charts, product information, meeting reports, project proposals, surveys, work environment reports and other data of relevance for the case studies.

However, it is in fact difficult to do exactly the same studies over again. The companies have developed, one company does not exist anymore, management has changed, and business development recipes have changed. But similar studies can be arranged using the same methodology.

The techniques, presented above, which I use to demonstrate and justify that the measures I chose to use, reflect the specific changes that I study not only improve validity, but also the reliability of the information. Building the case narratives up in a way that presents a chain of evidence from the initial research questions to the conclusions increases reliability as well as validity (Yin 2003) I also used multiple sources of evidence/triangulation to improve both validity and reliability. The question of reliability will be further commented on in Chapter 11 after the case study analyses.

I will now give some additional comments on action research, which was the applied methodology in the development phase of the ED 2000 and VC 2010 projects. As I consider action research a variant of a case study design, I will not repeat the general discussion about advantages and shortcomings of case study design above. In the following I will highlight specific characteristics of action research and what I find to be the most important aspects of action research as a method.

7.2.5 Additional and specific aspects of doing case studies through action research

By taking the traditional case study researcher role of the observer, I had little influence on the TKA implementation and business development in the Aker Stord case. In Yin's (2003) case study variant the researchers have no control over the situation. In action research the researchers share "control" with the other participants and plan the actions together (Greenwood & Levin 1998). In the regional development coalition, the local business network activities and the two action research case studies that I refer to in this book, my co-researchers and I were active partners in the development processes together with representatives from the local community, the networks and the companies.

I find action research to be a suitable case study research strategy in a situation where "how" and "why" questions are central issues. Action research can give a good understanding of the actual TQM and BPR processes when we cooperate very closely with managers, employees, facilitators and shop stewards in the development processes over time. Action research has been defined in numerous ways, and many different action research practices have been proposed (Greenwood & Levin 1998, Kemmis & Taggart 2000, Coughlan & Coghlan 2002, Reason & Bradbury 2006). However, some characteristics are common. According to Greenwood & Levin (1998) action research is composed of three elements:

- Research
- Participation
- Action

If one of these three elements is absent, it is not action research (Greenwood & Levin 1998). Research shall generate knowledge. In my case I want to gain knowledge for the further development of the theory of exploitation and exploration, among others by developing means that promote employee's contribution to innovation activities. Action research is a participatory process. Everyone involved takes on some responsibility, and there is a commitment to the democratization of the knowledge generation process. In my action research as many as practically possible were involved in the development process, contributed to the generation of new ideas, solutions and insights, and had influence on decision making. Action research is action oriented and aims to alter the initial situation of the participants. It promotes democratic social transformation through social research (Greenwoood & Levin 1998).

In line with Greenwood & Levin (1998:4), I define action research (AR) as: *"Social research carried out by a team encompassing professional action researchers and members of an organization or community seeking to improve their situation. AR promotes broad participation in the research process and supports action leading to a more just or satisfying situation for the stakeholders. Together, the researchers and the stakeholders define the problems to be examined, co-generate relevant knowledge about them, learn and execute social research techniques, take actions and interpret the results of the actions based on what they have learnt".*

Action research is constructive social science (Kalleberg 1992). Among the many variants of action research that exist, my approach is in line with a practice Greenwood & Levin (1998) have called "pragmatic action research": *"We consider ourselves participants in change processes where democratic rules guide decision making. We bring to the table certain skills and knowledge, and other actors do the same, bringing their own capacities and experiences to bear on the problems"* (Greenwood & Levin 1998:11).

The action is a democratic process, and the authors equate democracy with the creation of arenas for lively debate and decision-making. This is in line with the concept of "Democratic

Dialogue" (Gustavsen 1992), which is fundamental to my methodological and theoretical approach. Consequently the pragmatic action research approach is well suited to my purpose.

7.2.5.1 Action Research as a co-generative process

"Action research (AR) is a co-generative process through which professional researchers and interested members of a local organization, community, or a specially created organization collaborate to research, understand, and resolve problems of mutual interest. AR is a social process in which professional knowledge, local knowledge, process skills and democratic values are the basis for co-created knowledge and social change" (Greenwood & Levin 1998: 93).

Conventional social research separates the phenomenon or process being studied, the research activity itself and the application of research results. In action research these elements are linked together in the same process (Greenwood Levin 1998). It is *"research in action"* (Coughlan & Coghlan 2002:222). Members of the system that is being studied participate actively. Such participation contrasts with traditional research where members of the system being studied are objects of the study (Coughlan & Coghlan 2002). In line with the principles of democratic dialogue where all participants have the same status in the dialogue arenas, no knowledge is superior to other knowledge (Gustavsen 1992). This includes research and professional knowledge. The co-generative process is fundamental to pragmatic action research and is visualized in Figure 9.

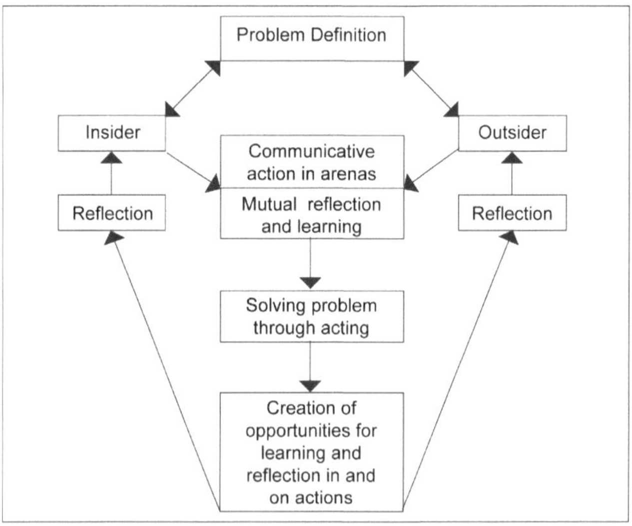

Figure 9 The co-generative action research model (Greenwood & Levin 1998:116)

In Figure 9 we see that action research starts out with a clarification of a research objective as a problem definition. In the action research variant "pragmatic action research" the research process is recommended to emerge from problems and demands outside academia (Greenwood & Levin 1998). At Aker Stord the problem was how to implement TQM in line with the company's industrial traditions. This was the "problem" that initiated my research activity that started in 1994. At Titania the problem was market challenges and how to engage employees in an organizational development process. At Serigstad the problem was related to productivity challenges and how to engage employees in developing new solutions.

Two main groups of actors participate in action research, insiders and outsiders. These are shown in the upper right and left side of Figure 9. The insiders are the owners of the problem, and are not any homogenous or egalitarian group. In my case we find employees, staff, shop stewards, managers and owners.

Insiders and outsiders cooperate in the development of "communicative actions in arenas" for learning processes, dialogue, idea generation and decision-making. Central in my theoretical approach to the further development of exploitation and exploration, including the additional tools that I have developed together with industry, are arenas for democratic dialogue and "mutual reflection and learning". Insiders and outsiders cooperate on the next phase in Figure 9, on "solving problems through acting".

How this was carried out in my case, will be described and analyzed in detail in the two action research case studies that will be presented later. In short, action researchers facilitated the development of new solutions and also participated in the development of these solutions, mostly by introducing methodology and asking critical questions. These solutions were implemented, and action researchers from RF/IRIS participated in the follow up phase mostly by data gathering and by coming up with new questions for further discussion and learning. An important role in my action research has been to promote the creation of opportunities for learning and reflection in and on action.

The reflection is done in close cooperation between outsiders and insiders in the arenas that were developed. Figure 9 seems to indicate that insiders and outsiders separate when they carry out reflection. They both do it together and they do it separately. In addition to reflecting jointly, insiders and outsiders reflect independently of each other. My reflection process when I write this book is an example of that. I have some discussions with people in industry when I write, but most of my reflection is done alone or together with colleagues from academia.

Figure 9 also shows that this co-generative process is a never-ending circle as long as the action research program is running. Reflection on learning and experiences continuously reshape the initial question in new problem definitions to be discussed and acted on in the same way.

7.2.5.2 My role as the "friendly outsider" in the co-generative process

The interaction between local expertise and outsiders' expert knowledge through a co-generative process where no knowledge is superior to other knowledge is a core feature of action research (Greenwood & Levin 1998). As a "friendly outsider" in a pragmatic action research approach I assumed several different roles. The most important ones are:

- Participate in the training of employees
- Support development processes
- Advise on the companies' running projects
- Import knowledge from outside
- Document, evaluate and report experiences
- Communicate knowledge

In other words, I took different roles in different phases of the development processes, as a teacher, a consultant, an action researcher, an observer, an analyst and a communicator. I will not go into detail on each of these roles. My colleagues' and my own role differed over time and depending on what was needed in the actual situation in order to get activities going. In some cases we were able to leave most of the activities to the companies. In other situations it was necessary to be heavily involved in training activities, planning and follow up.

7.2.5.3 Where does action research differ from consulting?

Gummesson (2000) presents four ways in which action research differs from consulting:

- Researchers are more rigorous in their inquiry and documentation. This book and the large amount of data that have been collected, analyzed and reported in ED 2000 and VC 2010 is one example of that.
- Researchers require theoretical justifications. E.g. the action research that my colleagues and I have conducted in ED 2000 and VC 2010 is founded on theory and contributes to the further development of theoretical propositions.
- Consultants work under tighter time and budget constraints. My research is part of a 50 year action research tradition, and has been carried out through a 5 year and a 10 year research program. However it does not stop here. The research continues in a new ten year program (VRI) that overlaps with, and represents a continuation of VC 2010.
- Action research is cyclical whereas consulting is frequently linear. Consultants engage, analyze, act and disengage. Action researchers, as in my case, gather data, feed it back to those concerned, analyze the data (as part of a co-generative process), plan action, take action together with those concerned, evaluate, which leads to further data gathering and so on.

Action research encounters the same validity and reliability challenges as described for case studies in general. Action researchers engage in both shaping and telling the story, and it can be an even greater challenge to give a valid, non-biased presentation compared to case study designs where researchers take the role of the distant observer. Coughlan & Coghlan (2002) find that the lack of impartiality on the part of the researcher to be the principal threat to validity for action research. How I met this challenge in my action research will be clarified in the presentation of my two action research case studies and further critically reflected on and summarized in Chapter 11.

To build scientific knowledge based on case studies is challenging and there are several pitfalls to be aware of. I will now discuss some of the most important criticism in relation to my methodological choice.

7.2.6 The case study – comments to common criticism

That case studies are valuable for hypothesis development is something both proponents and critics agree on. In my case I was in the initial phase of an action research project (Greenwood & Levin 1998, Coughlan & Coghlan 2002) with a development agenda anchored in international business development recipes. Contribution to the further development of the theory of balancing exploitation and exploration was on the agenda. Why, however, consider using a single case design, considering that this method has been criticized by many for several reasons (Abercombie, Hill & Turner 1984, Dogan & Pelassy 1990, Diamond 1996)? Flyvbjerg has summarized some of the criticism and what he calls the *"general wisdom"* about case study research in five, so-called "misunderstandings":

1) "General, theoretical (context-independent) knowledge is more valuable than concrete, practical (context-dependent) knowledge.
2) One cannot generalize on the basis of one individual case; therefore the case study cannot contribute to scientific development.
3) The case study is most useful for generating hypotheses; that is, in the first stage of a total research process, whereas other methods are more suitable for hypothesis testing and theory building.
4) The case study contains a bias toward verification, that is, a tendency to confirm the researcher's preconceived notions.
5) It is often difficult to summarize and develop general propositions and theories on the basis of specific case studies" (Flyvbjerg 2006:221).

I will use these "misunderstandings" to organize my discussion of these criticisms.

The first "misunderstanding" is the notion that theoretical context-independent knowledge is superior to concrete practical context-dependent knowledge. On the basis of his own research and literature studies Flyvbjerg (2007:221) argues: *"First, the case study produces the context dependent knowledge that research on learning shows to be necessary to allow people to develop from rule-based beginners to virtuoso experts... "*.

I find the case study approach well suited to produce the kind of knowledge that general theoretical and rule based knowledge can be built upon. Initially I wanted to learn more about TQM, the implementation processes in a Norwegian industrial context, how things worked and the reasons why. Alternatively, quantitative data could not give me the "expert" process knowledge that I needed when I started my research in ED 2000. Eventually, I wanted to use quantitative and qualitative methods complementary to increase external validity of my findings.

This leads me to the second of Flyvbjerg's "misunderstandings", the challenge of generalization and of hypothesis testing based on single case studies. My research makes me believe that social science knowledge can be valuable in contexts other than those where it is developed, but I do not believe in the idea that the transferability of knowledge from one setting to another is achieved by abstract, uncritical generalizations about that kind of knowledge. Transferring knowledge from one context to another seems to rely on an understanding of the contextual factors in the situation where the knowledge generation took place, and an understanding of the contextual factors of the situation where the knowledge is going to be applied. Neither do I believe that valid knowledge can only be produced by objective outsiders using formal quantitative methods that eliminate bias. That is not to say that the latter is not useful. In my case I use a quantitative study as a means to improve external validity of the findings from the Aker Stord case. Qualitative and quantitative methods meet different purposes, and can be used complementarily (Eisenhardt 1989, Yin 2003), as in my case.

However, can case studies alone be used as a basis for generalizations? Flyvbjerg (2006) rejects the idea that they cannot, among others by citing important and well known studies and scholars of science like Karl Popper (1959). It all depends on the case one is speaking of and how it is chosen. One example he refers to is the Luton case study – "*the affluent worker*" (Goldthorpe et al. 1968-69). In this case study the relationship between working class people obtaining middle class standards economically and class identity was tested. The study showed that the working class culture prevailed, and gave support to theories of class identity. The city of Luton was selected among others because of comparatively high wages and prosperity. It was not a representative case. It was an extreme case in Stake's (2000) terminology. So if class identification did not change here, it was considered less likely to change elsewhere were the wage level was lower.

Following this line of reasoning, both the Aker Stord case and my two action research case studies can be used to support generalizations. If for example company-wide employee participation in TQM and BPR does not work in a Norwegian industrial setting where ID holds a strong position and the mutual trust between the labor market parties is high, it is not very likely that this will be easy elsewhere. In other words, a proposition e.g. about ID as an important contextual factor for building exploitation and exploration capability synergistically, might be rejected if it did not turn out true in such a favorable setting regarding ID.

Another cited example is the science philosopher Karl Popper. A case study can be used for testing a theory. If just one observation does not fit with the proposition, it is "falsified" (Popper 1959). The theory must be revised or rejected. Accordingly Flyvbjerg concludes: *"One can often generalize on the basis of a single case, and the case study may be central to scientific development via generalization as supplement or alternative to other methods. But formal generalization is overvalued as source of scientific development, whereas the force of example is underestimated"* (Flyvbjerg 2007:228).

The Aker Stord case gives the insight necessary to evaluate theory and build theoretical propositions. By combining case studies with a survey, external validity is improved. I can hardly imagine that it could have been the other way around – starting with a survey. Among others, I did not have the insight necessary to ask the right questions. Beveridge may have a point: *"More discoveries have arisen from intense observation than from statistics applied to large groups"* (Beveridge 1985:95).

The third of Flyvbjerg's (2006) five "misunderstandings" is that case studies are most useful for generating hypotheses in the initial phases of a research process, whereas hypothesis testing and theory building are best carried out by other methods later in the process, derives from the above mentioned generalization misunderstanding. Flyvbjerg (2006) argues that case studies are useful for both generating and for testing hypothesis. As mentioned above, this is in line with Popper (1975) and also with Eckhoff (1975) who argues: *"Case studies are valuable at all stages of the theory-building process, but most valuable at that stage of theory-building where least value is generally attached to them: The stage at which candidate theories are tested"* (Eckhoff 1975:80).

The fourth of Flyvbjerg's (2007) "misunderstandings" is that case methodology maintains a bias towards verification (Diamond 1996). Case studies are often seen as less rigorous than quantitative methods, and leave more room for the researcher's subjective judgment. Campbell (1975) has shown that the opposite can be the case. He argues that a case study is as rigorous as quantitative research, but in another way. It can close in on real life situations and test views directly in relation to the phenomena as they unfold in practice (Campbell 1975). Geertz adds: *"The field itself is a powerful disciplinary force: assertive, demanding, even coercive"* (Geertz 1995:225).

According to Ragin (1992), the subjectivity problem and the bias towards verification are a challenge to all methods. He argues that the subjectivity can be significant e.g. when variables and questions for quantitative studies are being prepared and that this subjectivity will survive without being thoroughly corrected throughout the whole study. Compared to case studies, it is less likely that the objects being studied in surveys "*talk back*". The researchers are closer to the objects than in surveys (Ragin 1992). In line with Ragin (1992), Flyvbjerg (2006) concludes that case studies contain no greater bias towards verification of the researcher's perceived notions than other methods of inquiry, rather the contrary.

In the Aker Stord case shop stewards, facilitators, operators and managers at different levels followed the study very closely. This included planning, carrying out and result-analysis. I wanted employees to falsify and verify my findings. However, this created another challenge in relation to bias mentioned above. I got closer to the employees, it could be difficult to be critical, I might want them to succeed and I might be looking for results that favored my partners. I will return to this challenge when I discuss my experiences in Chapter 11.

Finally, a short comment on Flyvbjerg's fifth "misunderstanding" that it is often difficult to summarize and develop general propositions and theories on the basis of case studies. Many scholars (Goffman 1963, Nietzsche 1969, Rortye 1985, Peattie 2001, Flyvbjerg 2006) find it valuable to present the case in all its details for various reasons such as learning, the danger of losing essentials, and the possibility for readers to draw alternative conclusions. There is no contradiction between presenting the case in full and the development of propositions and theory. Researchers can do both and this is considered a strength of the approach (Eisenhardt 1989). The Aker Stord audit reports will be presented in full to give readers a good picture of the situation, the challenges and relevant solutions.

However, this strength of the case study approach can also represent one of its weaknesses. A case study typically contains rich but voluminous data. As regards theory building, the result of a case study can be a theory that is very rich in detail, but lacks the simplicity of overall perspective (Eisenhardt 1989). Furthermore, single case studies may not be sufficient to develop "theories in any grand sense" (Eisenhardt 1989:547). Here multiple studies might be needed. It may be more appropriate for developing more modest theories. Case studies can also be used as part of the accumulation needed to build more comprehensive theories (Eisenhardt 1989). E.g. the theory contributions from the Aker Stord case and the two action research cases are of a modest kind and can be seen as part of a cumulative process to the further development of the theory of exploitation and exploration.

7.3 Summary

In order to study the interaction between TQM and BPR I use two variants of case study design. In my initial study of Aker Stord I use a single case study design, in line with Yin (2003). In this case the researcher primarily takes the role as an outside observer. A key objective in this case is to describe and analyze an industrial company's efforts to combine TQM and BPR synergistically. For the development and testing of ways to combine TQM and BPR synergistically, I use an action research design in line with Greenwood & Levin (1998). In this design the action researcher is still an outsider, but shares the responsibility for the actions being studied with the insiders. I will now turn to real business life and the Aker Stord case.

8 Combining TQM and BPR synergistically, Aker Stord – a case study

When my cooperation with Aker Stord began in 1994, the company was one of Norway's leading suppliers of offshore floaters and deck structures for offshore installations. The company had about 1800 employees and used a considerable number of subcontractors. In peak periods more than 5000 persons worked inside the yard. The company was established in 1983 and is located at the island of Stord between Bergen and Stavanger at the west coast of Norway.

The Aker Group, that Aker Stord is part of, started their total quality journey in the autumn of 1993. The TQM initiate came from the concern, and was named "Total Kvalitet i Aker" (Total Quality in Aker) – "TKA". At Aker Stord it was given the name "Total Kvalitet på Aker Stord" (Total Quality at Aker Stord). In the following the TQM process at Aker Stord will be referred to as "TKA".

In this chapter I will first describe Aker's TQM approach "Total Kvalitet i Aker" (TKA), how it was introduced at Aker Stord and how it was organized. Aker Stord has long lasting traditions for industrial democracy (ID) and had a high focus on involving employees in TKA. Consequently, the company expressed very clearly that they expected employee participation to become an important issue in the cooperation with research. I will therefore pay special attention to how direct and indirect employee participation in the process was arranged.

Secondly, I will present the initial audit of employee participation in TKA that researchers from Rogaland Research carried out together with staff in 1996. This includes the observations and recommendations from the audit. I will present and comment on the EFQM Self Evaluation study at Aker Stord that was carried out at the same time as the first audit. Then follows a description of how BPR was introduced in 1997.

Furthermore, two follow-up audits of employee participation in TKA in the years 1997 and 1998/99 will be presented and commented on regarding important changes. The results from the three audits will be presented in full in order to give the reader an overview of how the situation evolved over time and in relation to important events, and to show the most important challenges that were identified and how the audit teams and the company proposed to meet these challenges.

The individual observations and recommendations will not be discussed and analyzed during the audit presentations, except for a few comments for clarification reasons. Important challenges, changes and actions will be analyzed in a section of its own after the audit presentations. In that section I will analyze and comment on Aker Stord's efforts to combine TQM and BPR synergistically in their specific ID context. A few selected observations, development directions and actions from the audits that relate to my theoretical discussions and my research objective will be picked out and commented on. As part of this analysis and to improve external validity I will compare the experiences from Aker Stord with results from a national survey on quality management. At the end of the chapter I will discuss the Aker Stord experiences in relation to my research objective and research question, and identify some shortcomings in existing practice that need to be dealt with in the further development of the theory and practice of combining forms of exploitation and exploration synergistically.

My methodological approach to the Aker Stord case study was discussed in Chapter 7. Some further details about the practical use of the different elements of the case study design in this specific setting will be given and shortly commented on as I proceed. The results and experiences from this case study became very important for the direction of my action research the following seven years and for my approach to combining TQM and BPR synergistically.

8.1 Start-up of TKA at Aker Stord and timeline

The TKA program succeeded other programs focusing on improvement at Aker Stord. All these programs have contributed to the development of the company, and many of the routines and activities that will be described below were started as part of previous development programs. In other words, these programs should not be seen as individual and isolated activities that are finalized when a new project starts, but rather as part of the ongoing development of the Aker Stord organization.

The TKA program was developed by the former Aker Group in collaboration with 3M, and Aker Stord developed its own variant of TKA. The process started as a result of the so called "Rationalization project" that was initiated in 1991 and aimed at quality improvements. In April 1993 "TKA at Aker Stord" was established, and a project manager was appointed. During the first year the Aker Stord TKA approach was discussed and an implementation plan was developed and assessed. A company-wide program for basic training in TKA was developed, and TKA materials were prepared. A system for planning and support of improvement projects was also developed and specific improvement projects were initiated.

The formal implementation decision was made by the top management in May 1994. During the following month, June 1994, the top management accomplished a basic TKA training program. Thereafter, TKA training was to be provided for everyone. At the turn of the year 1996/97, some 65% of Aker Stord's employees had been given basic TKA training. A fundamental principle of the implementation plan and this training program was that managers were responsible for the training of those who reported to them. In practice, the training was conducted in collaboration between line managers, the appointed TKA Project Manager and staff.

The organization of the TKA process at Aker Stord was simple. The only additional organizational element was the TKA project manager. It follows the basis organization. However, the dual organizational structure at Aker Stord, comprised of the basis organization and the building project organization, provided a challenge. The basis organization is as it appears on the formal organization chart. The building projects are organized as project units. The challenge is that the building projects are staffed by persons from the basis organization that report to line managers in the basis organization who they seldom see. Their daily work is managed by project managers.

Until autumn 1996, TKA had its own project manager who acted as an internal facilitator for the process and provided assistance to management and trade unions. When the TKA manager left the company in 1997, he was not replaced. His tasks were divided among several persons in the organization, and one person in the top management assumed ownership of the strategy process. TKA became an integrated and basic element of the further development of Aker Stord. Consequently, TKA was included as a basic element of Aker Stord's strategy and action plans. A time line for important events in the Aker Stord case study is given in Table 4.

1993	TKA introduced to Aker Stord. Project manager TKA appointed. TKA implementation plan developed. TKA training program and material developed.
1994	TKA plan decided on. Initial TKA training for top management. Initial status mapping by researchers
1994-1996	Training program for employees. 65% received training.
1996	First audit of employee participation in TKA. EFQM self evaluation. Program manager TKA left the company. Not replaced.
1997	New top manager. BPR introduced to Aker Stord. Aker Stord became part of the main production process in the concern. Aker Stord's position as an independent company reduced. Implementation plan for BPR developed and decided on. Three employees received BPR facilitator training. Start up of three BPR projects. Second audit of employee participation in TKA.
1998	Initial BPR projects finalized. New ones started
1998/99	Third audit of employee participation in TKA
1999	Case study cooperation with researchers finalized

Table 4 Time line for important events in the Aker Stord case study

All these events will be described below. I will now give a brief description of the TKA philosophy and strategy with a special focus on employee participation and its role in the TKA effort.

8.2 TKA: The basic philosophy and strategy. The importance of employee participation

In the Aker Group, TKA was an important means to realize the group's vision, business idea, strategic goals and basic values and to ensure that focus is constantly kept on customer needs and customer expectations. The foundations of TKA were laid down in the Aker document "Det handler om ..." (It concerns ...) (Aker Oil and Gas Technology 1994). The following is an extract from this document:

> "It concerns ..." is the foundation of the improvement process. It provides the direction in which we are moving, the values, personal attitudes and principles in which we believe, and the concepts of which we shall have a common understanding. The purpose is to create actions and favorable conditions for a continuous improvement process in all our establishments" (Aker Oil and Gas Technology 1994:3, my translation).

Total Quality in Aker (TKA) was defined as follows:

> *"Total Quality is a systematic process that builds up a customer and improvement-focused culture with participation from all employees.*
>
> *This means that each and every one of us – individually or in collaboration – always need to be on the look-out for improvements in our relationship with our customers and society in general.*
>
> *Continuous improvements must also apply to all our internal work processes and all our employees"* (Aker Oil and Gas Technology 1994:8, my translation).

"Five essentials" should exist if such improvement processes are to give results:

> *"We define quality as fully complying with the customer's expectations.*
>
> *We measure quality through customer satisfaction rather than self satisfaction.*
>
> *The target is to comply with expectations 100 % of the time.*
>
> *All employees are committed to achieve quality through prevention and improvement projects.*
>
> *The management's commitment and obligation drive the improvement effort"*
>
> (Aker Oil and Gas Technology 1994:9, my translation).

In order to achieve these five essentials and realize the TKA philosophy, a further "8 elements" are described in Aker Stord's manual "Total Quality in Aker" (Aker Stord 1994:55-103) under the following headings:

- *"Management of the improvement process*
- *Organization of the improvement process*
- *Training for the improvement process*
- *Understanding of the customer's expectations*
- *Measurement of compliance with the customer's expectations/measurement of improvements*

- *Communication about the improvement process*
- *Initiation of improvement projects and measures*
- *Annual plan for the improvement process"* (Aker Stord 1994:55-103, my translation)

These eight elements were seen as essential elements that had to be given substance and activities.

As it appears from this brief summary of Aker Stord's TKA philosophy and strategy, employee participation and commitment towards the process are of fundamental and basic importance. The process depends on this. This is also in line with the doctrine of the dominant schools of thought within the TQM tradition (Tønnessen 1992).

In order to further emphasize the importance of employee involvement, Aker operates with the concept "personal quality" as being essential for all other quality. Personal quality is defined as:

> *"Personal quality is the sum total of your attitudes, competence, ability to cooperate and reliability.*
>
> *Your personal quality emerges through your contribution to the value creation. Personal quality is a prerequisite for all other quality"* (Aker Oil and Gas Technology 1994:8. My translation).

In other words, the success of the TKA was considered to be dependent on competent employees who both wish to participate and have the ability to participate in the continuous improvement of every single function in the business, and on managers who realize the importance of such participation and actively facilitate it.

The company's need for an overview of how far they had reached in this area after a year's practical TKA effort was the background for conducting an extensive audit in 1996 at Aker Stord in cooperation with the RF researchers. This audit and the 1997 and 1998/99 audits constitute most of the data basis for the Aker Stord study, along with the previously mentioned status mapping and the EFQM self-evaluation. To carry out the Aker Stord case study, including the audit on employee participation in TKA, and as an element in the building of validity, I first needed a clear definition. I will now define and describe the use of the term employee participation in TKA and also the reasoning behind the chosen definition.

8.3 Interpreting "employee participation in TKA"

TKA is defined above. The definition is quite wide, and employee participation can in principle comprise most of what has happened at Aker Stord. In order to make the concept manageable, a more detailed interpretation rooted in practical reality had to be provided. I needed a clear definition. Hellevik (1994) distinguishes between a theoretical and an operational definition. I start out with a theoretical definition.

8.3.1 Theoretical definition

One way of limiting the topic could be to start by drawing a line between operational and development tasks (Pålshaugen 1991), and anchoring TKA in the development part. It may be reasonable to draw such a demarcation line in several industrial environments, where operation is the main task and development is something that occurs "outside" what personnel considers as their daily work. On the basis of preliminary discussions with Aker Stord personnel, I found that such a demarcation line could represent some problems in some staff environments, as for many staff personnel development is one of their main tasks and represents their "work". This division of terms could also prove problematic for purely TKA tactical reasons. One of the main challenges of the TKA process was to develop an organization where development is an integrated part of individual employees' day-to-day duties and integrated in the production of goods and services. Such a demarcation could further consolidate an attitude considered to be held by some of Aker Stord's operators, namely that "operation" is the work. Development is something they do when they are not working, or something that is someone else's responsibility.

Furthermore, if Aker Stord should be able to establish a practice with broad participation in improvement activities, they found it tactical to give work of this nature the status of "operation" and value creation. In addition, the aim was also to operate with an extended customer concept, where elevating the status of internal customers was considered important. A further emphasis on the demarcation between operations and development could play a part in cementing the low status of "internal customers".

Consequently, we decided to let the demarcation between the practical performances of specific work duties aimed at an external end customer and developing more appropriate ways of working, remain somewhat fluid. This, however, does not mean that the concepts of operation and development tasks cannot be useful in other contexts or that "employee participation in TKA" cannot be precisely defined. After some discussions with Aker Stord personnel, the theoretical definition of "Employee participation in TKA" became:

Employee participation in TKA is the participation of employees in specific TKA activities, their knowledge about these activities and their attitudes towards these activities.

Even if it could be said that TKA is more comprehensive than this and is meant to permeate the minds and conduct of individuals in their day-to-day work life, by such a definition we considered it possible to capture the commitment of employees to the most important activities that Aker Stord had initiated in order to improve such conduct and ways of thinking.

8.3.2 Operational definition

Participation in TKA concerns all employees. The target group for the audit was employees without line responsibility. Line managers were only involved in their capacity of being employees in line with the other employees in the improvement process, or in their role as prime movers and facilitators for employee participation in the TKA activities. Therefore, even if focus has been kept on employees without line responsibility, the line managers' TKA activities and attitudes were considered an important condition for employee participation in line with TQM philosophy (Juran 1989, Deming 1986).

On the basis of the theoretical definition, the study was limited to a set of TKA activities, which are listed and explained below, plus the employees' knowledge and attitudes towards these activities. However, no total overview of these activities existed at Aker Stord, and the information given as to what was included in TKA varied, depending on whom we were talking to. The overview had to be made, and represented what became the operational definition (Hellevik 1994) for the audit. Due to the comprehensive character of the topic, this definition became somewhat extensive and detailed by necessity.

In order to arrive at this overview (the operational definition) I initially differentiated between various stages of involvement in the process.

8.3.2.1 Three stages of participation in the TKA process

Employees were involved in the TKA process:

1) Through participation in TKA training and being given TKA information.
2) Through participation in problem identification and generation of ideas.
3) Through participation in performing the improvement activities, implementation and follow-up.

In this context, the stages were anchored in the target, notably improvement activities. First it had to be ensured that the employees were informed, were given competence and that arrangements were made to motivate for improvements. For the sake of brevity, I call this "information/training". Then challenges need to be identified and solutions outlined, hereafter called "idea generation". Thereafter the solutions needed to be implemented and followed up. I call this "Implementation/follow up". These three stages could be divided into several categories. In practice, the different ways of becoming involved are in many cases overlapping and take place in the same fora. This division is done on the basis of the TKA practice at Aker Stord at the time of the audit, and will be described in further detail later. Employee involvement can take place in two basically different ways (see Section 9.2).

8.3.2.2 Two ways to participate

- Direct participation
- Indirect participation

Regarding the second one, I could have differentiated between representatives appointed by the management to participate in certain activities/meetings/groups and representatives elected among the employees. I found this unnecessary. According to Aker Stord personnel it rarely happen that employee representatives are appointed by the management alone. In the event that the management asks employees to attend on behalf of the relevant work group, it is common practice to involve the unions in selecting representatives. Consequently, we decided to focus on indirect participation through representatives elected by the employees and disregard management-appointed representatives as a separate category in the further analysis.

8.3.2.3 Five organizational levels of participation in TKA

TKA participation can take place at various levels of the organization. Together with Aker Stord personnel we decided to divide the participation into five levels and limit the audit to evaluate TKA activities that had been planned. This means that we did not include the influence individual employees have on enterprise development and on self-initiated improvement initiatives taking place under their own control without involving others to any mentionable degree. We also agreed to disregard directly work-related initiatives suggested to management, union representatives and/or staff in areas that affect the employees' own and possibly other people's work situation unless this is included in planned TKA activities.

This is a delimitation that excluded parts of Akers Stord's total quality philosophy, which stated: "*This means that each and every one of us – individually or in collaboration – always needs to be on the look-out for improvements in our relationship with our customers and society in general*" (Aker Olje og Gass 1994:8 my translation). We decided nonetheless to limit the audit to make it easier to handle. An audit that also covered all minor and major improvements that took place outside the more formal TKA activities was considered too time consuming and comprehensive in relation to the objective of the audit.

This does not mean that the traditional ways of conducting improvements, such as taking initiatives without involving others or by proposing ideas to their own manager, union representative or staff outside the TKA activities were not considered equally important. Employees should continue to strive to further develop these forms in line with Aker's quality philosophy. A high degree of self-determination at work was an important development objective. At Aker Stord it was considered important that managers and employees cooperated well with each other and that problems were solved at the lowest possible level with the least possible use of resources. It was the planned and systematic approaches to working with improvement that was of interest for the audit. The activities that were included in the TKA process should however stimulate making arrangements for further development of the traditional ways of working with improvements together with management.

In the discussion of what to include, both Aker Stord personnel and the researchers considered autonomy at work and experience with initiating improvement initiatives and changes to their own way of performing the work could have positive effects on the individual employees' ability to participate constructively in TKA activities.

Despite the fact that some people at Aker Stord questioned the actual value of the TKA process, there was a general agreement among Aker Stord personnel that the above-mentioned traditional ways of working with improvement are insufficient. There will always be many issues that affect several people and need to be assessed by others and in relation to the enterprise's targets and strategies. TKA activities should be an opportunity to realize the possibilities of having employees participate more actively in a dialogue about and a choice of development initiatives that are of such a nature that they affect more people and more extensive change initiatives. It is the more systematic and dialogue-based part of the improvement effort that this audit focused primarily on. However, some very individual oriented tools like the suggestion box system were also included in the audit. I will now describe the five levels of employee participation in TKA that were used to design the audit.

Level 1. Employee

At the lowest organizational level we find individual employees. As mentioned above, in Aker Stord personal quality was considered a prerequisite for all other quality. Personal quality (the sum total of an individual's attitudes, competence, ability to cooperate and reliability) was something that managers and employees should develop collaboratively. In this context we focused the audit on two initiatives, the "Appraisal Interview" and the "Suggestion Box".

The appraisal interview is a means available to managers and employees to achieve a more formal and, ideally, a more evenly balanced dialogue about initiatives that may further the employees' professional and personal development in accordance with the enterprise's vision, goals, strategy and value basis.

The appraisal interview was not a TKA initiative, but was introduced before TKA. However in line with theory (Deming 1986) the TKA process starts in some areas in the enterprise and then gradually spreads outwards until the process and the philosophy behind it eventually constitute a central foundation for all the long and short term decisions that are taken at all levels of the enterprise. The appraisal interview, if it is conducted in accordance with the intentions, could be the most important means for planning systematically the further development of individual employees' personal quality through generating improvement initiatives and following them up. The appraisal interview could be viewed as the "improvement group" (manager and employee) for developing personal quality.

Also the suggestion box arrived before TKA, and is the means available to the employees for presenting suggestions for improvements on a current basis. The suggestion box is a well-structured system with detailed procedures for follow-up, feedback and rewards for the received suggestions. Most often the suggestions are provided by individuals, but also to some extent by groups of employees.

All suggestions are forwarded to a suggestion secretary who records them, informs the senders that their suggestions have been recorded and forwards them to local suggestion committees, and in some cases to local suggestion secretaries. For each suggestion, a recommendation is then prepared locally for the local line manager. If the suggestion proposed is implemented, the size of the monetary reward for the person making the suggestion is calculated according to set regulations. The local line manager may reject the suggestion, and then it is up to the person making the suggestion if he/she wants to appeal. Otherwise it is the manager's responsibility to ensure that suggestions that the committee has decided to implement and reward are realized.

Level 2. Assignment/job

Level 2 is made up by individual assignments/orders/jobs. Relevant types of improvement activities at this level are "pre-job meetings", "during job meetings" and "after job meetings". Best known and most used are "safe-job analyses" (SJA), which is a common practice in the oil and gas industry. Common for all of these meetings is that the work team in a systematic way, often with a process chart and lists of keywords, review the different phases of a job to be done in order to weed out weaknesses and plan improvements. Employees are given the opportunity to ask questions, point out problems and present suggestions for doing things better. An "after job meeting" is conducted to ensure that experiences are transferred from one job to the next.

The "improvement team" conducting the job being assessed in such a meeting is not a permanent improvement group, but consists basically of the people who are going to carry out the assignment. Such teams may also be representatively composed. In such cases local union representatives may meet on behalf of the employees.

Level 3. Organizational unit

Level 3 is the organizational unit where employees belong in the organization. As an "improvement group" this unit is of a more permanent nature, and is not dependent on individual jobs that come and go. Any changes to the composition of this group appear as a result of organizational changes.

At the time of the first audit, Aker Stord was divided into five main areas. The size and the further hierarchical division into sections, departments and groups varied between the different areas. A further division into sub-levels would have been possible, but we decided not to do that. In the analysis that follows, we did nonetheless touch onto the different levels and the importance of a continuous management commitment along the entire line.

At this level, quality circles are a common TQM practice. There are several variants of this practice, and various enterprises use different names for this kind of activities (Bergman & Klefsjø 1994). Aker Stord does not have quality circles in their traditional form, but stimulates its employees to present suggestions for improvements through brainstorming sessions and group conversations.

Improvement activities at this level will mainly be aimed at existing work practices in the department, and to a lesser degree at the improvements of individual tasks. Also here it will be possible to participate directly (for example through brainstorming sessions) or indirectly through elected representatives (e.g. in the Departmental Committee – DC).

Level 4. Process

In Chapter 4, process was defined as a set of continuous activities, which are meant to result in a particular product/service for a defined customer (internal or external). This term may be applied to all types of interlinked sets of activities/jobs that create value for a customer. Therefore, a process at Aker Stord should as a minimum have a process owner, a customer, consist of a set of activities that are interlinked and have a defined value creation for the customer as its aim (Aker Stord 1994).

Some processes involve few workers and take place mostly within the limits of the department or in the groups people belong to. Other processes may span over the entire enterprise. The difference between level 2 (assignment/job) and level 4 (process) is that the last-mentioned category often has a formal and a more permanent character while individual assignments/jobs can change considerably. However, a job can be defined as a process, and a process can encompass several jobs and vice versa. At main process level we are talking of processes that are so large that it may be difficult for individual employees to gain an overview of them. The process may affect several areas or departments, and can be subdivided into a hierarchy of sub-processes. Otherwise such cross-sector processes at a higher level have the same characteristics as processes of a lesser scope. There were no specific arrangements for direct employee participation in process innovation when our cooperation with Aker Stord started. Regarding indirect employee participation, process changes were discussed at existing arenas for employer-employee cooperation. Shop stewards and some individual employees were eventually invited to participate in process redesign tasks forces when BPR was introduced to the company.

Level 5. Company

Here, participation at enterprise level means participation in making overruling decisions that concern the entire business. These may be issues related to the business idea, vision, main objectives, overarching strategy, planning processes and company values. Employee participation in decision making processes at this level seems to be very limited or insignificant in most enterprises. If there is any participation, this often takes place through representative arrangements, e.g. board representation, union representatives' participation in work councils (WC), Work environment Committees (WEC), internal steering groups for development processes (such as for TQM processes) or corresponding bodies. With the exception of the WEC and the WC, union representatives are a minority in these bodies. At Aker Stord all these kinds of arrangements exist.

Therefore, at this level it appears as if employee participation in enterprise development depends on representative arrangements, possibly in combination with a process whereby employees are invited to participate through working in groups at different company levels.

8.3.2.4 Employee participation in general

TQM places significant emphasis on a broad and constructive dialogue between employees and managers on improvement. A broad participation in the selection of initiatives, competence building and a more decentralized control and decision-making structure is emphasized. External expert solutions are played down in favor of internal solutions based on employee insight and experience. If external experts are used, it is mainly in the role of external process facilitators, and to a lesser degree as experts on specific solutions.

What characterizes the TQM tradition is the considerable emphasis that is placed on group work in one form or another (Bergman & Klefsjø 1994). Many of the activities that are part of TQM may generally be viewed as various forms of improvement work in groups. This is an important starting point for the following analysis of the employees' conditions for actual participation in TKA. Several of the activities that have already been mentioned, from the appraisal interview to the WC and WEC at enterprise level, can be viewed within such a framework. However and as mentioned previously, not all ongoing improvement is formalized or taking place in groups. Think for example of the suggestion box system.

8.3.2.5 TKA activities at Aker Stord

In the following I will list the practical activities that we chose as our basis when assessing employee participation in TKA. The listing will follow the division into the three stages that were made above: 1) Information/training, 2) Idea generation and 3) Implementation/follow up. Again, I will point out that self-initiated individual improvement initiatives have been excluded from the analysis. On the other hand, a few agreement-based joint collaboration bodies have been included, as these have important roles in the TKA process. In particular, this applies to the Departmental Committee (DC) and the Works Council (WC). The TKA activities are listed below.

(1) Information/training

- Via the line in local meeting forums
- Written information
- Basic training in TKA
- Other training/information provided in TKA seminars and TKA activities
- TKA information in meetings where all employees are gathered

(2) Idea generation

- The suggestion box
- Appraisal interviews
- Group conversations
- Pre-job, during job and after job meetings

- Customer-supplier interviews
- The Departmental Committee (DC)
- Brainstorming sessions at different organizational levels
- Process improvement teams
- Audits
- The Work Environment Committee (WEC)
- The Works Council (WC)

(3) Implementation/follow up

- Improvement projects
- Implementation of initiatives/follow-up

(4) Summarizing the three stages

By summarizing the different activities that we are linking to the TKA process after the three stages of involvement and the five organizational levels, we get the following matrices for direct and indirect participation in Tables 5 and 6, respectively.

When summing up the TKA involvement in Table 5, we see that some activities are overlapping and that two boxes are empty. There is no direct employee participation in idea generation at the company level, and there is no direct participation in training/ information at process level. In other words, employees do not participate directly in idea generation at the upper company strategic level. Neither do they get any systematic training, information or ability to participate directly in process identification, evaluation and improvement. These two observations became important for my action research, and will be further discussed when I present my framework for a more synergistic combination of TQM and BPR and my action research case studies.

	Information/ training	**Idea generation**	**Implementation/ follow up**
Company	Written material. TKA meetings for all employees		Implementing improvement proposals
Processes		Process improvement teams. Customer-supplier interviews	Implementing improvement proposals
Department/ group	Info in group conversations. Info and TKA training in departmental/ group meetings and seminars. Written material	Group conversations. Brainstorming sessions. Customer-supplier interviews	Implementing improvement proposals

	Information/ training	Idea generation	Implementation/ follow up
Job/task	Pre-, during – and after job meeting	Pre-, during – and after job meetings	Implementing improvement proposals
Individual	Appraisal interviews. Suggestion box	Appraisal interviews. Suggestion box	Implementing improvement proposals

Table 5 Direct employee participation in TKA activities

Table 6 sums up the corresponding indirect participation in TKA activities. The activities where employee representatives participate as ordinary employees are not included in Table 6. Accordingly, the boxes at the individual level are empty. WC and WEC at company level and DC at department level do not take part in implementation processes but follow up on the activities in meetings. Employee representatives have no role in relation to pre-, during – and after-job meetings. In cases where employee representatives participate in the practical implementation of improvement proposals, it is in their role as ordinary capable employees to perform the necessary actions.

	Information/ training	Idea generation	Implementation/follow up
Company	WC, WEC	WC, WEC	WC, WEC
Processes	Process improvement teams	Process improvement teams Revisions/audits	WC, WEC, DC
Department/ group	DC. Customer– supplier interviews	DC. Customer–supplier interviews	DC
Job	Pre-, during – and after job meetings	Pre-, during – and after job meetings	
Individual			

Table 6 Indirect participation in TKA activities

Both the direct and the indirect participation tables above show that there were several arenas and TKA tools for direct and indirect employee participation at Aker Stord. The audit should evaluate how successful the company had been in implementing and using these means. Before I present the audits and the results, I will briefly describe how the DC is used as a coordinating body for many of these activities.

8.3.2.6 Departmental Committees (DC) as the coordination body for local improvement activities at Aker Stord

The dominating union, called "Fellesforbundet" (The General Workers Union), is part of the Norwegian Federation of Trade Unions (LO). The formal basis for DCs is found in the main agreement between the LO and NHO (LO & NHO 2006). In DCs, like in the Works Council, the parties shall co-operate *"to work for the most efficient production possible and for the maximum wellbeing of all who work at the enterprise"* (NHO & LO 2006: 47).

When the question of how to organize the TKA improvement activities was raised, the union was strongly in favor of using the arrangements for cooperation that already existed, and these committees were the union's choice. When Aker Stord first started their formal TKA process, they consequently found that the necessary structure and many routines for lower level improvement activities were almost in place.

The DCs in the production units at Aker Stord have the following composition: The head of the department, two foremen, one shop steward from one of the work groups, the safety delegates and the union's productivity shop steward (who participates in all the different departments' DC meetings). The DCs usually meet once a month.

Proposals to a DC may come from different places. They can come from the Works Council (WC), the Work Environment Committee (WEC) or the management group in the form of topics/issues that they want the departments to discuss and do something about. The topic of the month is one way of doing this. How to prevent falls from elevated places is an example of such a topic. Furthermore, proposals can come from group discussions, brainstorming sessions, from other departments and from individual persons such as supervisors, heads of departments, shop stewards and operators who report proposals to members of the DC. Input and ideas may also come from the customer-supplier interviews between the departments. In this way the DCs are connected to and integrated with many of the improvement processes that are taking place at other places and at other levels in the company. See Figure 10 for an overview of the DCs' role and link to other arenas in the TKA process.

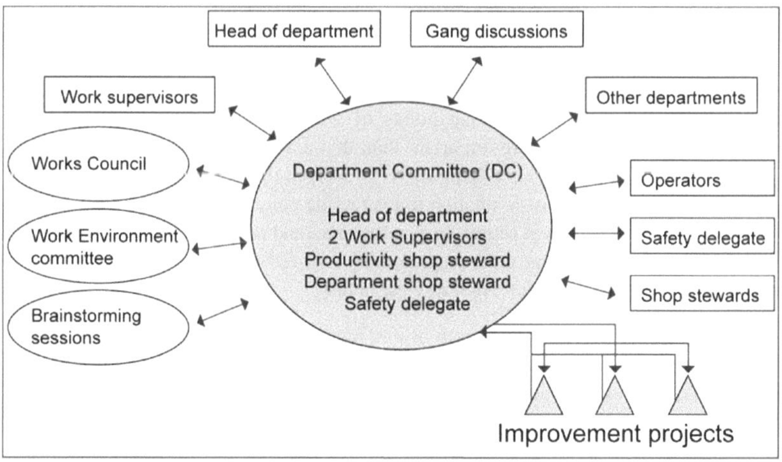

Figure 10 Departmental Committees as the arena for coordinating local improvement activities

8.4 The practical carrying out of the audits "Employee participation in TKA"

The first audit was specified as one of the initiatives in the annual plan for 1996. The practical carrying out of the audit was planned following the case study methodology (Yin 2003). In order to carry out the audit, an audit team consisting of three representatives from staff at Aker Stord and two representatives from RF was established. One of the representatives from Aker Stord was appointed audit leader.

First, a set of basic conditions and criteria for employee participation in TKA was developed in accordance with my research question and reflecting my initial case study propositions. This initial list was used as a guideline for the development of the interview guides for the audits. The list was mainly based on experience gained by Aker Stord personnel and the RF researchers, but also supported by theory (Howell & Dipboye 1986, Deming 1986, Juran 1989, Hammer & Champy 1993, Hammer 1996). The list was developed by the audit team in a seminar that was facilitated and prepared by the researchers, and further developed as we gained more experience and knowledge about employee participation in TKA through the three audits and the analysis of the Aker Stord case. The final version of this empirically based list is shown in the presentation of my framework for employee participation in Chapter 9.

Before the first audit the audit team also made a list of the information needed in order to evaluate employee participation in the identified TKA activities and a list of the personnel categories we needed to interview to ensure that the information we gained was as representative as possible. Based on this, three interview guides were prepared, one for employees, one for managers and one for union representatives. I made the first drafts, which were then discussed and revised by the audit team before each of the three audits. The interview guides contain open questions. The interviews of ordinary employees were conducted in groups consisting of two or three. This was done to provide the opportunity to discuss matters and extract contrasting points of view on which there was not complete agreement. I considered this more important than that any personal points of view were not expressed due to the presence of colleagues. However, my impression was that there was a great deal of openness and trust in relation to expressing points of views among employees at Aker Stord. The order book was satisfactory at the time, and people expressed a feeling of job security. Managers and central union representatives were interviewed individually. Some foremen and local shop stewards were interview in pairs.

All interviews were carried out by a team of one researcher and one company representative. The company representative operated as the leader of the team and asked the questions. The researcher asked follow up questions, took notes and taped all the interviews. Afterwards, all interviews were transcribed and printed out. A total of 75 persons were interviewed in the three audits, representing the different departments in the company, different levels and shop stewards.

In addition to the interviews, a considerable amount of documentation was collected in the protocol for all the three audits. The documentation contained meeting reports, presentations, statistics on improvement proposals and projects, statistics on TKA activities like customer–supplier meetings, statistics from the suggestion box and the organization chart.

The audit team condensed all the results into summaries, observations and improvement proposals. The reporting format followed the company's tradition of presenting audit results. All the informants and other shop stewards, staff personnel and managers were invited to half day seminars where the results were presented and discussed. Some minor changes were made. The results are summarized next.

8.5 Results from the first audit. Employee participation in TKA at Aker Stord in 1996

The presentation of the audit results is organized according to the stages in TKA participation used above: "information/training", "idea generation" and "implementation/follow up". The results are presented in full to give the overall picture of the status of employee participation in TKA in 1996. The different observations and recommendations from the audit will not be commented on and analyzed individually, but form a background for my analysis and theoretical discussion of the case study.

The recommendations represent important learning and identify strengths, weaknesses and improvement proposals for many of the tools and techniques that are being used. Much of this is reflected in my theoretical and methodological proposals for the further development of a synergistic approach to combining TQM and BPR. It is also reflected in my action research projects in the other participating companies.

8.5.1 A summary of the results of the 1996 audit

The audit revealed that virtually all employees had heard about TKA and had some idea of what TKA was. However, commitment to, participation in and attitude towards TKA varied considerably in the company.

The top management, most of the section managers, several heads of department and some employees and union representatives displayed an active commitment. Most head of departments and first-line managers/supervisors/foremen expressed little commitment to the process and many employees were outside the process.

The audit also revealed that the main part of the TKA effort had been directed towards the basis organization, whereas a growing percentage of the employees were organized in building projects.

The audit produced 22 observations that are described on the following pages. Based on these observations, the audit group developed 22 recommendations that will be presented afterwards.

8.5.2 Observations

On the basis of the data analysis, the audit group concluded as follows:

(1) Information/training

1) Most have heard about TKA and have an idea about what the process is meant to achieve.

An introduction to TKA has been given to 65% of the employees. These appear to have some insight into what the target of the process is. Emphasis is placed on becoming more effective and to achieve better cooperation and communication. Some of those who have not attended an introduction course have no knowledge of TKA.

2) The written information about TKA is adequate.

The TKA information that is disseminated via e-mail, in newsletters and on notice boards is experienced as considerable and adequate. However, the interest in reading it varies.

3) Individual understanding of what TKA is and which activities are included varies considerably among both managers and other employees.

At one extreme there are a few employees, union representatives and managers who believe they are working with TKA 100% of the time, and that the TKA idea is an important part of the way they are working. At the other extreme there are many who fail to see the connection between improvement projects, customer-supplier interviews and other improvement initiatives and the TKA process/training initiatives. Between these two extremes there is a great difference between how managers, union representatives and employees interpret the TKA process.

4) The attitude towards TKA varies.

Most managers above supervisor level say they are positive towards TKA. A few demonstrate that they also take it seriously through specific actions. Most of them just follow the trend and do what they feel is necessary. Many say they have insufficient time to follow up the process the way they should due to high work pressure.

Most of the union representatives are very positive towards TKA, and demonstrate this through involving themselves in the TKA activities. In particular, this applies to the Departmental Committee (DC) activities and improvement projects.

Some supervisors/first-line managers/foremen are negative about the process. Some are positive. Many supervisors are outside the process. Most employees have a very relaxed attitude towards the process or do not involve themselves at all.

5) There are three levels of commitment in relation to TKA. a) The actively committed b) Those who feel they must. c) Those that are outside the process.

Top management, most section managers, several heads of department, some employees and a few union representatives display a very strong commitment. Many supervisors hardly involve themselves in the process and many employees are outside the process.

6) Initially, the TKA training was too theoretical and not very specific. A seminar for DC members turned it into a more useful and action oriented direction.

About 65% of the employees participated in the introductory training. Few remember the content of the introductory training back in late 1994. Many fancy words were presented but little was said about what to do. Several of the managers who were to convey the content of the training to their co-workers felt a little silly when they presented the TKA transparencies.

Nonetheless, most of them understood the message of becoming more effective, and several started improvement projects and internal customer-supplier interviews after the introductory training. Some also followed up with refresher training the following year. A seminar in Haugesund, focused on improvement techniques for Departmental Committee (DC) members in the production area, was experienced as very good and action oriented, and very practical and relevant to the task that DCs should take in the TKA process.

7) Line management's role in disseminating TKA, setting requirements and demanding results varies greatly and seems essential for commitment to the TKA.

In the few departments where commitment to TKA is strong and the activity level high, such as "Prefabrication of Pipes" and "Welding Techniques", there is an identifiable management commitment along the line management up to area manager level. In areas where no results are demanded, either from the area manager, section manager or the head of department, the level of TKA activity is low.

Some heads of department report very little interest in the TKA activities among their own superiors. Among others, no one ever enquires about the status of the improvement projects.

(2) Idea generation

8) Conversations with the managing director are considered a very positive initiative.

Employees are given the opportunity to present points of view and ideas from the shop floor directly to the top manager. Ten employees are selected each month.

9) TKA is anchored in the basis organization, but the organization as a whole is now converting into a project organization. This is a great challenge to the future of the TKA process.

The anchoring in the basis organization leads to insufficient anchoring in the building projects that most people have gradually become involved in.

The projects conduct extensive TKA activities, such as customer-supplier interviews and experience transfer meetings that most people do not link to the TKA process.

There is some confusion concerning real responsibility, influence and involvement in TKA among line managers and project leaders. Who should do what?

Many experience appraisal interviews with their supervisor in the basis organization as meaningless in a situation where they are in the projects most of the time and therefore supervised by others.

10) For some it seems as if the attention around TKA has stopped up over the past six months.

Several report a higher activity level and greater attention around TKA in 1995. TKA training had stopped up, and many mention time pressure as an important cause. However, we see that activities such as improvement projects and customer-supplier interviews continue as before. Activities such as the EFQM evaluation and the three projects that were started after and as a consequence of the evaluation have gone unnoticed by most people. Those who are involved are reporting about high activity and commitment.

11) TKA tools, technical aids and transparencies that have been developed internally are considered to be very good.

"These technical aids are plentiful" (Supervisor, my translation).

12) Few use the employee appraisal interviews as an active tool to plan and follow-up individual development.

Members of staff appear to benefit more from employee appraisal interviews than operators. Most staff representatives feel that the employee appraisal interviews are useful, even if nothing happens afterwards. In some departments and for some of the staff, employee appraisal interviews are used as an active tool for planning their own development and further career in the organization.

The most important function of the employee appraisal interviews is to clear the air, and address issues that are unsuitable to address at other times of the year.

Many operators have insufficient trust in their supervisors to address problems that concern them.

Many consider the technical aids that have been developed to support employee appraisal interviews unsuitable. The key words on the form are not exactly relevant and many people avoid using them.

Only in exceptional cases do the employee appraisal interviews appear to be followed up during the year. Some people address last year's interview to see what has happened during the year.

13) Few are systematically transferring experience from one job to the next. Experience transfer is considered a good practice where this has been done.

Few have participated in the pre-job meetings where those who are going to participate in the work are systematically reviewing the various phases of the job to be done. So far, the experience transfer meetings in the latest projects are described as positive. Most people feel that this practice will provide increased effectiveness. In major projects where experience transfer has been practiced, this is considered very useful and important.

14) Few conduct brainstorming as a basis for improvements in their own department or workplace. Those who have attempted this have gained very positive experience.

Members of staff have mainly tried this practice in external training/course contexts. The "Welding Techniques" department has practiced idea generation (brainstorming) in its development meetings. The operator groups that have conducted brainstorming sessions in, for example the department Prefabrication of Pipes, report good results. Almost everyone regards this type of initiatives to be very useful.

15) Work group conversations are mostly experienced as one-way communication.

The attempts to involve the operators in presenting and discussing suggestions for improvements in work group conversations have not been very successful, except for in a few departments. This also applies to an arrangement conducted by some Departmental Committees (DC) called the "topic of the month". Each month one topic for discussion is selected for discussion in the group meetings. The intention is to encourage employees to forward improvement proposals to the DC. Some proposals do emerge. In exceptional cases, a good discussion may start about issues that concern people. However, many meetings are marked by insufficient time to think of new good ideas, which is often the last item on the meeting agenda. Many foremen/supervisors feel responsible for the production job that has to be done and are concerned with getting started again instead of spending time in meetings.

16) Internal customer-supplier interviews are a very successful initiative.

Internal customer-supplier interviews have eased cooperation and communication between the departments. Many proposed initiatives from these conversations have resulted in increased efficiency. It has become easier to propose improvements and discuss how colleagues in other departments can ease their own work situation. The initiative has definitely led to a change of attitude in the organization. This has happened although many people have not in fact participated in such interviews. All have heard about them, and are describing them as positive.

The extent to which people use the existing system and routines for registration of the initiatives that emerge from them varies. In other words, the scope of this activity appears to be considerably wider than what is being registered. Those who use the routines to the full and register everything believe that the routines are very good. Those who do not use the routines often explain this by lack of time.

17) The organization is insufficiently focused on systematic process improvement through identifying, assessing and improving important, value adding processes.

Process diagrams/flowcharts are hardly used. Few have participated in activities where the topic has been to identify important value-adding processes, describe them, assess the course of the work as a process, assess the different phases of the process and come up with improvements.

18) Operators view the suggestion box system as useful. Staff personnel do not.

The operators consider the initiative as a useful tool. The system ensures that their suggested initiatives are given proper consideration. Managers and staff appear to share the same opinion of the significance that the suggestion box has for those who work on the shop floor. The operators produce a number of good suggestions that could otherwise have been stopped at an earlier stage in the line of command. With few exceptions, the follow-up and feedback on the suggestions that arrive are experienced as good. However, at some places the suggestions have not been followed up, which has been de-motivating.

Several staff groups believe that the suggestion box is superfluous, partly because development and improvement is an integrated part of their job.

(3) Implementation/follow up

19) In parts of the organization the DC and corresponding meeting places function as a very efficient body for assessing and for following up suggested improvements.

In departmental committees (DC) such as those of Prefabrication of Pipes and Welding Techniques all suggestions are evaluated, the further fate of the suggestions is decided on, and the projects are reported to these meetings. DCs appear to have a great deal of trust, and many believe that more issues should be considered there. In contrast the meeting arenas for other staff groups function more as information meetings than as bodies for assessing and follow up improvements.

20) The trade union's involvement and attitude towards TKA are hardly perceptible for many employees. Nonetheless, several union representatives make a significant contribution and fill important roles in the local improvement activities.

The trade unions have signaled that everyone should do their best but otherwise they have no observable roles in the TKA process for many employees.

Several group representatives and the area safety delegates are very active and contribute positively in the DC, in improvement projects, and some also in customer-supplier interviews.

Union representatives believe it is important that they take part in the process. As union representatives they have no specific roles in TKA apart from being representatives in DC, WEC and WC.

TKA does not address topics of particular relevance to the trade unions. Nor have the trade unions made any mark on the content of the TKA process in a way that has been noticed in the organization.

21) The number of improvement projects that has been started in the organization is considerable. The scope and the follow-up of the projects vary from one department to another.

There are reports of several successful projects, such as for instance a new budgeting system.

However, some departments report poor progress in the projects and report that managers and managers' managers are not demanding results. In some departments, managers' managers have never asked why some projects are never finished.

Many consider the projects that can be solved within the framework of the DC as the most successful. Often major projects are taken out of the department and then prove more problematic to carry out.

22) Departments with a high TKA activity report considerable improvement with regard to efficiency, communication and cooperation.

Departments such as Prefabrication of Pipes and Welding Techniques have spent a lot of time on improvement activities and are reporting considerable efficiency and quality improvements. These departments have no doubt whatsoever of the effort having been worthwhile.

In departments with low TKA activity the main reasons given are shortage of time and scant interest from the line management.

Some further comments on the TKA training and information provided to employees are given below as a background for a better understanding of the recommendations given after the audit.

8.5.2.1 Direct information from own manager

Here, practice varied greatly from one manager to the next. At one extreme, several operators felt that their manager had never informed them about TKA. At the other extreme, there were managers who used TKA as a basis for most of what they did, and who actively conveyed the message to their co-workers on a day-to day basis. The department Welding Techniques was an example of the latter.

8.5.2.2 Introductory training in TKA for all employees

By the end of 1996, 65% of all employees had received introductory training in TKA. All managers and salaried staff and 550 of the operators had been given this training, and the planning of courses for the rest of the operators was well on the way. Line managers at different levels were responsible for training their own staff, supported by their managers and staff. The training of subordinate employees followed more or less the same pattern in the whole company.

The salaried staff had held a one-day seminar, where one representative of the top management introduced TKA. The local line managers, who had already taken part in TKA training for managers, addressed central quality concepts, the five essentials and the eight elements as referred to above. Finally, there was group work aimed at identifying specific improvement initiatives for the actual department/ group. Then local managers were given the responsibility for following up these suggestions for improvements.

The operators attended a three-hour seminar. First, a meeting was held for each department, where a representative from the top management introduced TKA and the reason for this effort. Thereafter, the work continued in the work groups in which the Engineering Manager, Production Manager and the Works Manager took part. However, it was the foremen who trained their own people. The operators were given a compressed version of the arrangement that was made for the salaried staff with an emphasis on the DC's place in the process. There was no group work in this seminar.

Otherwise, the audit uncovered the following about the introductory training. The training seminars were considered too theoretical and many did not remember the content a year after. In spite of that, those who had attended the courses felt that they had understood the main message about becoming more efficient and the importance of better collaboration and communication. Some of those who had not received this training said they did not know what TKA was. Nonetheless, everyone appeared to have some knowledge of some of the TKA activities that had been implemented, although several did not view them as being part of the TKA process. Several departments started improvement projects and customer-supplier interviews after this introductory training. Therefore, although the course did not significantly help to develop individual people's understanding of TKA itself, it appears to have had an informing, motivating and mobilizing effect on some of the TKA activities.

8.5.3 Recommendations from the 1996 audit "Employee Participation in TKA"

The recommendations that were given based on the observations above are organized according to the stages in the TKA participation used above: "information/training", "idea generation" and "implementation/follow up".

The introduction training in TKA should continue and be provided to the 35% of the employees who had not received training yet. The audit team found that at least four general conditions had to be met if this type of training was to have the intended effect with regard to the willingness, ability and opportunity to take part in TKA:

Before asking line managers to motivate and train their subordinates in TKA, it is necessary to make sure that they have sufficient knowledge and insight in the process and methods.

- The content of the introductory training must be better adapted to the various target groups. At operator level, the TKA terms must be linked to specific improvement techniques and methods such as customer-supplier interviews, how to present suggestions for improvements, etc.

- The training seminars should provide an opportunity to work with specific improvement initiatives that concern people on an everyday basis as part of the seminar. This means that, when the seminar is finished, a useful piece of work has been done, which is to be followed up/implemented when returning to work.
- Arrangements must be made to ensure that what has been learnt in these seminars is put into practice afterwards. In practice, this means that managers must be aware that, if anything is to happen after the seminars, local improvement competence must be available. To provide support, local facilitators are needed and routines must be established for handling and follow-up improvement suggestions.

On the basis of these considerations, six specific recommendations were given for the category "information/training", thirteen were given for "idea-generation" and three for "implementation\follow up". These are presented below.

(1) Information/training

Recommendation 1:

The training course content must be better adapted to the target group. At operator level the TKA terms must be directly related to specific improvement techniques and methods, such as the customer-supplier interviews, the suggestion box system, the opportunity for presenting improvement suggestions in the group conversations, the role of the DCs and the improvement projects.

Recommendation 2:

Before the first-line managers are asked to train others, it is necessary to make sure that they have been given sufficient background to provide this training. The teaching material that is used must be adjusted according to recommendation 1.

Recommendation 3:

Practical training in improvement techniques should be offered to everyone who has a role as facilitator, project leader or prime mover in the TKA effort. A plan should be drawn up for how and in which forms this can be achieved in the different departments.

Recommendation 4:

Practical action should be tied in with all training, and the practical initiatives that are implemented should be monitored through the established meeting structure.

Recommendation 5:

The information that is issued in the organization about TKA and improvement activities should make better connections between the process and the different activities so as to make it is easier for employees to discover the relation between them.

Recommendation 6:

The future role of supervisors in the organization and the required skills must be clarified. Further development in personnel management and strategy is required to enable them to perform their present role.

(2) Idea generation

Recommendation 7:

All who will be conducting employee appraisal interviews must be given training in this through leadership training that is otherwise arranged in the organization or in the form of a separate mini-course/seminar.

Recommendation 8:

Better and more flexible tools/schematics for conducting employee appraisal interviews must be developed to make it easier to adapt them to the various target groups.

Recommendation 9:

It is necessary to make sure that the result of the employee appraisal interviews and the agreements that are made are followed up through training and actions.

Recommendation 10:

To ensure that the employee appraisal interviews become a means for the development of personal quality, the interviews must result in specific development targets and improvement activities for individual employees.

Recommendation 11:

The practice of systematic transfer of experience from one project to another that has been implemented in the large projects must be further developed systematically and distributed around the organization both with regard to type of activity and the involvement of relevant personnel. The practice should vary from "pre-job meetings" for small but demanding projects to a series of well prepared meetings for major projects.

Recommendation 12:

Each department/project should arrange a brainstorming session to which individual employees are invited to assess work practices in their own department/group and present suggestions for improvements. This could take place within the framework of half a working day and preferably by expanding an existing meeting.

Recommendation 13:

Group conversations and similar staff meetings should be strengthened as meeting arenas in order to encourage suggestions for initiatives. With some exceptions, many of these meetings are currently characterized by one-way communication.

Recommendation 14:

The customer-supplier interview practice should be expanded to all parts of the company. As well as using it between departments, and between departments and projects, it is recommended that this practice is further widened to also be used within departments, and that all employees should have participated in such an interview over a two-year period. These interviews could be held either between individuals or between groups in a customer-supplier relationship. It is necessary to make sure that the interviews are followed up.

Recommendation 15:

The customer-supplier interviews that are conducted and the initiatives that are agreed on after an interview should be recorded in the TKA system. At present, many fail to do so. Such a practice would ease the follow-up of the initiatives and provide a better overview of the total TKA activity and the results.

Recommendation 16:

The practice and methodology of process management should be further developed. Important value-creating processes should be identified, described, assessed, improved and followed up at various levels and in different parts of the organization. It is best to start at company level and begin with the processes that are easiest to describe.

Recommendation 17:

The change towards a higher degree of project organization must have consequences for the implementation of TKA in the organization. Project managers must be given responsibilities and roles within the TKA process. Training, TKA tools and improvement initiatives must to a greater extent be directed towards and rooted in the project groups.

Recommendation 18:

The union representatives' role in the promotion of TKA, improvement tools and improvement initiatives could be better utilized. TKA strategy and action plans should to a greater extent be discussed and formulated in collaboration with the employee representatives in the Works Council (WC). This would, among others, give the employee representatives more security and acknowledgement for their collaboration with the line on the implementation of TKA.

Recommendation 19:

As inspiration and a source of ideas, Aker Stord should to a far greater extent document and disseminate information about successful TKA practice and improvement projects from the departments that have succeeded to the parts of Aker Stord that have not properly implemented TKA.

(3) Implementation/follow up

Recommendation 20:

Line managers at different levels must demand reporting of the status of planning and implementation of improvement activities. The Aker Stord QA Department should also be given a central position in follow-up and feedback.

Recommendation 21:

Better and more specific criteria for awarding suggestions through the suggestion box system must be drawn up, along with better routines for decision making, realization and following up of initiatives.

Recommendation 22:

The practice that the DC and corresponding departmental staff meetings function as a follow-up body for suggested improvements should be expanded. Successful practice at the departments Prefabrication of Pipes and Welding techniques should be used as a source of inspiration for others to further develop their practice.

There was a general agreement to follow up all these recommendations both in the half day seminars where all informants, union representatives, managers and staff were invited and in the top management group. Now what happened? This became an important issue for the 1997 audit of employee participation in TKA. I will first describe the EFQM self evaluation process and the introduction BPR.

8.6 The Aker Stord EFQM Self Evaluation

This evaluation was planned without any involvement from the researchers, but I was invited to participate and to take notes from the discussion. I had been trained as an EFQM assessor earlier, had facilitated EFQM evaluations in other companies and was familiar with the EFQM self evaluation tools. I could give the company some assistance, and it was a good opportunity for data collection and for me to get better acquainted with Aker Stord.

The EFQM evaluation was carried out in accordance with the EFQM model (see Chapter 3). I will briefly show how the evaluation fitted into the TKA process, present the mandate for the evaluation, how the EFQM evaluation team worked, the results, the recommendations and how the main conclusions related to and contributed to increase the quality of the Aker Stord case study.

The self evaluation was initially planned to be the tool that should be used to "measure" the status of the TQM process at Aker Stord. The audit presented above came in addition.

The mandate of the EFQM evaluation was to:

- Analyze facts (strong/weak aspects) that emerged through the evaluation process. Recommend initiatives.

- Suggest an information strategy concerning the results of the self evaluation (who, what and how).

The recommendations should form the basis for:

- Evaluating any reprioritizations in the Annual Plan for the improvement process for 1997.
- Input to Aker Stord's Strategic Plan.
- Input to the Organizational Development Plan for Aker Stord.

The evaluation process started with a meeting in which the top management and the leader of the largest trade union participated. At this meeting the EFQM model, the self evaluation process and expectations from both the individual participants and the process were described. The participants were given homework and carried out their evaluation of the company on the performance elements individually according to the EFQM method.

The next meeting was a consensus meeting at which the individual evaluations were discussed and agreed on. In this meeting the managers and the trade union representative worked in five groups representing the whole company. Each group had five participants. The TKA facilitator facilitated the process assisted by me.

The results of the consensus meeting were further analyzed and summarized by a task force of five middle managers and the facilitator. The task force reported the results and recommendations to the management group with the CEO as responsible. The management group discussed the recommendations and then decided on further action.

The EFQM self evaluation task force concluded to the following strong/weak aspects:

- The methods of the TKA approach are good, but the ability to carry them out is lacking
- Complicated/diffuse regulations result in inconsistent management
- Competence is not emphasized in Aker Stord's strategies
- Training is limited to trades with external requirements
- We excel in training apprentices and certifying welders
- Aker Stord lacks process improvement competence

In addition, the results from the managers' scoring on the nine EFQM elements described in Chapter 3 showed that "people results" and "processes" achieved the lowest scores. These results and conclusions made the task force come up with the following recommendations:

- Revise plans and strategies, "to be managed by TKA"
- Re-engineering of personnel policy, get rid of rules and acquire competence
- Establish basic understanding of process improvement techniques

On this background the management concluded to start up the following projects in their 24.05.96 meeting:

- Define governing documents and structure, define values, and make TKA color governing documents and values
- Reengineer the personnel strategy
- Build up basic understanding of process improvement techniques

The second and the third were process improvement projects, which I will comment on in the next section.

8.7 Introducing BPR at Aker Stord

An organization's ability to identify, evaluate and improve its processes is important in the EFQM model (EFQM 2003). Both the first TKA audit and especially the EFQM self-evaluation made the company aware of its shortcomings in this area, and motivated it to improve its understanding and skills in process management. The same audit and self-evaluation also made the company aware of the large potentials that a better involvement of employees in TKA and development activities could represent.

The audit and the EFQM self evaluation happened at the same time as the Aker Group started to process-orient the whole organization, among others with the help of Cap Gemini as its main consulting partner. The Aker Group wanted to define its business processes and develop a common operating model.

Aker Stord was at that time a very independent company within the Aker Group and wanted to develop its own BPR approach. Several consultants were contacted. The international consulting company Ernst & Young was chosen. Ernst & Young had developed the concept "BPR Norden" together with the consultant Bjørn Erik Willoch (Ernst & Young & Willoch 1996). Aker Stord bought the concept with some modifications. In order to use it, some internal facilitators needed to be certified and three Aker Stord employees received a six months' training. As part of the training, a practical re-engineering job was done and a manual for BPR in Aker Stord was developed.

In the course of the process of choosing the concept, the company found that the concept had some shortcomings in the area of employee participation. BPR Norden was very much top-down. After some discussions, important modifications and adaptations to Aker Stord's needs were made before the company concluded on how to run their process centering journey. Aker Stord wanted that the employees who were affected by BPR changes should participate in the change process. The result of this training and introductory phase was:

- An introductory process renewal handbook describing what Aker Stord means by process orientation and how process innovation projects should be conducted.
- Training material based on the introductory handbook and a set of slides.

- Three facilitators with an in-depth knowledge of the method. The task of the facilitators was to provide guidance to and support process innovation groups.
- A book of advice for facilitators giving practical tips on how to conduct projects.
- The BPR-Norden method, which describes the implementation of the project in detail. The facilitators used the documentation.
- A new work process for procurement of personnel ready for implementation.

Thus, two of the three projects that were decided on after the EFQM evaluation were realized during the BPR training period. The company built up its process-reengineering competence, and the proposed project "Reengineering personnel strategy" developed a new process for contracting external personnel – "Procurement Personnel". In the Aker Stord BPR handbook we find a clear link to TKA, for example in the "Introduction": *"In the spring of 1996, Aker Stord was evaluated by use of the EFQM self evaluation methodology. Process is an important enabler in the EFQM model. As well as gaining a low "score" in this measuring area, we acknowledged that we lacked an agreed understanding of the term "process" in our enterprise. The top management appointed a task force, which was given the mandate to "establish a basic understanding of process techniques" at Aker Stord. The task force decided to work together with Ernst & Young Management Consulting"* (Bjelland et al. 1997: 3, my translation).

That BPR should be a part of TKA is even more clearly documented when the handbook defines BPR: *"BPR means Business Process Reengineering and is a tool that is meant to help us get a step further in our TKA effort. The improvement tools we apply in TKA handle our traditional improvement projects, whereas the BPR tool is intended to act as a means to carry out radical changes to comprehensive work processes in our enterprise. In TKA we need both these work processes"* (Bjelland et al. 1997: 4, my translation).

This connection is illustrated in the Aker Stord BPR handbook (Bjelland et al. 1997: 4, my translation) and presented in Figure 10, which depicts various types of improvement projects in a holistic model.

In Figure 10 we find a combining of the TQM and the BPR ways of thinking. With reference to my theory section, both concepts are customer and process oriented. Development is based on TKA, processes are central and we find Aker Stord's version of the PDSA cycle (Deming 1986) "See – Think – Plan – Do – See" at the bottom.

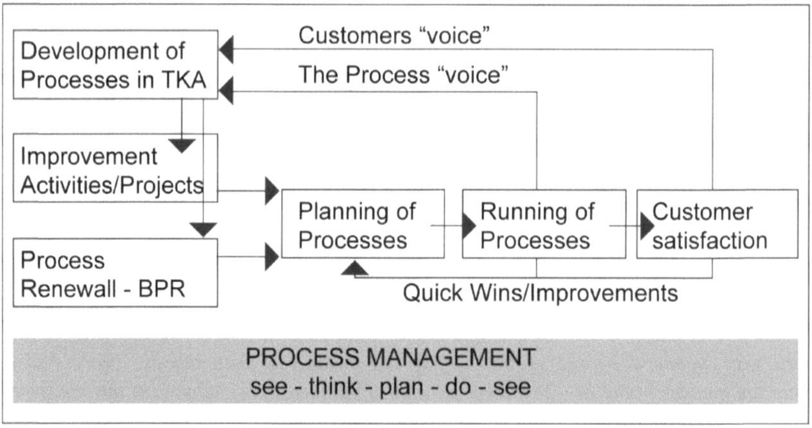

Figure 10 BPR as a TKA tool for radical changes to comprehensive work processes

Furthermore it is stated that Aker Stord shall be characterized as: *"A company with a focus on the value chain, having simple, secure and self driven work processes, where demanding customers and process owners continuously challenge"* (Bjelland et al. 1997: 4, my translation).

The process orientation that is presented in the handbook is in line with the fundamentals of BPR described in Chapter 4, and will not be repeated here. However, I will give some further comments regarding the link to TKA, employee participation in BPR and the way this new approach should change the organization.

Having defined process, the handbook states: *"Process orientation focuses on the enterprise's value chain, which consists of processes. The management's responsibility will be to clarify/monitor and develop the processes on a continuous basis, to ensure that their performance, stability and products are correct. Employees in individual processes are involved in their control, and they are also invited to continuously improve the process"* (Bjelland et al. 1997: 6, my translation).

In other words, employees shall take part both in the management and the continuous development of the company's processes. This kind of participation was an important adaptation of "BPR Norden" (Ernst & Young & Willoch 1996) to Aker Stord. The company did not want the top-down approach of BPR Norden, and the TKA focus on continuous improvement should be important.

The handbook was very clear about how this new approach should challenge the organization:

"Traditionally, we have been focusing on the value creation and the use of resources within various organizational units. We have clarified and removed grey areas towards adjacent

departments/functions through customer-supplier interviews and instructions/ functional descriptions. The result and follow-up of targets is strongly tied in with individual organizational units in the line organization. At Aker Stord process thinking has traditionally been aimed at areas of responsibility, and viewing processes across departments and enterprises has not been easily achieved. To a great extent, the processes meet the vertical organization's barriers when the process extends from one department to another. This leads to breaches of the chain, waiting time and duplicated work" (Bjelland et al. 1997:6, my translation).

Figure 11 illustrates how the vertical processes cross the department-borders A, B, C and D. The lines in the vertical barriers symbolize the breaches in the value chain when the processes meet department-borders. In 1996 it was the responsibility of individual managers to operate, develop and improve their own processes, but they were not responsible for seeing their own area as part of the whole process. Aker Stord realized that if they should achieve radical improvements, they needed to improve and renew their processes across the organizational units. The company's challenge was to see the individual sub-processes in a wider context.

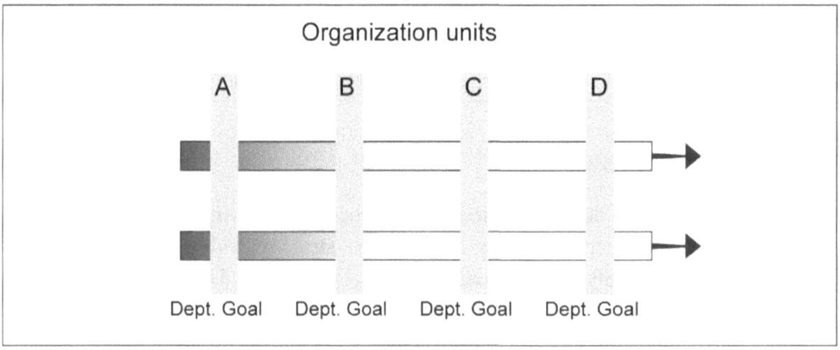

Figure 11 Processes meeting departmental barriers

Through process orientation, they wanted to seek control of the processes across organizational units. The company increased the focus on results from and development of major, holistic processes. A new approach was developed in line with Figure 3 from Chapter 4. In this approach the processes were in focus. In Figure 3, there is no breach in the vertical processes when processes cross department-borders such as demonstrated for Aker Stord in Figure 11.

Aker Stord developed its process hierarchy in the BPR way I described in Chapter 4. Main processes were built up by sub-processes, which in turn are built up by activities.

Aker Stord chose to divide its main processes into three categories:

- Direct value-creating processes
- Support processes

- Strategy and management processes

This is illustrated in the Figure 12.

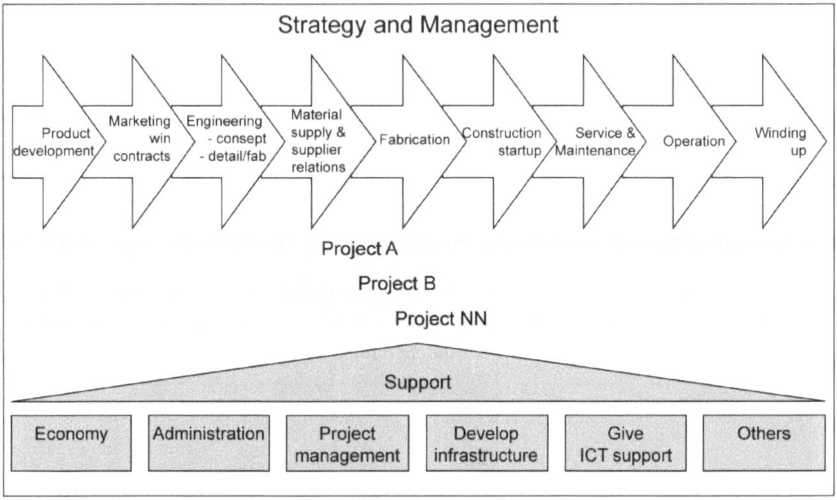

Figure 12 Main processes at Aker Stord

The process chart above defines which main processes are included in Aker Stord's overarching process landscape. On top we find the strategy and management processes. Below we find the company's main value creating processes. Then we see the building projects, and at the bottom of the chart the support processes "Economy", "Administration", etc. A process owner is appointed for each main process and made responsible for ensuring that the process is efficient and optimal at any time.

Each main process is divided into sub-processes that are handled by line managers, also called process managers. The process owners along with their process managers (the process management team) monitor the process from an efficiency and stability perspective. Process targets are drawn up for the process and followed up by the process owner. Experience transfer is the responsibility of the process owner and shall take place in cooperation with the process manager. The process management team ensures that the process is continuously improved.

Aker Stord found that if the company should succeed with process thinking, the traditional "box thinking" and sub-optimization within organizational units must be removed. This means that:

- Main processes and sub-processes are identified and described for Aker Stord
- Process owners are appointed

- The top management establishes and updates an improvement agenda on an annual basis (annual TKA plan)
- The improvement project requires the implementation of radical change methods as part of the TKA tool box

Once the company has decided to re-engineer a process, the reengineering process and methodology prescribed in the handbook are used in accordance with the BPR way described in Chapter 4 on BPR. As illustrated in Figure 13, the reengineering process at Aker Stord has the four main phases:

- Mobilize BPR team
- Map and quantify existing processes
- Establish a new/improved process
- Implement a new process

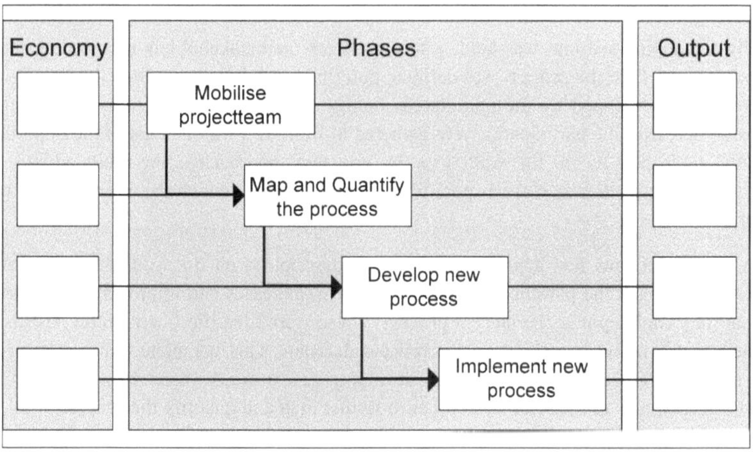

Figure 13 The four phases of a BPR project at Aker Stord

The management selects a project from the enterprise's improvement agenda. The process owner establishes the project mandate, selects the project participants and follows up the progress of the project. A process team is chosen and mobilized. When selecting the team, the line management needs to be involved to ensure that correct resources are provided 100% to the project. Persons who are working in the process on a daily basis and customers of the process ought to be part of the team. When selecting team members, also persons should be included who through their input could influence the team to think untraditionally. The selected persons must be given sufficient time to carry out the project. The BPR team should consist of between 5-8 persons.

When the BPR team has been established, the team's facilitator will train the team. A facilitator will actively oversee the daily progress of the team. After the establishment of the BPR team, the next phase, the mapping and quantification of the selected process is started. The result of the work is an "image" of the present situation (what is actually happening) including the present weaknesses and improvement opportunities of the process. The actual mapping takes place in a mapping seminar in which the BPR team and the stakeholders of the process participate. Subsequently, the BPR team undertakes a more extensive mapping of the process by conducting interviews wherever more information is needed. Also the required time, cost and quality are quantified, and the various events that emerge in the process are identified.

As mentioned, the initial facilitator training included the pilot: "Procurement Personnel". This pilot was carried out in accordance with the handbook. To demonstrate how the new BPR methodology was used in practice, I will summarize the pilot next.

8.7.1 The BPR project "Procurement Personnel"

Initially a mapping seminar was held, where managers and stakeholders met to describe the present process. First the process was defined, described and delimited. Then the workflow of the process was mapped by the participants noting on post-its the activities in the defined process. Thereafter the participants were gathered in front of a wall and given the opportunity to fasten their post-its on the wall. One by one they approached the wall, adding their descriptions of activities as close as possible to the sequence in which the activities take place in real life.

When the participants had approved the process description on the wall, the leader of the seminar encouraged the participants to identify any weaknesses and improvement opportunities that they could spot in the present process. These were identified, written on post-its and fastened on the relevant activity on the process diagram. This complete "image" forms the basis for the BPR team's further work to develop a new process. Interviews were then held with the participants in the seminar in order to further map and quantify the process.

In order to determine the new process, the customer was contacted to present requirements/wishes/targets for the new process. On the basis of a quantification of the present process and the customer's wishes, an analysis was performed, which described the gap between the present and the future process. Then phase three, the design of a new process was started by creating a vision and preparing an overarching description of the new process. Results from the gap analysis in the preceding phase were used to create the new vision. The handbook states that a good vision is characterized by participants recognizing their own ideas and the vision as radical and feasible. At the same time the vision must be neutral as to who does what in the new process.

Having completed the mapping and quantification of the selected process, the work started to find out what the new process should look like. During this phase, input that had emerged during the mapping seminar and the interviews were used.

Individual team members thought of new solutions based on the material that had emerged during the previous phases. Independently, the BPR team members decided to draw/outline their own solutions, which were gathered into a joint solution.

The BPR team arrived at a vision that was introduced in a vision seminar attended by the participants of the mapping seminar. The intention was that those who were affected by the changes should be able to influence the design of a potential new work flow. It must still be possible to evaluate opportunities for improvement and threats, and to influence the design. The vision was presented in three stages:

- The first time very superficially, to introduce the participants to the vision.
- The second presentation was made more in-depth while the participants were only required to listen and note any comments on a sheet of paper. After the introduction, the participants noted their comments on post-its, which were fastened in the correct positions on the wall.
- During the third presentation of the vision, the participants gave their inputs, and discussions were allowed.

The company's experience from this vision seminar was clearly positive. The group had the opportunity to give their inputs to problem areas as well as to future opportunities, which gave them ownership of the new process. The fact that the BPR team had considered new solutions in advance was favorable. And, as the team had made a draft, it was easier to explain and build faith in the new process.

A decision making basis was then prepared, which consisted of:

- The vision, as it was concluded during the vision seminar, presented in individual elements/phases with a brief description of the present situation, and a possible solution
- An evaluation of consequences in terms of competence and resources required, plus the use of technology
- A cost/benefit analysis
- An overall implementation plan

It was then decided by the management/process owner that the new process was to be implemented. By approving it, the process owner assumed responsibility for the implementation of the new process. A detailed plan was drawn up and the necessary activities initiated. The BPR team supported the work by developing detailed solutions, but it was primarily those who were meant to work on the new process that performed the detailed work of clarifying and establishing the new form of working within the set framework.

After the first pilot, two other radical process redesign projects were carried out by applying the methodology described in the handbook. These were Prefabrication of Pipes and Prefabrication of steel. I will give a short summary of the Prefabrication of Pipes case in relation to the 1998/99 audit in Section 8.8.2. This case, which will be referred to several places in my data-analysis, had an important influence on the development of my change approach in my action research cases.

Now, what happened over time? Did the company succeed with combining TQM and BPR effectively in the same development process? How was employee participation affected by the changes? To answer these questions I will turn to the other two audits on employee participation in TKA.

8.8 Time series – repeating the audit "Employee Participation in TKA in 1997 and 1998/9

Most people that I talked to as a researcher considered the three above mentioned re-engineering projects to be successful. Informants told me that when they bid for a new large job for the Valhall oil field their offer could be reduced by 15 % compared to calculations based on earlier calculation figures from the so-called "blue book". Aker Stord won the contract. I was also told that Prefabrication of Pipes carried out a job on the Sleipner Drilling Module with a time consumption of 44% of what was usual based on calculations with the basis in the "blue book". There were no doubts in the company whether the changes resulting from the re-engineering projects were successful economically.

As referred to in Chapter 7, I use "time series analyses" (Yin 2003) both to increase the quality of the case study but also to describe and analyze what happened over time. The audit that was carried out in 1996 was repeated in 1997 and 1998/9. The interview guides that were used for the three audits had to be expanded each time because new TKA and BPR activities had been introduced. However, they were basically built on the same principles. All interview guides were filed in the case study protocol. Translated versions of the 1997 interview guides are enclosed in Appendix 2. I will first present the results from the 1997 audit and important changes that took place since 1996, and then go into the 1998/99 audit. After that I will discuss the results and the changes in relation to my theoretical discussions and research objective.

8.8.1 The 1997 audit on employee participation in TKA. Important changes.

Like the 1996 audit, the results are presented under the main headings 1) Information/ training, 2) Idea generation and 3) Implementation/follow up. In addition important changes, important projects and a summary are given. The audit team did not come up with a list of recommendations as in the 1996 audit. However, a conclusion based on the findings was made. Before presenting the results, I will give a brief description of the situation in the company at that time and how different employee categories experienced it. This setting is important for understanding the concluding points. Most of the information is based on interviews.

8.8.1.1 The situation at Aker Stord in 1997

At the time of the second audit, very few people at Aker Stord had a clear idea of what the improvement process actually entailed. The year 1997 brought many changes, new managers, organizational changes and a closer integration with and clearer distribution of work in the Aker Group. A number of terms, such as "TKA", "the reorganization process", "the improvement process", BPR, "AS 2000" (Aker Stord 2000) and "SPIN" ("Store prosjekter i Norge" – Large projects in Norway) were bandied about. The latter was a Group-wide project that placed Aker Stord as an assembly workshop in the Group's value chain, rather than as a fully integrated supplier. Very few people had the full picture.

The audit team's understanding was that the improvement process was to be the umbrella term and that the methods of approach that had been developed through the TKA were to be Aker Stord's fundamental way of working with development. This also meant that the organization was to develop by means of a process-based rather than a function-based method of operation. During this process, central mechanisms and techniques from the BPR concept had been introduced. In other words, in 1997 the methodological basis for the improvement process had been expanded into a combined TQM and BPR approach.

Employees

The employees' knowledge of the improvement process was characterized by a series of terms linked to the Aker Stord reorganization process. Some talked about TKA, others about the reorganization process, about the financial targets in AS 2000 (Aker Stord 2000), or about earning money. What they all seemed to share was a fragmented picture of what was actually going on. They seemed to have understood some aspects of the ongoing process without being able to see the connection between the different development activities and the terms. What people seemed best informed about were, quite naturally, the issues that affected themselves. But also this knowledge seemed to vary considerably, both between and within departments. Salaried staff appeared not to know any more about the improvement processes than the operators did. If anything, it seemed to be the other way around. The following statement from one of the informants serves to illustrate how many employees experienced the situation: *"Perhaps they should call all of it TKA, but they give it different names all the time, so it is impossible to keep up with it"* (Salaried staff, my translation).

Shop stewards

Interviews of shop stewards revealed that they were well acquainted with what was going on. They knew about the objectives of SPIN (Large Projects in Norway), Aker Stord's future role in SPIN and most of the improvement projects in SPIN. They appeared to have a clear idea of the connection between this and the activities in AS 2000. They appeared satisfied with the information they had been given, they got what they needed when seeking information, and did not feel that anything was kept hidden for them.

On the other hand, they felt that things were progressing a little too fast and that much of the information reached them somewhat late. They felt that they had less opportunity to influence than before. The most important decisions were now being made in a SPIN context, and they were unsure as to what influence shop stewards had at SPIN level. However, a number of SPIN-related meeting places for shop stewards were about to be established. In 1996, the local shop stewards appeared to have a greater influence on developments at Aker Stord than in 1997. The changes in the TKA process were less dramatic and were kept more or less within a given framework. Now the framework had been changed radically and the numerous minor changes/improvements were pushed more into the background. All these changes made it difficult for the local union shop stewards to involve themselves in everything.

Shop stewards who were members of the local union board had been thoroughly informed in committee meetings by management representatives. They did not remember every detail, but appeared to be informed about matters that concerned the departments they represented.

Managers

The managers' information about the process depended on how far out in the organization they were placed. Top level managers appeared to be very well acquainted with the ongoing activities. However, numerous people, employees, shop stewards and high level managers referred to two processes. They often talked of "the other process". Therefore, even if top management appeared clearly to have integrated TKA and the process direction (as some of them called it) into one and the same improvement process, they had failed to convey this to the organization.

The three projects from the EFQM evaluation

The three projects that were started after the EFQM self evaluation referred to earlier seemed unknown to the employees who had not participated in them, although central shop stewards had been informed of these projects. One of the shop stewards that we interviewed knew them well through his own participation, but other shop stewards had no or hardly any knowledge of the processes. High level managers had thorough knowledge of the processes. Several of them had taken part in the EFQM evaluation, and had been given roles related to the three projects. First-line managers had little knowledge of the processes, and some had never heard of them.

Now, this was a short background that may ease the understanding of the changes that took place and the results from our second audit. I will now present the results from the 1997 audit and then summarize the changes from 1996 to 1997.

8.8.1.2 Observations

TKA information and training

Basic training in TKA/improvement processes

During the corresponding audit in 1996, we found that 65% of the employees had been given basic training in TKA. At that time it was a target that all employees should receive this kind of training, and several of our informants expressed expectations of such training. The HR Department had developed a revised scheme, which had still not been brought into use. The TKA training had stopped up. The remaining 35% of the employees never received any TKA training.

Few of those who had participated in this basic training in 1996 remembered what it was all about. They remembered that it contained a lot of fancy words and pretty slides, but the training scheme contained few references as to how to act. Nonetheless, several expressed that terms like personal quality, internal customers and quality were given increased attention and that the number of improvement activities in several departments increased after the training seminars. Some of the departments started improvement activities for the first time after the seminars. Some people also reported that the training took a more practical direction after an evaluation of the first training seminars in 1996. In spite of that, employees considered it too theoretical.

Several also mentioned that the intention was to train 300 so-called central persons in process understanding in order to facilitate an efficient transformation to a process-centered organization, and to teach people process improvement. This did not happen. Attempts at process organization were made in the departments by reviewing the work methods being used and to see whether they could be better structured. But many felt they lacked the tools to achieve that kind of changes. However, the reengineering changes from the three projects mentioned above were carried out.

Departmental follow-up TKA seminars

All the departmental follow-up TKA seminars that were conducted in cooperation with the DCs and the HR Department were considered useful and positive by the informants. The seminars had taught them a simple technique for idea-generation, which had been given various names such as the yellow post-its method and brainstorming. They had also worked on specific issues that were rooted in people's everyday working life, and in virtually all cases there had been some follow-up of the initiatives taken up in the seminars. On the other hand, disappointment was voiced that not more of the initiatives suggested in the seminars had been followed up.

In the Prefabrication of Pipes department improvement groups with a broad operator representation had been appointed to prepare suggestions for initiatives to the DC. The works manager prepared a list of all the suggestions that had been tabled and posted the list on the notice board. This list was used as a basis for selecting improvement projects in the DC and other initiatives. The department had also started to involve its internal customers, such as the "Painting department" in its DC seminars in order to get their opinions. The shop stewards appeared satisfied with this practice.

Improvement techniques

Brainstorming (the yellow post-its method) was the technique that most people felt to be the most useful one, and that they felt easiest to handle. This applies to employees that had participated, to shop stewards and managers alike.

Information about the improvement process from superiors

Several employees and shop stewards referred to the general meetings, to which all were invited, as important sources of information. Beyond this, there appeared to be a difference between operators and salaried staff. The operators had gained most of their information through the group discussions and corresponding meetings. In these meetings it was not the foremen that informed them, but representatives from the management. Apart from reading out newsletters to employees, the foremen appeared to give limited information about the improvement process. Salaried staff appeared to have gained more of their information directly from their immediate superiors, who informed them about the process and activities in department meetings or similar. But also here the quality of the information appeared to vary considerably. Some reported about very little and superficial information whereas others were more satisfied. Again Prefabrication of Pipes was held up by the operators as a good example, where the works manager took part in most of the group discussions and informed them of the various elements of the improvement process. Also "Procurement/Storehouse" was an operator environment that was very satisfied with the information they had obtained through management representatives who had visited the various units and given information.

Otherwise, it was claimed that managers interpreted and selected differently when they gave information out in the organization. The information received by employees from their managers varied from one department to the other.

Members of Aker Stord's top management group were very satisfied with the information they had received, and they had received it repeatedly in various contexts. Managers at the next level were also satisfied with the information that the top managers provided. However, the information received by managers at the level below appeared to vary considerably. One of them reported that information about the improvement process had been good for a long period, but had stopped this autumn.

Written information about the improvement process

A total review of the written information that had been sent to the employees' homes (SPIN-off and Puls (concern magazines), the Aker Stord paper and a newsletters) gave very good insight into the overruling objectives of SPIN, the improvement projects and the most important improvement activities that had been initiated at Aker Stord. However, this was presented over an extensive period of time without placing the various activities in a holistic framework under a joint heading. In addition, most employees reported they had read most of what they received quite superficially, and this applied in particular to the two first-mentioned magazines. People tended to read what was most relevant to their own situation. Otherwise, people felt that the internal information provided by Aker Stord was very good as long as people took the trouble to read it, including the small bits of information that were provided about the improvement process on a continuous basis. Information posted on the notice board appeared to be read.

Top level managers in the organization reported that they read more of the written material that was issued than others. They were very satisfied with the information provided, and had often received the information in management meetings before it came in a printed format.

Information from shop stewards

Information given to employees from local shop stewards was sparse. It was assumed that local shop stewards informed of what they knew, but that they knew very little. Shop stewards who served on the local union board received good information. Central shop stewards provided broad information to their members when special situations arose. Nonetheless, the shop steward felt that the main responsibility for providing information about the improvement process ought to lie with the line management, but that also they needed to inform, and particularly about the processes they were involved in.

Information about the improvement process in the building projects

Informants from the building projects describe the information as either the same or less compared to 1996. Most reported less information, and commented that the information that was given in the building projects was more job related.

Changes in the information about the improvement process from 1996 to 1997

Those who felt they had been given poor information felt that the information was just as poor in 1997 as in 1996. However, most people reported that they were better informed in 1997 and that they now could see that more things were happening. Others were of the opinion that compared with what was happening now, they were being less informed than before and several said that too many things had not been clarified.

Participation in problem identification and idea generation

Appraisal interviews

The stated objective was to hold annual appraisal interviews. Apparently, practice varied in this respect, from those who reported having had two appraisal interviews in 1997 due to a change of manager, to those who reported that it was several years since the last interview. The main impression was that most had dropped the appraisal interviews in 1997 due to the reorganization process and lack of time.

Also from a content point of view, the interviews varied considerably. Central topics were well-being, cooperation, conduct, courses and future work duties. Some took the opportunity to clear the air and address issues that concerned them and which they considered important. Many said that courses and future work duties were addressed. But apart from follow-up concerning the necessary certificates and courses, the follow-up of the interviews was reported as poor. In this context, several mentioned that in the present situation it was very difficult to plan. An agreement may have been reached, but due to all the reorganizations and the projects, the work situation may have changed completely shortly after the interview, and what had been agreed on became irrelevant. In particular in the salaried staff environments the work situation appeared to have changed for many compared to 1996 due to reorganizations.

There was a wide difference between people's attitude towards the appraisal interviews. Some felt they were pure nonsense; others had a great deal of faith in them. Most people appeared to have landed somewhere in between and felt it was all right. Some felt it was all right to have some schematics to be thought through in advance, while others felt that the schematics were unsuitable. Otherwise, the operators in particular mentioned that the prerequisite for gaining something from an appraisal interview was having trust in the work supervisor. Many were afraid to express themselves or their opinions.

There was no consensus among high level managers about the benefit of appraisal interviews. In parts of the organization they were not in use, while others used them very seriously. Most high level managers felt it was a useful tool but also that the form was too formalized and not quite relevant. Leaders in the basis organization pointed out that it was often a little artificial for them to run appraisal interviews with people who were placed in projects and whom they rarely saw. On the other hand, project managers mentioned that it did not feel natural for them to run appraisal interviews, as people were in the project for short periods only.

Pre-job meetings

Many project meetings in some salaried staff environments were reported to have the character of reviewing pending work tasks and receiving input from the employees who participated. But it was not general practice at Aker Stord to involve employees in this type of meetings, although large building projects were mentioned as exceptions. However, good experience and a positive attitude towards this kind of meetings were reported both from managers and employees who had participated in the meetings. The meetings provided an opportunity to weed out problems and misunderstandings before starting the work. It became easier to plan one's own work in relation to that of others, and it became possible to reduce some of the time of waiting for orders from the foreman. The respondents felt that the time spent in these meetings was quickly worked in again as soon as the work started.

Brainstorming sessions in own department

It was not a general practice to gather all employees in the departments to run a brainstorming session in order to extract improvement initiatives. However, in some departments both operators and staff had tried it and had obtained a number of suggestions for initiatives. Nevertheless, many of them experienced that the follow-up failed when they returned to everyday life. Then again there were good examples of cases where groups had been established to follow up some of the suggestions (for example in Prefabrication of Pipes), and where action plans had been developed to implement the initiatives.

Most managers, employees and shop stewards had a very positive attitude to such brainstorming sessions. They felt they generated results and wanted to participate in them. In spite of this, many of them reported that it had been a little difficult to follow up the issues they had had on their action plan lately due to the reorganization process. Their attention was diverted to other parts of the organization and the changes that were taking place there. Therefore, the action plans needed to be revised when things had settled a little. Still, one of the high level managers believed that it was too time consuming to involve all employees in such brainstorming sessions.

Group discussions and department meetings

Group discussions and department meetings lasted from 30 minutes to two hours. The frequency varied. Some staff groups held weekly meetings. Most operator groups were meant to hold monthly meetings, but the meetings could be less frequent, and particularly in the projects. They had also encountered problems holding the meetings in the parts of the organization that had undergone major changes. The quality and content of the meetings varied. In the group discussions it was common for the foreman and invited persons from for example the HR Department, the management or the HSE Department to first provide information and then ask for input from the employees, preferably linked to the topic of the month.

There were however reports of a great deal of variation concerning the degree to which the meetings were coordinated with DC meetings, whether the content of newsletters was addressed and whether input was received from the employees and included in the minutes or directly dealt with or conveyed to the works manager. Prefabrication of Pipes, for example, reported that some group discussions functioned very well. The employees were active with regard to presenting suggestions for improvements, there was a great deal of trust in the foreman and the works manager was very willing to follow up improvements suggested. A lot of ideas from employees were realized. Other places reported a far lower activity level. In one department, where people had problems addressing issues in group discussions with so many people gathered, the work supervisor addressed topics that he had picked up through conversations with people in smaller groups between the meetings, and managed to start the discussion in this way.

This kind of supervisor engagement had also been made to work in some staff environments, for example in "Method Engineering". In this department they had managed to receive numerous suggestions from the employees, which had been prioritized and taken into an action plan. Some meetings also addressed particular topics, which were determined from one meeting to the next. In general, it appeared as if a great number of initiatives were started, but that people were not properly informed of what had in fact been implemented. To many people it seemed as if little was happening.

Among the staff it seemed as if it was easier to address suggested improvements in a day to day context rather than in the department meetings. But also here there was great variety as to whether these meetings were one-way communications or if the communication went both ways. People reported that very much depended on the manager's engagement and ability to involve people.

Departmental committees (DC) as bodies for managing the improvement activities

Also here, practice and opinions varied considerably. Most interviewees felt that the DC was a suitable body for managing local improvement activities and referred to successful experiences at, for example, Prefabrication of Pipes.

Shop stewards in the workshop believed the DC to be one of the most important forums in the enterprise. They felt that this body had to be strengthened and that an effort had to be done to increase the commitment of the managers and the employees. There were too many good improvement suggestions that came to nothing.

In some departments there were problems committing people to present suggestions both individually and in group discussions. Often only a small number of people (those serving on the DC) became involved. The DCs had their own budget for implementing improvement initiatives.

Customer-supplier interviews

Few employees and shop steward had participated in customer-supplier interviews. Those who had participated reported positive experiences. They believed that more such interviews could have been held to solve a number of problems that had existed for a long time between departments.

Managers also pointed out that far fewer customer-supplier interviews had been held recently. At the same time, the managers disagreed on the benefit of running such interviews. Some felt they had to wait with such interviews until the new organization was in place, as the interviews could have a cementing effect on the organization, and that there was no time for this now. Others felt that it was an important mechanism, as at present there was a lot of confusion/friction between new units concerning the division of work and tasks that had to be clarified.

The suggestion box

Everyone appeared to know the suggestion box arrangement. It was not much used by members of the staff as they felt it was a natural part of their job to present improvements. Furthermore, for the above reason, it was more difficult for them to receive the full reward. It seemed more natural to the staff to address suggestions for improvements there and then, either in department meetings, in the projects or with their own supervisor.

The suggestion box was easily accessible. The rewards were popular, although many did not always understand why some were awarded prizes and others were not and why some received large prizes and others smaller prizes. It also served as a channel to get things done if going via the line management had failed to give results. Some complained about a long processing time. Others felt that the scheme led to less discussion in the groups about improvements because people kept their suggestions to themselves due to the potential prize. In other environments ideas were openly discussed without any fear of others stealing them. The opinions were many and the experiences differed. But most interviewees felt that the scheme was positive and useful, although very many pointed out that it is very unfortunate to award prizes for suggestions and then not implement them. Everyone knew of such examples and believed they undermined the scheme.

Some stated that many operators were reluctant to put their thoughts on paper. Some felt it was easier to address issues in group discussions.

Registration and overview of improvement suggestions

There were good statistics on suggestions that had arrived through the suggestion box system, and it was the operator environments that delivered most suggestions per head. Prefabrication of Pipes had been on top over the past two years. The staff had far fewer suggestions. Compared to the period January-November 1996, the total number of suggestions had fallen from 650 to 491 during the same period in 1997, with a marked fall after February. The share of awarded suggestions had also fallen from 39.1% in 1996 to 34.8% in 1997.

As regards the statistics on suggestions that were received from other places such as group discussions, customer-supplier interviews and department meetings, the system that had been established for registration of improvement suggestions was used to a highly varying degree. A number of departments did not use it at all. Some used it only for especially important initiatives. Some managers said they did not know whether they had access, and others said they had problems with the computer system. It was therefore not possible to gain a reliable overview of the other suggestion and improvement activity at Aker Stord in 1997, although it was reported that there were improvement groups and improvement projects going on in most departments.

Many believed there had been fewer suggestions for improvements in 1997. The reason for this was that they were now more concerned with implementing the centrally initiated changes they had been ordered to carry out than with focusing on locally initiated improvement activities. Many employees felt that the managers followed this up less well than in 1996, and put less emphasis on it. There were fewer requests from the management for results from local improvement activities than before.

Reports on undesired events (RUH)

This practice was not covered in the previous audit, but was included in the 1997 version on request from the management. Also here the practice differed between members of staff and the operators. Everyone agreed that it was a very important and useful tool for the operators, but some staff members felt that it was not relevant for them, and that they had no injuries or near misses to report. The operators felt it was an important tool that helped to uncover problems. It was also understood as a means of making things happen if there were problematic conditions at some location or other. It was also stated that there was some under-reporting at some locations, as people did not like to write about their own mistakes and hesitated to report on other people. Otherwise, many felt that the follow-up of RUH in the form of initiatives was good, as people also would start immediate initiatives when something happened. RUH was regarded as important in getting people to focus on matters that could threaten their own or other people's safety.

The Managing Director's monthly staff meetings

Most people had heard of these and felt they were a very positive arrangement and a good way for the top manager to keep abreast of what was going on. People felt that they were listened to. Some believed that such an arrangement ought to have continued when the company experienced a change of managing director in 1997. Others believed that it could be perceived as a form of mistrust in the line management, and that they should not be necessary. Regardless of this, most people believed that it was important for the top manager to meet the employees in various ways, and most of them reported a high degree of trust in both the previous and the new top manager.

The opportunities for obtaining suggestions for improvements in the projects

The members of staff believed they had good opportunities for obtaining improvement suggestions in relation to matters in the projects they were working on. The operators reported that it was a little more difficult here than in other places, as time pressure was greater. Focus was placed on the specific work tasks and matters were not facilitated in the same way as in the basis organization.

Implementation/follow up

Participation in improvement projects

Virtually all interviewees had participated in the implementation of improvement projects. Some of the projects were quite extensive, but most of them concerned minor improvements in individual departments. There were also some examples of projects in which two or more departments collaborated. Projects were normally organized in the form of minor groups with a project manager in charge. Projects that were run in the departments would normally have a link to the local DC. As mentioned, when proposals were discussed in the DC, the DC would in some situations appoint project groups that reported back to the DC.

How the participants in the projects were chosen varied according to the scope and nature of the project. Often, it was the head of department or the work supervisor who picked out the participants based on criteria such as competence, the extent to which participants were affected by the project and that several shifts should be represented. Some people felt that the selection process was a bit arbitrary. In major projects it was often a criterion that there should be representatives from both management and the operators. Most often people were appointed in agreement with shop stewards.

The application of improvement techniques varied greatly from one group to the next, and large differences between the various departments were also reported. The most common technique was yellow post-its, particularly during phases where it was important to obtain alternative initiatives and solutions. But also other techniques were mentioned. Some process improvement teams reported that they had used the brown paper methodology that had been developed as part of the BPR introduction and training. This had been used to map the work processes and identify improvement initiatives. However, several people stated that the improvement work in the groups took place without using any particular techniques. Instead they discussed until they found solutions that the project group could agree on.

Assessment of the implementation of the improvement projects

Most employees considered improvement projects, both incremental and radical, as an efficient and positive method of work. Some employees felt that improvement projects could have been given even higher attention, while others were more skeptical and had mixed experience of this kind of project work. Among others, it was stated that these projects took a lot of time, which increased the work pressure of those who were involved. This was expressed as one of the reasons why some employees did not want to take part in improvement projects. In the initial phases of the projects, people were committed and had a lot of good ideas, but as time went on, people stopped attending, and the efforts resulted in fewer results.

Most employees felt that they were taken seriously during the improvement projects, and that everyone who took part experienced having part of the responsibility for the result. All who wished to were allowed to present their points of view, although some were more active than others.

Also the managers more or less agreed that improvement projects were a constructive method of working, although they saw that the company had an improvement potential in this area. It was considered positive that several persons were involved, although they felt that there should not be too many projects ongoing at any one time, which tended to happen in some departments.

It was also pointed out that change projects concentrated more strongly on processes now compared to previous years. The interviews indicated that the process thinking was about to fasten itself with some but not all of the managers. The tendency was that the higher up in the organization the managers were placed, the more likely it was that they were thinking about processes in connection with the improvement projects. This could mean that process thinking was about to be spread downwards from above, but also that it could be traced back to a few managers at the top of the organization.

The shop stewards reported to have the same experiences with the implementation of the improvement projects as the rest of the personnel. Normally, shop stewards had no particular role in the improvement projects, and most of them felt that this was quite correct. Their most important role was to participate indirectly through representation in the DCs, where the shop stewards felt they had an important function in coordinating the improvement work. However, some of the shop stewards stated that they ought to be represented in company-wide projects at higher levels of the organization. They also provided examples of this happening in practice in some cases.

Evaluation of the follow-up of improvement projects and their consequences

Some employees viewed the follow-up of improvement projects as a weakness. In some cases the follow-up was relatively arbitrary and there was a lack of clear routines. However, the follow-up that was conducted by some of the DCs was reported to function well.

The employees reported that the improvement projects had basically provided positive consequences for the employees. This was because they had led to better and more efficient work methods while creating improvements in the work environment and general well being as well. Some employees disagreed and pointed out that the improvement work in itself required a lot of resources and led to more pressure on the employees.

The way projects were followed up by the managers varied. Some of the managers undertook direct follow-up and registration of the improvement projects, while others said that the reporting was done to the DC. Several managers believed that the follow-up was insufficient and that this was due to lack of available time. The audit team was left with an impression that the follow-up part of the improvement work was a weak point. There was no uniform practice, which may be due to lack of routines or because the routines were not known.

The managers believed that they had an important task in promoting an improvement practice among their subordinates. The key was to involve people, to listen to their arguments and encourage them to present proposals for improvements. The extent to which they succeeded in this seemed to vary a lot. Some of the managers appeared to be satisfied with their efforts in this connection, while others acknowledged that they could do better. Also here time was pointed out as a limiting factor. The time barrier is reported to be even more prevalent in the projects. In addition to unclear links to the basis organization, this had negative consequences for the improvement work.

In relation to the shop stewards, the audit team emphasized obtaining points of view on the consequences the improvement process had for the trade unions and their participation in bodies such as the Departmental Committees, the Work Environment Committee and the Works Council. In general, the shop stewards felt that the improvement process had not weakened the trade unions' importance in the company. On the contrary, many felt that the trade unions' position had been strengthened and that cooperation with the management had improved. It was also pointed out that the shop stewards had been given several new tasks, which was basically positive but could result in an increased workload for some.

For the joint cooperative bodies in the enterprise the improvement process was regarded as having provided positive consequences. Most people believed that the DC had become a more active body where the shop stewards could exercise influence on the improvement work. Only one of the shop stewards we interviewed believed that the Works Council had no influence on the improvement process, and that this body had been weakened. With regard to the consequences of the improvement process in general, the shop stewards expressed approximately the same points of view as the employees.

The importance of the improvement process

Most of the interviewees believed that the improvement process had changed everyday-life at Aker Stord significantly. Generally, these changes were described in positive terms, although the picture is not entirely uniform. Some pointed out that it led to an increased tempo and stress in the work situation, whereas a small minority said that the improvement process had not had any significant consequences for everyday life in the company.

With respect to changes in people's involvement from 1996 to 1997, few clear points of view were offered. Most felt that this was difficult to assess. Among the managers it was mentioned that AS 2000 had involved a lot of people and that this was positive. However, several said that these changes appeared a lot clearer if viewed in a 10-year perspective. Compared to the situation 10 years ago, they felt that employees were much more involved now, and that the improvement work now had a better system. Several felt that the improvement process had positive effects on people's attitudes towards delivering quality at the right time, while some stated that time pressure resulted in the quality not being all that much better.

On the whole, everyone felt that the improvement process was something that had come to stay. Several felt however that it was a problem that the process was constantly given new names. The fact that a new concept (TKA) had been introduced with a great deal of management involvement only to be changed for a new one a couple of years after confused many people. It also gave the skeptics good arguments. Others said that these kinds of changes were a natural part of the development work. The main impression was nonetheless that most people considered it an advantage if the improvement process could be given a name that lasted year after year.

Consequences of the reorganization process

At the time we conducted the audit, the organization was marked by a great deal of uncertainty as a result of the reorganization process in the Aker Group. Most of the interviewees had experienced this uncertainty. The audit attempted to find out whether this situation in one way or another became significant to the improvement work and people's commitment. But there was no clear picture. Some of the informants from all three categories believed they had observed a clear tendency of people becoming more passive as a result of the reorganization process. *"People have lost their spark"*, as one of the foreman said.

On the other hand, others pointed out that some employees had displayed a stronger commitment during this period. Most people nonetheless reckoned that the general commitment would improve when the heaviest reorganization had been completed and the situation had become normalized.

There was also a clear tendency of the informants hoping and believing that the reorganization process that was meant to turn Aker Stord into an assembly workshop in the group's value chain would strengthen Aker Stord both in the market and within the Aker Group. This was an understanding that existed among managers, employees and shop stewards alike.

Most people also felt that the reorganization process could have a negative effect on the shop stewards' influence. The fact that more important decisions had now been centralized in the Aker management in Oslo was often expressed as significant in this respect. Some also felt that the shop stewards had difficulties in following this kind of change processes. The shop stewards themselves felt challenged in coordinating the trade union work between the different levels in the organization, and particularly between the Aker Group and Aker Stord.

All the employees and shop stewards we interviewed felt that their opportunity to influence their own work situation had been strengthened as a result of the improvement process. They felt more involved. Some pointed out that they wished to increase their opportunity to influence even more. In general, the employees felt that the managers had become more positive with regard to listening to the suggestions and ideas presented by the employees. People seemed generally motivated to participate, but lack of time was the most important obstacle to employees becoming even more involved. Many expressed that it is important to set aside time for this, and that the management should clearly signal that time can be spent on improvement work.

On the question to managers whether TKA had any effect on their personal development as a manager, the replies can be divided into two. About half of the interviewed believed that TKA had not had any significant effect. The other half believed that they developed positively as a result of TKA. In particular, this group stressed that they had become better at involving the employees, and that the development work was now progressing more because of increased employee engagement. One of the managers also said that TKA had made him more humble vis-à-vis the employees.

The improvement process and democratization

All the interviewed shop stewards believed that the improvement process had promoted democratization in the enterprise. It had increased co-determination, and more people took part in decision making. They also felt that the improvement activities had resulted in more people becoming more motivated to involve themselves in Aker Stord's development.

8.8.1.3 Changes from 1996 to 1997 summarized

"I'm thinking TKA, has it died out? Why is no one talking about it anymore? A new director and a new slogan, or what is this? This is probably not noticed in everyday life. People are concerned with doing a good job, TKA or no TKA. As I see it, the dangerous part here is that we go out with a lot of hullabaloo declaring that now we have TKA, and from now on things are going to be done in such and such a way, and that this is important and this is how we'll do things from now on. They say yes, all right. This is probably just another short-lived idea. And then they are right, in a way. They are given yet another argument to be negative. It would have been an advantage to use the same terms. They recognize things, and that is important". This quotation from a first-line manager (my translation) represents a common viewpoint. The changes from 1996 to 1997 were summarized by the audit team as follows:

- The term "TKA" has now become less alive and is less used at Aker Stord.
- The employees receive more information and there is more activity in total in the enterprise.
- Viewed in relation to the activity level, people report a poorer holistic understanding of the improvement process.
- More decisions about the improvement process are taken centrally in the Aker Group.
- The use of appraisal interviews is reduced.
- The use of customer-supplier interviews is reduced.
- The scope of local improvement activity in the departments appears to be reduced.
- The enterprise's overview of local improvement activity is significantly reduced.
- Training related to the improvement process is significantly reduced.
- Greater uncertainty among managers and employees about what is most important to prioritize in the improvement process. A more "wait and see" attitude.

8.8.1.4 The 1997 audit conclusions

No recommendations were prepared after the 1997 audit, but the audit team concluded as follows:

- Most people believe in the reorganization that is taking place, and express trust in the management
- At the same time many people feel great uncertainty
- The objective and content of the improvement process is unclear. Many feel that there are two processes on the go, *"TKA and the other process"* (first-line manager, my translation). Most people lack an overall understanding
- Direct information from the top management is felt to be most useful
- Training related to the improvement process has stopped. The process itself continues

- The local improvement activity has weakened due to greater attention to a few large radical reengineering projects and major reorganizations
- Some TKA activities that took a long time to build up, and which have gradually started to work, are no longer given priority
- The shop stewards' influence on decisions related to the improvement process in Aker Stord is reduced
- The DC has a strong status as a body for coordinating local improvement activities
- Aker Stord has succeeded in obtaining several suggestions for improvements from various parts of the organization. The follow-up of these could be better
- Aker Stord has a large and unexploited improvement potential for achieving a more efficient integration of the best of their own traditions and the strategy for the local improvement activity and TQM, BPR/process direction, and the opportunities that the changed general conditions in the Aker Group represent
- Few are conducting pre-job meetings or similar experience transfer meetings Experiences are good where the meetings are held
- The improvement process is rooted in the basis organization. With the exception of the project "Total work improvement", the process is not all that visible in the projects (Steffensen et al. 1997)

Important changes had taken place at Aker Stord in 1997 and many of the people we interviewed believed that the local participatory improvement activities would be better followed up when things settled and come back to normal. What happened over time? Our last audit was carried out in the period December 1998 – January 1999.

Before I present the audit itself, I will shortly describe important organizational changes that happened after 1997 due to the BPR efforts in the Aker Concern. One of the most radical changes came as a result of the BPR projects "The future workplace" and the "Prefabrication of Pipes reengineering project". These changes took place over a long period of time and continued for several years after the 1998/99 audit had been completed. For editorial and continuity reasons and in order to show the results of the changes that started in 1997, also changes that took place after the third audit are briefly described below.

8.8.2 Important changes – the BPR project "The Future Workplace" and the Prefabrication of Pipes reengineering project"

The situation facing Aker Stord in 1997 was a decline in the market for large integrated offshore platforms. More and more offshore fields were being developed by the use of underwater installations, the produced petroleum was transported in pipes and constantly more of the processing was done onshore. This intensified the competition for building and assembling the few large top-sides that were decided to build. The offshore modules became smaller and thereby opening up for new competing suppliers.

Aker Stord lost some large bids and the order situation became critical. Something had to be done to improve competitiveness. The people at Aker Stord also realized that the change from being an independent company to becoming an assembly plant in the Aker value chain created some new challenges. Most of the production planning and documentation was now prepared at engineering departments in Oslo and Stavanger. This created some efficiency problems: *"Yes, I get my products, my bills, material lists and all procurement planning from Oslo. There are some communication challenges, then, there is the distance, and they have many people without any background from production. You get products that you are not pleased with, and there are lots of things that you do not handle as easy as you did before……. We did not have anybody we could run over to and discuss problems, production schedules, drawings etc."* (installation manager, my translation).

Based on input from many sources, including experiences from offshore construction, top management pursued the idea of establishing operations centers/central control rooms where the personnel in projects could handle challenges, move technical clarifications away from the construction units, place technical personnel close to the construction area to handle remaining clarifications and improve productivity by improving the work planning and preparations. Top management decided to use experiences from the first BPR pilot to work out the details and plan the practical implementation. One of the trained BPR facilitators was given the responsibility to facilitate the process. A project group composed of staff members and managers, and a steering committee representing top management, were appointed. This project was part of the process of transforming Aker Stord from a primarily functional oriented company to a process oriented organization. Based on the BPR methodology from the first BPR pilot, the main principles for the future work organization for building projects were developed. These are presented in Figure 14.

The most important structural differences compared to the former organization were that the number of managers in the construction organization was considerably reduced. Two layers of managers in the line of command disappeared, the discipline managers and the foremen. A new role, the installation manager appeared, and partly self-managed operator teams consisting of 3-6 employees reporting to the installation manager were established. The former foremen, who earlier supervised and controlled the daily work of the operators, were substituted by the operator teams and installations managers, who were now responsible for several operator teams. Operation centers were established, first at Prefabrication of Pipes and Prefabrication of Steel, and eventually for all disciplines. These operation centers are manned by technical personnel who are responsible for preparing documents and drawings for the operator teams, detail plans, logistics, field engineering, HSE and mechanical completion. Operators can turn to these operation centers to discuss or clarify technical issues, plans, get information, report the need for adjustments etc. Computers were also placed at the production locations and operators can now order materials, report on progress, on man-hours and other functions that earlier were done by foremen.

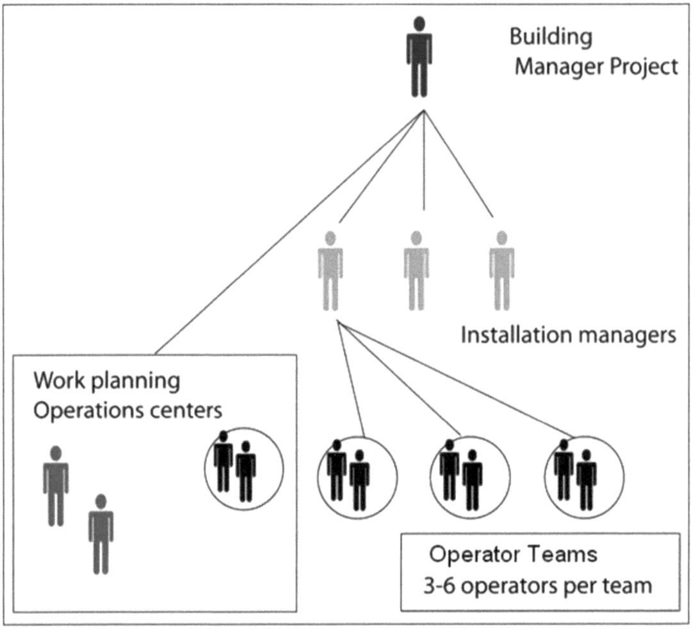

Figure 14 Principles for the new organization for building projects (my translation of internal Aker Stord slides).

The operators can also pick up drawings, plans and documentation on the computers themselves without having to go to see the technical staff in the operation centers or ask managers. The change principles for operators were summed up as follows:

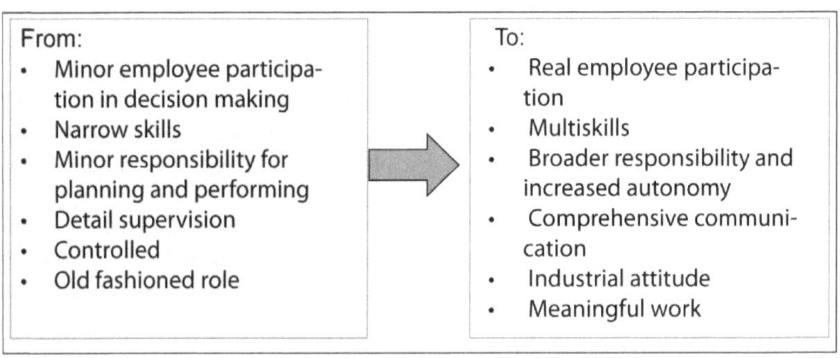

Figure 15 Change principles for the operator role (my translation of internal Aker Stord slides)

Corresponding principles were developed for the installation managers, who were the first-line managers:

From:

- Expert
- Daily operations
- Supervision of individual employees
- Detail supervision
- Directing
- Dominance and control
- Authoritative management
- Old fashioned role

To:

- Facilitator and advisor
- Adaptation and development of process
- Management of operator teams
- Borderline coordination
- Dialogue and cooperation
- Facilitate for employee autonomy
- Key role in change processes

Figure 16 Change principles for the role of first-line managers

Managers should now focus more on facilitating and advising and not on direct control and supervision of individual employees in line with basic principles of "The New World of Work" (Hammer & Champy 1993). The operator teams are responsible for a specific and defined process and the products the team delivers. The members of the team cooperate on planning and performing the tasks. The installation manager is responsible for the product that the discipline delivers.

When the principles for the new organization were presented for the top management by the project team in 1998, it was decided to test out and evaluate this radical new way of organizing the building work in two pilots. Prefabrication of Pipes and Prefabrication of Steel were chosen. However, it took more than two years before the pilots were formally started. In the meantime the company started building the operations rooms and started using them gradually from late 1998. I will now give some further details about how the development of the detailed solutions to the new organization was carried out at Prefabrication of Pipes, and how employees were involved in forming this radical new organizational innovation.

The change process at Prefabrication of Pipes was facilitated by one of the trained BPR facilitators, and the brown wallpaper methodology from the first BPR pilot was used. The work started up with a kick off meeting in November 1999 where 35 managers, shop stewards and employees at Prefabrication of Pipes met. The employees that met were picked out in cooperation with the union. The question was "How do we want our jobs to look like in the future?"

The kick of meeting was a combination of presentations and group work. The plans for the new work organization were presented and the group work was arranged as brainstorming sessions where ideas were written down on yellow post-its. A considerable number of ideas were presented and arranged under different headings such as HSE, work planning, flexible working hours and job content. These ideas were gathered and used as inputs to and guidelines for the further development process. They were also used as a basis for developing a vision for the future work organization at Prefabrication of Pipes. The development of the vision was facilitated by the facilitator and developed by one work planner and two operators. The vision was approved by the management of Prefabrication of Pipes and presented to all employees.

The brown wall paper methodology was used all the time, not only for analysis and planning, but also as a means of communication the process to all employees. The methodology was used both to develop the whole Prefabrication of Pipes process and also for planning different sub-processes and work-processes. The work was done in minor work groups composed of managers, shop stewards and operators that had been appointed in cooperation with the union. 25-30 operators, managers and shop stewards participated in these work groups (Aker Stord 2000). All the process charts, pictures and proposals that were developed by the groups were presented for the employees in workshops. Here employees were asked to give comments, especially in relation to weaknesses, risks and opportunities. In addition, members of the work groups continuously presented and discussed ideas with colleagues and received feedback and ideas. Furthermore, the TKA practice of idea-generation in team meetings continued. Many proposals from employees on how to further develop the new future organization also came up through this channel and were treated in the DC along with other ideas.

To support the brown wallpaper methodology, several of the TKA techniques that ensured broad employee participation that the employees were familiar with were utilized. In addition to team meetings and a comprehensive use of workshops, examples are yellow post-its, brainstorming, customer-supplier interviews and stakeholder analyses to ensure smooth interfaces between disciplines and functions. Employees participated in all these. The brown wallpaper methodology itself was developed in a way that ensured employee participation. Consequently, Prefabrication of Pipes was one of the departments where managers, employees and shop stewards had some experience with elements of the new work organization before the formal pilot started. In fact they had started to use an early version of Operations Rooms from late 1998, without fully implementing the new organization.

The organization of the work at Prefabrication of Pipes was developed according to the principles for the new future organization. The introduction of the new organization, facilities and equipment came gradually from 1998 and onwards, Most was in fact in place before the formal pilot started up in the first half of January 2001. The pilot was managed by one of the trained BPR facilitators, and the project group was composed of two installation managers and four operators, who were selected in cooperation with the union. The mandate for the pilot clearly expressed that detailed planning should be performed in a way that ensured employee participation (Gandrud 2001). The most important products from the formal pilot were procedures needed to make the new organization work, checklists and an evaluation of managers, operators and shop stewards viewpoints and experiences with the new organization.

An evaluation of the new organization was carried out by me and another action researcher from IRIS in December 2000/January 2001, and is summed up by Gandrud (2001). A considerable number of viewpoints were given. For example, the operators had now started working without being supervised by a foreman. They help each other, cooperate, plan, find solutions to challenges coming up, they report progress, order materials and other actions that make the production go smoother than before. Some report that the coordination and communication between disciplines are better, that they discuss challenges, support each other and that there are fewer conflicts than before. The job packages are now larger and more complete, which they find very positive. Many have received more training, and get to know more about other disciplines. They find the work more challenging, they experience more freedom, and autonomy has increased. Some find it easy to get support when they need it, others express that they sometimes find it difficult to know where to go when they have questions. Many find that the work planning and preparations are not good enough, that equipment and materials are not in place when they need it, and that many routines are not yet good enough or well established. Some also express that they find that the IT equipment can be too complicated for beginners. The main conclusion is that a considerable number of improvements and further development were needed. However, the majority saw the potentials of the new organization, that the production now runs smoother, and expects the arrangements to be further developed.

It was a radical change for the company and for the employees affected. The opinions expressed during the interviews that we carried out varied. A former foreman was not very satisfied: *"I had to get up at night and see what was written on the wall. I did not quite remember. I am bitter. I feel like stabbed from behind. I was not one of the selected"* (my translation). A few former foreman and some other managers were appointed installation managers. The majority were offered operator positions and consequently a salary reduction.

A member of staff gave the following comment: *"What is exciting about this is that operators are now being defined as customers, they enter the room to express demands, they are being asked to express what they want. Earlier they have been told, told, told, but all of a sudden people start asking what they want, so the process is thereby process oriented, customer and product is in focus"* (my translation).

An operator expressed it as follows: *"Today people go directly to work. We do not sit on a box and wait for the foreman to come and tell us that we can continue the work we did yesterday. People consider this change very positive. People now feel that it is their responsibility to get some work to do. Before, that was the foreman's problem"* (my translation).

Another operator said: *"I mean that it is good to have a job from A-Å. Earlier, a smaller job that was not finished by the end of the day, was often forgotten, until somebody asked. Now the team owns the whole job. We know what is on hold and what is coming"* (my translation).

The change gave more team work both inside the operator teams and between them: *"Now we are more active in searching for information and learning. We discuss with other skilled workers in other teams*" (operator, my translation). *"Before, we worked most of the time alone. If I had questions, I went to the foreman. Now, I have to address challenges and ask horizontally"* (operator, my translation). *"We have meetings in the morning – the team. If one team have problems, the operator teams gather. If we have serious problems, we go to the operations center"* (operator, my translation).

The pilot was finished by the second half of February 2001. The formal pilot and evaluation period was very short because the company needed the experience as a basis for planning for a new large contract that was coming up, the Valhall top deck. In sum, the organizational changes were considered promising, and top management decided that this was going to be the future organization for all building projects. Top management was so convinced that the new organizational model was heading Aker Stord in the right direction, that they used the estimated productivity improvements from one of the projects that Prefabrication of Pipes had finished as a basis for all disciplines when they calculated a new large bid. The Grane Drilling Module was completed by Prefabrication of Pipes, utilizing the new organization, with a 44% productivity increase compared to estimates based on earlier jobs and experiences (The blue book). *"We won the contract*" (member of staff, my translation).

The new organization soon became daily practice for all building projects for productivity, learning and work environment reasons. Now, BPR/process innovation became the dominant recipe for development, both at Aker Stord and in the Aker Group as a whole. This had consequences for both the division of work and the cooperation between the enterprises in the Aker Group, as well as for the internal organization in the individual companies. The whole Aker Group became process oriented. One of the consequences for Aker Stord was that the company lost some of its independence and became more integrated in the Aker value chain as the assembly plant in the group's main business process. At the same time Aker Stord got a new top manager who clearly gave priority to BPR. In other words, important changes took place. However, even if BPR and radical process change had now become an integrated part of TKA, most people still referred to the changes that took place as the improvement process. Consequently, when I refer to the improvement process at Aker Stord, it is an abbreviation and a company expression that includes both continuous improvement and radical process change. Now back to 1998/99 and the third audit.

8.8.3 The 1998/99 audit on employee participation in TKA and important changes

Like the 1996 and the 1997 audits the results from the 1998/99 audit are presented under the main headings 1) Information/training, 2) Idea generation and 3) Implementation/ follow up. The audits were basically carried out in the same way, but because of changes in the company over time, changes in the improvement activities and also in the way the company wanted the results to be presented, the way I present the results differs a little for the three audits. I still use the expressions "the improvement process" and "improvement activities" the way it was used at Aker Stord, namely including both incremental and radical change.

8.8.3.1 Information/training

Direct information from the top management was well received

Direct information to employees from the line managers had been reduced. The employees were quite happy with top managers visiting the departments and with general meetings to give information about improvement activities. Otherwise, employees reported that managers at all levels spent less time and resources on informing the employees directly about the improvement activities than they used to do. The reason offered for this is that the scope of the written and electronic dissemination of information had increased. However, the information provided by superiors varied considerably and was selective in relation to what they felt was most relevant. Previous audits showed that the direct information from managers, which had now been reduced, was highly valued.

The shop stewards did not manage to inform everybody

The local union committee, members of the WC and full time shop stewards was well informed. Beyond this, what the shop stewards gave information about appeared arbitrary. Most of the information went directly when they were talking to individuals, but they *"couldn't reach everybody"* (shop steward, my translation).

Information provided via e-mail was good, but did not reach most of the employees at Aker Stord.

A lot of good information is provided via e-mail and the company newspaper to managers and salaried staff. Most of the operators had no access to e-mail, but many work supervisors were good at posting the most relevant information on the notice boards. There were nonetheless many operators who did not read it. In other words, it could not be expected that everything that had been sent out via e-mail would reach or be read by the operators.

There was a lot of good information on the Intranet. Few read it. Few of our informants visited the intranet to gather information themselves. The few who did know what to look for found a lot of relevant and good information.

Internal newspapers functioned well as a dissemination channel

Very few read information provided by the Aker Group. Top managers reported that they read it. What is issued by Aker Stord (Aker Stord Avisa, System Posten) was characterized as good and informative, and was read by many. Those who had received project related information letters were very happy with such initiatives.

Employees who worked in the projects were better informed and involved in the improvement activities now than previously

The 1997 audit pointed out that the information about the improvement process and activities was poorer in the projects than in the basis organization. This had changed somewhat over the past year, as a result of many of the improvement initiatives now having been introduced in the projects and of people becoming more a part of what was going on. Also, more of the information was received electronically, which made it easier for those working outside the basis organization to stay informed. Otherwise the situation was as before. People working in the projects received most information about what was relevant for the project they were working on.

Different opinions

Salaried staff, central shop stewards and managers felt that the amount of information had increased and was sufficient. Operators wanted more oral information through line management. Most felt that sufficient information was being spread, both in a written and electronic format. Managers, salaried staff and central shop stewards had ample access to information. Many work supervisors did not read it and did not pass it on.

8.8.3.2 Participation in problem identification and idea generation

The number of appraisal interviews is considerably reduced

The attitude towards the scheme varies greatly. Salaried staff is positive, but operators are lukewarm.

The appraisal interviews were a voluntary arrangement, but managers had been instructed to offer employees annual interviews. Due to large reorganizations and a high tempo in 1998, the use of appraisal interviews appeared to have been considerably reduced in 1998 compared to 1997. Many said they had plans of reintroducing them in 1999. Both the use of appraisal interviews and the attitude towards them varied considerably throughout Aker Stord. Most managers and salaried staff considered it a useful tool. Work supervisors and operators were less positive. Many felt that the schematics were unsuitable, that the whole thing was too structured and that a more informal approach was actually being practiced, more or less in the form of chats. The most important aspect appeared to be to have an opportunity to talk freely about matters that concerned them.

The appraisal interviews appeared to function as a tool for planning and following up development and improvement objectives for some people, but this did not apply to the wider majority. One matter that made this difficult was the rapid changes that took place both in the Aker Group and in Aker Stord at the time. Frequently, new managers joined the company, there were frequent personnel moves between the projects, and many had little contact with their managers in the basis organization. Issues and wishes that were addressed in the appraisal interview often became irrelevant shortly afterwards.

Employees who had participated in idea-generation activities pleased

The employees that had participated in brainstorming sessions, customer-supplier interviews, the brown paper method, pre-job meetings and experience transfer meetings were very satisfied with the arrangements.

Brainstorming sessions, use of post-its, customer-supplier interviews (often without using the recommended schematics), the brown paper method used in the process redesign projects, pre-job discussions and experience transfer meetings were highlighted as examples of change techniques the informants valued.

The use of pre-job discussions varied as it was not considered relevant to many tasks. Some had used them during the preparation before complicated operations that included a lot of uncertainties. In those events they were often linked to risk assessments (safe-job analysis, and hazard and operability study). All who had participated in these meetings characterized them as useful tools and believed they should be used more often. For example, there were good experiences of pre-job discussions during the planning of maritime operations.

Experience transfer meetings between the projects were characterized as very useful initiatives. Information was provided and pitfalls to be avoided were pointed out. These meetings are often arranged in connection with risk assessments or in connection with the start-up of new projects.

Brainstorming is an improvement technique that most people had good experience with whether it had been used in annual departmental seminars, in teambuilding sessions or in projects. Everyone appeared to have obtained many good and relevant suggestions for improvements. However, most people expressed disappointment with the follow-up in line with previous years' audit results. Some initiatives had been implemented, but which ones were chosen seemed a little arbitrary and dependent on what individual managers decided. One of the positive exceptions was Prefabrication of Pipes, which over several years had been successful with a serious and systematic follow-up of the suggestions that had been produced at among others the department's annual DC seminars.

Customer-supplier interviews is a tool that all the interviewed persons appeared to have good experience with, and which had helped to bring about a number of improvements and also to kindle people's awareness and change their attitudes. But it appears that far fewer interviews were conducted now than in the previous year. *"There is no longer any pressure exerted to conduct these interviews"* (first-line manager, my translation). Most interviewees had conducted the interviews far less formally and without the recommended schematics.

The quality and frequency of department meetings/group discussions varied greatly

Most of the meetings appeared to be information meetings/one way communication.

The informants believed that few suggestions were obtained for improvements in these meetings although it was an objective at Aker Stord to make this happen. First-line managers did not always encourage this. For example, some employees who had presented suggestions for improvements in a group discussion had been asked to place them in the suggestion box instead. However, there were positive exceptions such as Prefabrication of Pipes.

The DCs function well in stable departments but are not adapted to a process and project oriented organization

DCs are the most important body for co-ordination, organization and follow-up of local improvement activities in the operator environments. But how well the DCs functioned varied considerably as referred to in the two other audits. As in previous years, it was Prefabrication of Pipes that was referred to as the star example, and which several other DCs attempted to learn from. Here a work practice had been developed that appeared very functional and which both managers and employees were very satisfied with. And, what is very important, they had managed to keep up the pressure on the improvement activities over a number of years, including the BPR period when radical process redesign was on top of the change agenda.

The DC is anchored in the basis organization. One of the challenges was that many departments had people working in the projects for lengthy periods, which left few people in the basis organization. The projects did not have a body that corresponded to the DC, which appeared to function well in stable departments, nor did Aker Stord have a joint cooperation body that was adapted to the process and project oriented organization the company had moved more closely towards recently.

There is less use of the suggestion box. The processing period is considered too long

The suggestion box was considered a useful initiative both by operators and managers. However, it was not used in the staff environments. The managers felt it was their job to ensure development, and they did not receive the same rewards as the operators. There were also complaints against the processing time being too long and that this had become worse over the past year. The number of received suggestions had fallen considerably over the past couple of years. In 1996, 712 proposals had been received, in 1997 it was 529 and in 1998 the number had fallen to 463. There were also complaints that a number of suggestions had been rewarded prizes, but had not been implemented.

RUH was considered a good and important routine, as it contributed to specific improvements and increased safety awareness

There was consensus among the informants about the importance of reporting undesired events (RUH). Several people also felt that they could see the value of reporting in the form of implemented initiatives. The safety awareness in Aker Stord appeared to be high. Several departments appeared to have sound and quick routines for following up RUHs. As for the suggestion box, also the RUH is not a TKA initiative, but the company wanted it included in the audit.

Participation in implementation and follow-up of improvement projects

Are there too many things on the go?

Several people pointed out that there were too many initiatives on the go to be able to follow up everything properly. New and comprehensive processes were initiated and took up so much time and attention that it was impossible to finish what was already going on in a proper manner. Whether it was possible to complete things was reported to depend very much on the managers' "pressure" on getting things completed. This was a challenge that was pointed out by both managers and shop stewards.

At that time Aker Stord representatives were expected to participate in development activities at both Aker Group and company levels. The objective for Aker Stord was to have three major company-wide improvement projects going on at any time. These were mainly to be anchored in whatever was given priority in the Aker Group. Several managers and salaried staff participated in Aker Group improvement processes outside Aker Stord in addition to the three internal projects.

The shop stewards expressed satisfaction about being invited to participate in all these activities. But it also worried them that things happened so fast and that so many things happened at the same time. There was no time to become familiar enough with it all.

The BPR projects had found a good methodology but some of the follow-up failed

The interviewees who had participated in Aker Stord's BPR projects were very satisfied with the approach and methodology. Reference was among others made to the brown wall paper method and the fact that so many persons from various positions were involved during the design phase. However, some said that they felt controlled during the group work and that they felt the end result was given in advance. They also reported that they felt the follow-up did not quite tally with the planning. Several expressed disappointment as not more had been implemented.

The importance of the reorganization and improvement process. Important changes

The reorganization process in the Aker Group and the limitation of Aker Stord's role in the value chain to assembly activities were important conditions for the further development of TKA. In general, the impressions from the interviews were that people felt that everyday life at Aker Stord had changed, but what had changed varied a lot. The most evident change was that people had to spend more time travelling, and that this commuting was perceived as very negative. The increased travel activity was linked to the change processes in the group and many were critical towards the centralization that had taken place: *"The centralization has led to our enterprise becoming a smithy in the countryside"* (first-line manager, my translation). A larger proportion of the installation work was now done on different geographically spread sites compared to the preceding years where most of the work was carried out inside Aker Stord's yards. Several pointed out that the employees' loyalty to Aker Stord had been weakened due to the increased travelling. A positive aspect was that people who work a lot in the field experienced that they learned a lot from working in different environments and that this knowledge benefited Aker Stord.

People felt uncertainty in relation to their job situation

Another important consequence of the changes that had taken place is that they created a lot of uncertainty among the employees. This applied to all levels in the enterprise. The uncertainty was in particular related to what would happen to Aker Stord in the future, and people were worried about losing their jobs. Some felt that the uncertainty adversely affected their motivation for TKA, while others stated that they were just as motivated as before or even more so.

Process orientation as an organizational principle has not "set" in the whole organization

In the interviews, some employees expressed that they tended not to feel that the enterprise had become more process oriented. In general, some felt that no great changes had taken place regarding the way the actual practical work was done: *"It is only on paper and not in practice"* (first-line manager, my translation). Some felt that there had been changes, and that this had resulted in changes in their relation to colleagues. Some departments appeared to be harder "hit" than others. There were fewer management levels, but some people reported that the work was organized and carried out more or less as before. They had, on the other hand, become more aware of the changes that had occurred as a consequence of the Aker Group's operating model, and of Aker Stord's new role in the group. The process orientation was most visible at a superior level and in the parts of the organization where radical changes had been implemented. The knowledge of what was meant by process orientation appeared to be scant: *"Process-control and process-organization as a principle have not set in Aker Stord"* (middle manager, my translation).

Increased autonomy at work, reduced autonomy for Aker Stord

Many of the interviewees felt that their work situation had become easier because the procedures had become better and fewer. They felt they had a relative high degree of freedom in their work and believed they had been given a greater autonomy. *"Before, we used to have an inspector at our back at all times"* (operator, my translation). The enterprise itself appeared to have lost some of its autonomy through the centralization in the Aker Group, while several of the employees experienced a higher degree of autonomy in their work situation. No one reported less autonomy.

Quality has been maintained although the tempo and pressure have increased

Although the employees reported an increased pressure of work through narrower time margins, most of them felt that this had not affected the quality of what was produced. This also had something to do with how quality was defined, or as one of them said: *"The question is: what is the main philosophy for the Åsgard Project? Is it to be good enough, or Rolls Royce quality?"* (first-line manager, my translation).

Strong trust in the shop stewards

Employees at all levels felt that the shop stewards did a good job in relation to the improvement processes. Most of the interviewees felt that the shop stewards' influence was as before or that it had been strengthened, but that the tempo of the changes had made it difficult for them for keep up. The shop stewards felt that they were involved in a lot of things and that this made their work situation difficult. They were also a little worried about all the time, attention and resources that the change processes required, and that this could affect the job that had to be done. High expectations combined with a high work pressure made it more difficult to recruit new shop stewards.

People were more afraid to talk

In the interviews, several people said that employees had become more afraid of voicing their opinions and doubts than before, perhaps from fear of not "being included" any longer: *"Either do people participate in the change processes taking place, or they do not"* (Operator, my translation). Furthermore, many of the processes were being managed from Oslo. Expressing one's opinions was more risky than it used to be.

8.8.3.3 Changes from 1997-1998/99 – summary

Those who know how to use e-mail and the internet are well informed. The others are less informed than they used to be

The interviews showed that the employees' general knowledge of the improvement process and the improvement activities in SPIN and with Aker Stord varied depending on their access to and competence with personal computers. Most of the informants reported that they were poorly informed of the change processes.

The top management was well informed of the improvement activities that were going on in the Aker Group and at Aker Stord. Most of them participated in or had participated in central activities both at Group and company level.

Middle managers were also well informed, and some had taken part in improvement activities both locally and in SPIN.

Salaried staff appeared to have good access to information, among others through e-mail. But they were very selective with regard to what they felt important to be informed about. Full time shop stewards, members of the local union board and in the Works Council were well informed about the improvement activities, both in the Aker Group and locally. Several also participated in the improvement activities either directly or indirectly through steering groups or reference groups.

The work supervisors' information about processes and activities varied considerably. Most of them appeared to know about the things that concerned them most, but were poorly informed about other activities. Some had access to e-mail and printed out newsletters. Others did not.

Local shop stewards had little information about most of the improvement activities and in which contexts they were included in SPIN. On the other hand, they were well informed about the activities that concerned them the most.

The operators were poorly informed of the improvement process and activities. Most of them did not have access to e-mail and read written information very selectively. Few had participated in specific improvement activities.

The volume of distributed information about TKA and the reorganization process had increased considerably over the past year, while oral dissemination via line management had been reduced correspondingly. Those who mastered, were happy to use, and had easy access to e-mail and internet were the winners. The rest received less information than before. It had become easier to keep informed for those who were stationed outside the Aker Stord yards.

Some experienced little transition from functional orientation to process orientation

A process direction as a governing principle had not settled at Aker Stord as a whole.

The process orientation was more visible at corporate level. There was less talk about TKA, although some of the initiatives were still running.

More management controlled change-processes, less participation from below

It was said that focus was now placed on management controlled change processes that were initiated at corporate level. In total there were fewer suggestions for improvements and initiatives from below. This concerned for example appraisal interviews, local improvement projects, brainstorming sessions, local development meetings, the suggestion box and customer-supplier interviews. Viewed as a whole, the employees experienced that the improvement process had become more top-down compared with the situation 2-3 years before.

DC as an arena for local improvement activities is well suited under stable conditions but is not adapted to a process and project-oriented organization

The experiences gained at Aker Stord indicated that a degree of stability and relatively clear departmental boundaries are important success criteria for a well-functioning DC. In large sections of Aker Stord these criteria were missing and the DC was not functioning as intended.

8.9 The challenges of combining TQM and BPR synergistically– empirical analysis

The experienced relationship between TQM and BPR at Aker Stord and the main findings will now be analyzed in relation to my research question. In my theoretical discussion I found that TQM and BPR have some important conflicting characteristic and some important reciprocally reinforcing characteristics. I concluded that in theory it seemed advantageous to use the kind of thinking that TQM and BPR represent together in the same development process in the same system, at the same time and on an ongoing basis. Consequently a main conclusion from the theoretical discussion was that a synergistic approach to combining TQM (exploitation characteristics) and BPR (exploration characteristics) seems appropriate. However, an important question is whether this is reasonable in practice?

8.9.1 TQM and BPR. Trade-off or reciprocally supportive?

Both the 1997 and the 1998/99 audits showed that there was a great deal of uncertainty among leaders, shop stewards and operators as to which initiatives and mechanisms were important right now. Was it the first TKA activities that promoted participation in continuous improvement that was important at the moment or was it the radical process redesign activities? There was also uncertainty related to the connection between TKA and BPR. In the interviews, people kept referring to TKA and "the other process". The company had failed to portray this as a change process with varying mechanisms. What were given most attention in 1998/99, were the BPR projects that had been initiated both at Aker Group level and internally at Aker Stord. Most people had a remote relation to the processes that were rooted in the concern, but they noted that these had consequences for their everyday life. It became a period marked by uncertainty.

Many managers were in 1998/99 more concerned with implementing the centrally initiated changes they had been ordered to carry out than with focusing on locally initiated innovation activities. Many employees felt that the managers followed this up less well than in 1996, and focused less on it. There were fewer requests from management for results from local improvement activities now than before.

This uncertainty manifested itself in many ways. First, there was less use of appraisal interviews. The same applied to customer-supplier interviews. Then, the scope of local improvement activity in the departments and in the DC diminished together with the enterprise's overview of the ongoing activity. Statistics of the improvement initiatives were no longer kept. Furthermore, the number of suggestions in the suggestion box had fallen. Also TKA training ceased.

Much of this reduction in local improvement activities was due to changes in departments and new reporting routines, but it was quite clear that the managers no longer asked for nor initiated many of the initial TKA activities. Many of the initiatives that it had taken a long time to build up and which worked in the initial TKA phase were no longer considered important.

Interesting from an EFQM perspective is the following paradox. In line with the EFQM model (EFQM 2003) the EFQM self evaluation group recommended that the TKA process at Aker Stord needed tools and practices for identifying, evaluating and renewing business and work processes. This should make the TKA process at Aker Stord more successful, stronger and more complete. What happened?

In fact, BPR resulted in a weakening of the continuous improvement activities in the company, which is fundamental to TQM. This is line with and supports March's (1991) trade-off position: *"exploration undermines exploitation"* (March 1995:433). The company pursued a predominantly synergistic approach to combining forms of exploitation (TQM) and forms of exploration (BPR). However, they ended up with a predominantly antithetical approach with the two alternating in time in line with Imai (1986).

A synergistic/dual approach (Sutcliffe et al 2000, Boer et al 2006) to the balancing between control/exploitation and exploration seems appropriate in theory, but difficult to accomplish in practice. However, I find many examples that show that the initial phase of TKA, which focused on TQM/Continuous Improvement, provided a fundament and supported the more radical BPR activities. E.g. customer-supplier interviews gave employees a better process understanding, awareness of and overview of the internal supply chain, and thereby contributed to prepare the ground for exploration. The Works Council gained experiences with the following up of improvement projects from the initial TKA phase and was motivated constantly to take on new and larger projects. People became familiar with the post-it technique through brainstorming sessions. This technique was extensively used in BPR projects. People who had participated in incremental change processes became motivated for participating in more comprehensive and radical projects. The department Prefabrication of Pipes had a well functioning legitimate practice for dealing with improvement projects that was also used during the period of radical process redesign projects.

8.9.1.1 The importance of organizational learning

The department Prefabrication of Pipes was a "fast learner" (March 1991) of the TKA codes. The department was considered good at continuously improving existing procedures, routines and ways of work. Before BPR was introduced, these improvements were predominantly incremental changes in existing practices. In March's terminology, the exploitation capability could be considered as good. According to March (1991), it should be expected that fast learners contribute less to the exploration of new possibilities than slow learners when a new concept (BPR) with new codes is introduced. March claims that exploration of new alternatives reduces the speed with which skills at existing ones are improved, and that improvements in competencies at existing procedures make experimentation with others less attractive. According to March, learning the organizational code and imitation inhibit experimentation.

This seems logical if exploitation goes at the cost of exploration. However, this was not the case at Prefabrication of Pipes. The fast learners of how to continuously improve existing practices also became the fast learners of experimenting with new radical alternatives. Prefabrication of Pipes was the local star case of the TQM/Continuous Improvement phase of TKA, and became one of the local star cases in the BPR phase. This department managed to continue the TQM practice at the same time as they reengineered their department and their work processes radically. In fact the total number of registered improvement projects decreased as for the rest of the company. However the projects became larger and more comprehensive and were followed up by minor projects.

This experience invites the question: Under which conditions are March right and when is March wrong about the connection between slow learners and exploration capability. On the one hand the situation was that some of the codes from the TQM concept could be used in the BPR efforts. For example both focused on customer value and some of the quality techniques could be used in the reengineering projects. On the other hand there were also fundamental differences in the codes like the radicality, the speed of change, employee involvement and expert orientation in BPR. Prefabrication of Pipes succeeded with the introduction of radically new work processes and practices very quickly, and with outstanding production results.

Sutcliffe et al. (2000) take another position than March (1991), which appears to be more in line with the experiences from Prefabrication of Pipes. They claim that as individuals gain control over key task behaviors and exercise discretion in performing those behaviors, they develop a sense of competence. As a sense of competence increases, individuals are better able to respond effectively in unfamiliar and/or changing situations, and effective action subsequently reinforces that sense of competence. The authors further argue that these kinds of processes at the individual level are similar at group and organization level, and that collective beliefs can have a positive effect on performance. They hold that as organizational units better understand their capabilities, competencies and identity, collective beliefs about a unit's capability for action can be important for facilitating resilience and for achieving synergies between exploitation (control) and exploration.

Prefabrication of Pipes was well known for their good results, proud of it and knew that they had a capability for action. According to Sutcliffe et al. (2000) better trained and multifunctional groups develop better sensing and coping capabilities. The reason for this is that as the capabilities for action increase, work groups that perceive many possibilities for action may be better able to grasp variations in their environment. The more a group sees in a situation, the greater the likelihood that it will see specific changes that need to be made. Prefabrication of Pipes very soon adapted to the new BPR codes, and came up with outstanding results in a short period of time. Aker Stord personnel recognized this. The most impressive example of BPR success they could produce was the remarkable increase in productivity at Prefabrication of Pipes. This department profited from and participated in the reengineering projects at the same time as they continued their continuous improvement activities.

These findings and considerations suggest that an exploitation oriented development approach constitutes a good foundation for a process exploration oriented development approach. This is not in line with March's predominantly trade-off positions, but in line with several other studies (e.g. Sutcliffe et al 2000, Cardinal 2001, Greve 2007, Isobe et al 2007, Yalcinkaya 2007). Thus, the Prefabrication of Pipes experiences support the theoretical position that forms of exploitation can enhance forms of exploration, at least in the radical process innovation area.

Now, why did this department manage to develop a mostly synergistic approach? The top manager was the same and gave the same signals to all departments and employees. The Aker Group and the company strategy was the same for all departments. Many other departments were in the same situation regarding level of competence/ professional skills, geographical position, involvement in change processes, being part of a production value chain, using the same structure for two party cooperation and employee participation, etc. The commitment of the department manager seemed to be crucial.

8.9.1.2 The importance of leadership

Prefabrication of Pipes had a manager who was very dedicated to and active in both the TQM/Continuous Improvement part of TKA and the new BPR approach. Even if the new managing director did not focus on TQM/Continuous Innovation so much anymore, the local manager of Prefabrication of Pipes did. He kept asking for improvement proposals, he continued to follow up the improvement activities, continued to ask for results and kept up the practice of using the DC as the coordination body for the local improvement activities. The BPR process was in fact supported by this practice. The radical change in the pipe logistics affected this department very much. Among others, the number of personnel was reduced to the half, and a need for many minor changes followed the more radical and comprehensive ones. In other words, the manager facilitated change in a way that made forms of exploitation capability support forms of process exploration capability. It is also part of the story that the manager of Prefabrication of Pipes was very often invited to give presentations of the way they worked both in internal meetings, in IfS network meetings and in other meetings outside the company. He was well-known for this both inside and outside the company.

Many other managers reduced their TQM/Continuous Improvement focus as BPR became more important. This underlines the importance of management focus in these kinds of processes (Deming 1986, Juran 1989, Yukl 2006). Neither I nor the company representatives that I cooperated with during the audits could come up with any other good reasons for this development in Prefabrication of Pipes. Again, the skills and competence levels were not very different from those of other departments. The same development directions were given to all departments from top management. Union membership and arrangements for two-party cooperation were similar to other departments. Most departments were located in the same area, and the composition of the workforce did not differ very much regarding background, gender or age.

For the rest of the company the management's and the new top manager's focus on process renewal resulted in a neglect of continuous improvement and the TQM practice that had been given a high priority in TKA the preceding years. The new managing director practiced and prioritized process renewal both in words and in action. The continuous improvement efforts more or less belonged to the previous CEO, even if TKA was planned to continue as before. This not only weakened the company's continuous improvement capability, but also employee participation in improvement activities, which is fundamental to TQM.

Regarding the latter, another important finding from the Aker Stord case is the company's strong focus on employee participation in innovation processes and the fact that employee participation was considered to be an important enabler for TKA success. Prefabrication of Pipes' success both in the initial TQM phase of TKA and in the predominantly BPR phase is an example of that. Prefabrication of Pipes managed to ensure an extensive participation of employees and their representatives in both phases. In addition to leadership focus, employee participation was considered to be the other important reason explaining this departments' success.

8.9.2 Employee participation in TKA

My theoretical discussion showed that TQM is based on direct employee participation in Continuous Improvement. It also showed that TQM is built on consensus thinking, and has no recipes for the inclusion of trade unions and indirect employee participation.

Aker Stord introduced or further developed several tools for direct employee participation as part of their TKA effort; e.g. the employee appraisal system, before-job meetings, safe-job analysis, team conversations, brainstorming/the post-it method and customer-supplier interviews. Aker Stord succeeded with the use of these tools, but not in the whole system, not at the same time and not on an ongoing basis. We found that some departments had success with some of the tools, other departments did not succeed with the same tools and some did not use them at all.

8.9.2.1 Factors explaining employee participation in TKA

A decisive factor for obtaining a high degree of employee participation in TKA is management focus. Interviews with managers, shop stewards and operators all pointed in that direction, which also corresponds with the quality literature referred to in Chapter 3 (Deming 1986, Juran 1989, Tushman et al 2006). We were presented stories about how the managers of departments such as Prefabrication of Pipes followed up the TKA initiatives systematically with training, idea-generating activities, change projects, employee involvement, asked for results and showed personal commitment. Other managers did less. So, comparable departments in the same line of reporting showed very different results. The managers who succeeded with TKA were also visible in internal magazines, where they promoted the improvement process and presented results and good practice.

Two other important factors explaining variation in TKA participation were degree of stability and organization. In units where conditions were unstable and changing, it was difficult to keep focus on the improvement process. Also project-organized units had difficulties focusing on other matters than the actual projects they were working on. In the latter case this was mostly due to the distance to line management, lack of information and lack of formal arrangements for organizing employee participation the improvement process. These three important factors for explaining variation in employee participation in TKA are summed up in Figure 17.

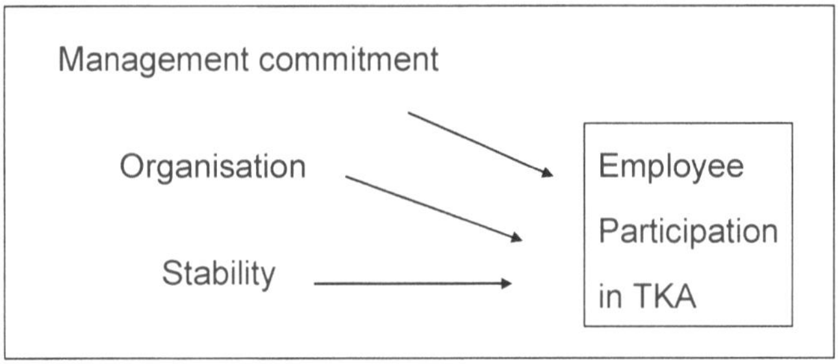

Figure 17. Important factors explaining variation in employee participation in TKA activities

8.9.2.2 TKA and arenas for company-union cooperation

Theoretically and also in practice direct employee participation in business development in TQM appears to be limited to lower level/socio-technical participation. Direct employee participation at higher/business strategy level was neither found in theory nor in the Aker Stord case. E.g. people did not participate directly in the development of Aker Stord's vision, main goals, strategy or values. The highest level where direct participation was practiced was in (some of) the departments, where people were invited to participate in brainstorming sessions and idea generation.

I further found that when Aker Stord started its TQM journey, the company deviated from the TQM textbooks and stuck to their long lasting tradition for company-union participation in business development when they decided how to organize the TKA process. This finding was unexpected, and built on mutual trust between the parties. The only change made in the organization structure when TQM was introduced was the appointment of an internal TKA facilitator/project leader. Except for that, the company and the union decided that the TKA implementation should use the existing arrangements for company-union cooperation as a basis for managing, coordinating and organizing the process. No extraordinary steering committee and no specific TKA development organization were created.

The most important effects of this arrangement were that the company's existing practice of indirect employee participation became a means to TKA implementation, and that the "opposing partner" to company management became the cooperating partner in the TKA process. TKA was introduced without major conflicts and with union support and contributions. Furthermore, the company did not need to establish new formal arrangements that would consume time and management attention. It could profit from a well-established practice, routines that work and were trusted.

The literature (Juran 1989) recommends the establishment of formal arrangement outside the ordinary organization in order to show visibility, priority and secure focus. However, at Aker Stord the top manager demonstrated high commitment to the process, demanded the same from his direct reports and gave the process considerable place and focus in management meetings. It was the line organization that ran the TKA process and not a committee outside the formal organization. TKA was integrated in the daily and long-term running of the business.

The 1996 audit showed that this arrangement made TKA and indirect employee participation work reciprocally supportive. E.g. at company level the Works Council functioned as the "steering committee" for TKA, eventually including BPR and important radical process reengineering projects. These projects gave the Works Council more interesting challenges and tasks. The council became more important for both parties.

Another example is the choice of the Departmental Committee (DC) as the coordination body for local improvement activities. This practice gave the TKA process a tool for coordinating local improvement activities, which was anchored in the main agreement between the labor market parties and considered legitimate by most employees. At the same time this strengthened the DC, and gave it more important tasks. Also, this finding was not expected. However, like the TKA tools for direct employee participation, this arrangement was not successful in all departments and appeared to depend on the success factors presented above, where management commitment was decisive.

As part of the TKA process Aker Stord succeeded in obtaining a considerable number of improvement proposals from the employees in some of their departments. Several of these proposals are being implemented through immediate decisions taken by the manager in charge. Proposals of a more extensive character are usually presented to the DC. The DC discusses the improvement proposals and decides their further destiny. In many cases, the DC decides whether to implement the proposed actions or not. In other cases, a small project group is established to manage the improvement project and given a mandate by the DC, and reports back to the DC. The DC may also decide to send the proposal to other departments, to the company's work environment committee or to other places depending on the actual issue.

The audit interviews indicate that this is an efficient way of coordinating local improvement activities. The ownership and responsibility for the solutions are shared. It is considered a very legitimate way of decision making, and employees are secured a fair treatment of their proposals. E.g. the system makes it difficult for first-line supervisors to stop a proposal from an employee, if they don't like it. Many considered the projects that had been performed within the framework of the DCs as some of the most successful.

However, many operators, particularly in large departments, have little knowledge of what is going on in the DC. Other DCs at Aker Stord have managed to make this practice function well for many years. Although many operators at Aker Stord show little interest in TKA in general, the company have managed to create interest and an attitude towards caring about what happens at their own workplace. The DC enjoys a great deal of trust from both managers and operators, and they have managed to achieve documentable gains with regard to both effectiveness and work environment through this scheme.

Consequently, a main finding in the Aker Stord case is that central elements in the Norwegian tradition for employer-employee cooperation, under conditions presented above and summarized in Figure 18, can be combined with central elements in the TQM tradition, with a reciprocal re-enforcement as a result. The activity in many of the DC meetings has increased and has become more constructive. Shop stewards, other DC members and employees have been trained to use problem-solving techniques, and have improved their ability to participate in business development and thereby in shaping their future work place.

The DC practice at Aker Stord also illustrates how the company has managed to combine direct employee participation, through brainstorming sessions, group discussions and individual initiatives, with indirect participation in DCs in an appropriate manner. It has been possible to mobilize employees to participate in improvement activities at grass root level, and the shop stewards have been given several improvement proposals that employees have prioritized for the shop stewards to follow up, among others in the DC. Many of the DCs have gained an increase in activity and have become a more important body for both managers and operators. The DC practice also shows that the enterprise has managed very well to connect local improvement activities at department level with the various activities and improvement arenas at other levels in the company.

Regarding TKA, the DC practice offered the concept valuable cooperation arenas for dialogue between parties that sometimes have conflicting interests. Bringing conflicting views to the surface, creating a climate for understanding and listening to the background for different views and then compromising, may reduce costly tense and frustration in the organization. It may become easier to join forces for common goals and prioritized innovation initiatives.

This way of organizing the innovation activities has given the shop stewards a very active role in the TKA process. Not only do shop stewards encourage employees to participate in the improvement activities. They also keep an eye on supervisors and remind them of their responsibility for their part of the improvement process. Shop stewards now spend more of their time and talents on business development than before. These experiences also show that the shop stewards' attitude to and support for the improvement process have a great impact on local activity.

Like the improvement process in general, there were three basic conditions that were important for the success of the DCs. These are management commitment, stability and organization. Wherever the DCs functioned, there were visible managers who were clearly aware that this was important. They followed up, imposed requirements and participated actively.

Regarding stability, TKA initiatives seem to require time to mature in order to function. The departments that did manage to get TKA to function had a higher degree of stability than those that failed. This makes me question the appropriateness of this kind of agreed schemes in enterprises with a high degree of process and project organization.

The DC is anchored in the basis organization, but many employees spend most of their time in projects, reporting to a project manager. A high pressure on process improvement creates constant changes in the basis organization. People and managers are on the move. Prefabrication of Pipes was considered to be about the closest we can come to a stable workshop at Aker Stord. The company's installation departments and the projects did not attempt or, even if they did, failed to succeed with the kind of cooperation arrangements that require stability, like a DC.

It appears that the DC, which is anchored in the main agreement between the NHO and LO, is not as easily adaptable to project – and process-oriented enterprises as it is for stable hierarchical and functional oriented organizations. The practical arrangements for cooperation between the two parties formulated in the main agreement are based on the existence of departments as fundamental and important units in companies. Not so much on process – and project-organization. This problem is a challenge for Aker Stord and other companies that process and project orient their organizations.

8.9.2.3 What to choose – The individual-oriented suggestion box system or the collective-oriented DC system?

For many years Aker Stord has had a suggestion box system that has not been linked to the total quality process activities to any great extent. The suggestion box system allows individual employees to suggest improvements. In principle, it could be said that this is a system that competes with the DC, where suggestions initially are addressed in group discussions, and thereafter brought before the DC. A high activity level in one scheme could have led to a low activity in the other, but this has not been the case. For many years Prefabrication of Pipes has been the department that generated most suggestions in the suggestion box per head. The proposals from this department increased in both schemes at the same time. The two schemes had a mutually strengthening effect through creating a positive attitude to employee participation in enterprise development. Encouraging results from one improvement proposal scheme seem to motivate and trains people for using other schemes.

8.9.2.4 The change of focus from TQM to BPR

During the introduction of BPR at Aker Stord it was emphasized to secure broad participation in the internal BPR projects by addressing them in the WC, by establishing steering and reference groups for the projects where shop stewards participated, and by involving the operators when running the projects. But, even in a company with traditions for company-wide employee participation, where both the union and individual employees had been invited to participate in the reengineering processes, the audits documented several obstacles to real employee participation in Aker Stord's process renewal practice.

The company, which in 1996 could show many good examples of moves in the direction of industrial democracy, now seemed to move in the opposite direction. The change processes were rushed. Information came late, and there was little time for reflection, discussion, alternative solutions and participation in general. Few individuals were involved in the processes. Some informants even expressed a feeling of being used as hostages to change processes already decided on by management (Gandrud et al. 1997).

Many operators and shop stewards felt that things were going too fast and that participation became less real. Several reported that they felt the most important issues had already been decided on in advance. Shop stewards reported of less influence and participation in things that happened in the enterprise. Many were scarcely able to keep informed of what was happening.

So, many of the tools for employee participation that were established for a hierarchical and function oriented company both before TKA and as a part of TKA got less focus as the company gradually became more process oriented and project organized. New and adjusted tools and practice for employee participation adjusted and suited to process orienting were not developed, implemented and spread at the same speed as these changes came along. Also this was an important finding from the Aker Stord case.

8.9.2.5 Autonomy

The possibilities for employee participation in dialogue based decision making was reduced towards the end of the four year case study period. What about autonomy?

Autonomy was not an issue in the audits. However, in the 1998/99 audit many interviewees reported that autonomy had increased. There were many reasons for that. The reengineering and reorganization projects had reduced the number of managers at all levels. The number of foremen had been reduced and their role was changing towards facilitating rather than supervising. There was less control by others in the work situation. Many operators had been trained to do some of the quality control on their own. There were more computers available at the shop floor. Operators could look at production plans and drawings, order materials and do more of the work planning and reporting than before.

I find this observation interesting in a BPR perspective. A better work environment and enriched jobs was important objectives in the Aker Stord way of introducing BPR, in the way BPR was integrated in TKA and also visible goals in the BPR projects. Aker Stord adopted important elements from Hammer & Champy's ideas of a "New World of Work" (Hammer 1996) such as: *"Work units change – from functional departments to process teams, jobs change – from simple tasks to multidimensional work, people's roles change – from controlled to empowered, job preparations change – from training to education, managers change – from supervisors to coaches, organizational structure changes – from hierarchical to flat, executives change – from score keepers to leaders"* (Hammer & Champy 1993: 65-82). Now people reported that this was happening. The reengineering efforts at Aker Stord headed the company in such a direction. Workers were heading in the direction of "process professionals" (Hammer 1996) who know their processes and not only isolated tasks. The reengineering projects had changed some complex processes with simple tasks to simple processes with complex tasks.

Consequently, direct employee participation in decisions concerning own tasks had increased, while direct employee participation through formalized channels and arenas had decreased, due to more process orienting, project organization and also the fact that more decisions concerning Aker Stord where taken at Concern level.

What can we learn from the Aker Stord experiences? Is Aker Stord a unique case regarding employee participation in innovation or is this common for other Norwegian companies? I will now compare employee participation in TKA with a national survey of quality management. On the basis of these analyses, shortcomings of existing theory and practice will be highlighted and form the basis for the development of my practical approach to combining TQM and BPR characteristics synergistically in a Norwegian industrial setting.

8.9.2.6 Aker Stord's participatory TKA practice in relation to quality management practice in Norway– a national survey

Direct employee participation in improvements is important in TQM processes but not that important in BPR. In TQM and BPR publications, trade unions and indirect arrangements for employee participation are hardly mentioned at all.

Norwegian traditions, laws, regulations and agreements promote both direct participation and indirect participation. At the initial phase of the start up of ED 2000 at Rogaland Research, two sets of fundamental questions were discussed among the participating researchers regarding indirect employee participation:

- Do strong unions, initially grounded on conflicting interests between the work place parties, represent a threat to the implementation of TQM processes? Or can they represent a resource?
- Will an extensive practice of direct employee participation, as part of a TQM implementation, reduce the importance of trade unions? Or will such a practice strengthen their power base? Will any of the existing bodies for cooperation between the two parties become superfluous, or can their importance be increased?

Furthermore, when the researchers first met with the trade unions that we cooperated with in the other IfS network companies it was an issue whether they should play an active role in TQM processes, or avoid taking their share of the responsibility for processes and solutions. BPR was not discussed with the shop stewards in the initial phase of our research. Neither was it on the agenda at Aker Stord at that time. However and with reference to Chapter 3, prevailing models for TQM include radical change, even if continuous incremental improvements dominate TQM processes. So, conceptually radical change was included, but not explicitly.

Aker Stord involved the unions in their TQM efforts in accordance with the company's own tradition for company-union cooperation. The experiences from using Departmental Committees as the coordination body for local improvement activities indicated that this could be a good idea. It increased both improvement activities and local union activity in many departments. Committed shop stewards and active unions eased the TQM implementation process. At the same time arrangements and bodies for cooperation became more meaningful, constructive and attractive for both parties.

At this early phase of the ED 2000 program the research group at RF discussed the challenges related to indirect employee participation in TQM considerably, and decided to look at the experiences in other Norwegian companies. Did other Norwegian companies co-operate with the unions when they implemented quality management? In that case, how common was this practice? What about direct employee participation? How common was that? How was the relationship between direct and indirect employee participation? Did some companies practice one of them, others both? Did the two forms of employee participation work reciprocally supportive in innovation activities like in the Aker Stord case, or the opposite? Did Norwegian companies follow the textbook and focus on direct employee participation only, or did they follow Norwegian traditions for cooperation between the labor market parties?

To answer these kinds of questions and also other questions related to quality management, the research group decided to carry out a national Norwegian survey on quality management together with the University College of Stavanger (at present the University of Stavanger). This decision was made for several reasons. Most important, we needed to learn more about how quality management was practiced in Norway before we started our action research together with the companies in the Industry Network in Sunnhordland. We needed to benchmark the Aker Stord experiences and future research results from the participating companies, and we wanted to increase the external validity of our research by combining qualitative and quantitative methods. At that time few Norwegian companies had any experience with BPR. BPR was not included in the survey.

We invited the Norwegian Association for Quality and Management (NFKL) as a partner in order to make sure that the survey reached companies that practiced quality management. We also believed that this cooperation would increase the interest for our study and thereby the number of companies interested in answering our questionnaire.

In this survey we decided to use the term quality management instead of TQM. The reason for this was lack of detailed knowledge about how the different companies practiced their quality management. This was a methodological challenge we needed to be aware of when we analyzed the data. However, NFK is a member of EFQM and promotes the EFQM model. Through conversations with NFK and after having participated in many of their activities I found it reasonable to believe that most of their member companies had at least some basic knowledge about TQM and the EFQM model, and practiced accordingly.

The questionnaire was formed on the basis of the Baldrigde Quality Award Model (Hart & Bogan 1992) and the European Quality Award Model (EFQM 2003). We also used early experiences from the Aker Stord case study to formulate questions related to direct and indirect employee participation in quality management. The questionnaire can be found in Appendix 3. See also Hansen & Sun (1997). The survey was carried out in the period December 1996 – January 1997, and was sent to the 900 member companies of the Norwegian Quality Association. We received 331 responses, which is 37% of the questionnaires that were sent out. Of these, 57% came from production companies.

Some of the results that are presented below have been published previously (Hansen 1997, Hansen & Sun 1997, Sun 1997).

8.9.2.7 Direct employee participation in quality management

Direct participation in improvement projects

One of the questions was: *"To what extent do employees participate in the carrying out of improvement projects"* (Hansen et al. 1997, my translation). 4% replied "very low", 18% replied low, 42% "medium", 31% "high", and 5% very high. (Hansen 1997).

This means that in 1997 22% of the responding companies that were members of the Norwegian Quality Association had little direct employee participation when they carried out improvement projects, 36 % involved their employees and 42% positioned themselves somewhere in between. Only 5% of the companies followed the TQM textbooks and practiced a very high degree of employee participation in quality projects. As the companies in the survey were members of the Norwegian Association for Quality and Management and, therefore, more likely to emphasize quality improvements more than other Norwegian companies, the findings cannot be generalized to represent Norwegian industry as a whole.

In terms of direct employee participation, we see that Aker Stord is in line with the large majority of the companies in the survey that replied "low" or "medium" participation. These numbers represent all the companies that answered, regardless of their length of experience with quality management. But what happens with experience? Do the degree of participation increase, decrease or remain the same?

Length of experience with quality management and direct participation in improvement projects

We did this survey only once and do not have longitudinal data. Consequently, we could not follow the development in the companies that replied over time. We divided the responding companies into four groups according to the length of experience they had with quality management to see if the length of experience correlated with the degree of participation.

When we exclude the companies that did not take any strong position and reported medium participation, we found that 58% that reported less than one year of experience with quality management reported high or very high direct participation. In the group with 1-3 years of experience, 59% reported high or very high. In the group that had 3-5 years of experience, 60% reported high or very high and in the group with more than 5 years of experience with quality management 75% reported high or very high. Thus, direct employee participation in quality projects seem to be higher in companies with the most quality management experience.

8.9.2.8 Indirect employee participation in quality management

Indirect participation in improvement projects

We were also interested to see whether Norwegian companies involved arenas for indirect employee participation in quality management like Aker Stord did. Looking back at the quality management textbooks referred to in my theory section, indirect employee participation was not a subject at all. Consequently, we should expect little indirect employee participation. One of our questions was: *"To what extent do existing cooperative bodies like the Works Council, Departmental Committees and the Work environment Committee participate in the planning and implementation of quality improvement"* (Hansen et al. 1997, my translation).

We found that 16% reported "very little" indirect participation through these arenas, 34% reported "little", 31% reported "medium", 16% "high" and 2% "very high" (Hansen 1997).

If we compare these results with the same figures for direct employee participation, we see that 18% practiced "high" or "very high" indirect employee participation in improvement projects through the above mentioned cooperative bodies, while 38% reported "high" or "very high" direct participation in improvement projects. Apparently, many companies follow authors like Ishikawa (1985), Imai (1986), Deming (1986), Juran (1989) and Feigenbaum (1991) and focus on direct participation in their quality processes.

However, 18% of the companies deviated from the textbooks and practiced indirect employee participation in quality management. This may indicate that they did as Aker Stord and followed their tradition of company-union cooperation in business development also for quality management. Aker Stord was in line with this minority.

Length of experience with quality management and indirect participation in improvement projects

Around 5% of the companies in the group with less than one year of quality management experience reported a high or a very high degree of indirect employee participation in improvement projects. For the group with 1-3 years of quality management experience, the figure was 21%. Companies with 3-5 years of experience reported 36% and for companies with more than 5 years of experience the figure was 33%. Remember, though, that we excluded the 31% of the companies that did not take any strong position and reported medium.

So, indirect employee participation in quality projects is higher in companies with some years of experience with quality management compared to companies with little experience. Practicing a high degree of indirect employee participation in TKA from the start, Aker Stord belong to the 5% minority in the group with less than 1 year of quality management experience that reported a high degree of indirect participation.

We also questioned top management's attitude towards trade unions' participation in quality improvement: *"To what degree does top management mean that trade union participation in quality improvement is important?"* (Hansen et al. 2007, my translation).

In the group with less than one year of experience, 21% reported a high degree. In the group with 1-3 years of experience, 29% reported a high degree. In the group with 3-5 years of experience the figure was 36% and in companies with more than 5 years of quality management experience 47% of top management held the position that the importance of trade union participation in quality improvement was high.

In other words, the more experience these Norwegian companies have with quality management, the more top management values trade union participation. Aker Stord had long lasting traditions for company-union cooperation in business development and valued trade union participation in TKA from the very start.

8.9.2.9 The relationship between indirect and direct employee participation in quality management

At Aker Stord we found that direct participation in TKA did not threaten or make indirect participation superfluous. Rather the contrary, the two forms of employee participation both promoted the TQM process and supported each other. Could this finding be generalized, and what did it look like in other Norwegian companies?

The survey data showed that in the companies that reported very little or little indirect employee participation in improvement projects, 66 % of the companies reported high or very high direct employee participation. This makes logical sense as TQM is based on direct participation. However, of the companies that reported high or very high indirect employee participation in improvement projects, 91% also reported high or very high direct employee participation. This result is in line with the Aker Stord finding that direct and indirect employee participation in quality management may indeed reinforce each other in a Norwegian setting.

The survey also showed a positive correlation between employee participation and the companies' reported quality results (Hansen 1997). Consequently it is reasonable to assume that employee participation supports quality improvements in Norwegian companies.

8.9.2.10 Summarizing the national survey

Like Aker Stord, more than half of the companies that started up their quality journey less than one year before answering our survey, reported a high degree of direct employee participation in improvement projects. In companies with more than five years of experience about 75% reported a high degree of direct employee participation.

The picture for indirect participation was a bit different. Only 5% of the quality management starters did like Aker Stord and practiced a high degree of indirect employee participation in improvement projects. In companies with more than 5 years of quality management experience every third company reported a high degree of indirect participation. These results may indicate that many companies start their quality journey in line with influential scholars like Ishikawa (1985), Imai (1986), Deming (1986), Juran (1989) and Feigenbaum (1991) by focusing on direct, rather than indirect, employee participation. Even if we did not do time series analyses, it is reasonable to believe that after a few years of experience, many companies with traditions for company-union cooperation realize that indirect employee participation is advantageous in this area, as well as in other business development areas.

8.9.3 Summarizing the Aker Stord experiences and way forward

Do the experiences from Aker Stord as a whole support the view of BPR as an alternative to TQM? Aker Stord encountered great challenges when the company tried to combine TQM and BPR synergistically. Did the company experience this development because the TQM concept and the BPR concept worked against each other so that they excluded each other conceptually? I found no strong convincing evidence for that. Was the reason for this development at Aker Stord a change in top management and new priorities? Yes, that seems to have contributed.

Did we experience this development because of Aker Stord's new role in the value chain in the concern? Their role in the value chain became more visible, and their status as an independent company was reduced. More decisions that affected the company were taken at concern level. Yes, that seems to have contributed, too. E.g. the company, managers and shop stewards were expected to take part in the Aker Group's common reengineering efforts and spent a lot of time and energy on Aker Group activities that could otherwise have been spent on TKA at Aker Stord.

Did we experience this development because the company did not have the capacity to increase its ability to continuous improvement and its ability to radical change at the same time? Yes, that seems to have contributed as well. The interviews indicated that it was very demanding for the company to pay attention to both continuous improvement and radical process innovation in the same system at the same time and on an ongoing basis both at company level and at concern level.

So, all in all several factors seem to have influenced the development at Aker Stord. My empirical data does not permit me to rank the importance of different factors, to exclude any, or to present a complete list of factors. That was not intended, but it is obvious from the interview data that all these factors contributed. The challenges that Aker Stord encountered when management introduced BPR in a TQM process are in line with March's (1991) theory of the balancing between exploitation and exploration as a trade-off. BPR was introduced at the cost of TQM. Seen in that background this development should not come as a surprise. Furthermore, a successful TQM process depends on management commitment and visibility all the way from top management and down to first-line managers (Juran 1989). Top management in the concern, at Aker Stord and many middle and first-line managers stopped asking for continuous improvement activities. They focused on BPR. This development should not come as a surprise either.

An interesting question is whether Aker Stord might have managed to increase both its exploitation and its process exploration capability at the same time if an awareness of and a recipe for combining TQM and BPR synergistically, including adequate tools and methods had been available and practically operational at the time. Anyway, Aker Stord's and particularly Prefabrication of Pipes experiences inspired me as an action researcher to start developing such a recipe together with industry and test it as a contribution to the further theoretical and methodological development of the theory and practice of combining exploitation and exploration characteristics.

At the time of the first audit, the other action researchers at RF and I had started the TQM research program in the other companies in the industry network, and were in need of a better approach that combined continuous improvement and radical process innovation. Our ambition in this setting was to find a better approach, tools and ways to combine characteristics of TQM and BPR in a Norwegian industrial context. Neither my theoretical discussion nor my empirical case study Aker Stord made it possible for me to conclude that a synergistic combination of TQM and BPR was an unwise or an impossible mission. On the contrary, I still believed that efficiency and innovation could grow simultaneously through a development in this direction. This would, however not be an easy or straightforward task. Before proceeding, I will now briefly sum up the main findings from the Aker Stord case and discussions, the survey on quality management in Norway and important explanations. Based on that, I will briefly conclude and propose some ideas for further development.

8.9.3.1 Main findings and conclusions from the Aker Stord case study discussion briefly summarized

Main findings 1) TKA and direct employee participation:

- Lower level shop floor direct employee participation supports TKA.
- Many of Aker Stord's existing methods for employee participation support the introduction of TQM.
- Strategy level direct employee participation is not an issue.

Explanation:

Direct employee participation is a fundamental element in TQM. As for Aker Stord, many Norwegian companies practice direct employee participation as part of their quality management approach. The TQM concept focuses on shop floor level employee participation and has developed many methods to support that. TQM does not focus employee participation at strategy level, and has not developed adequate methods.

Conclusion and way forward:

Existing methods for direct employee participation should be considered as means to support TQM and continuous improvement. Methods to improve direct employee participation at company strategy level need to be developed, tested and evaluated.

Main findings 2) TKA and indirect employee participation:

- Indirect employee participation supports TKA.
- Aker Stord's arrangements for indirect employee participation support TQM, but the TQM implementation was not equally successful in all parts of the company.

Explanation:

TQM is founded on consensus thinking, and indirect employee participation is not an issue in TQM. However, Many Norwegian companies practice two-party business development cooperation and many arrangements and agreements supporting that have been developed. Aker Stord and many other companies arrange their quality management approaches in accordance with these. The Aker Stord experiences show that these arrangements and agreements form a legitimate, ready and good structure for identifying, prioritizing and following up improvement proposals. The experiences also show that they did not function successfully in all parts of the organization. Three important reasons for that were lack of management commitment, lack of stability and a project oriented way of arranging work in parts of the organization.

Conclusion and way forward:

In a Norwegian ID setting, formal agreements and arrangements for two-party business development cooperation can and should be used to promote TQM and continuous improvement. These arrangements should be further developed to support TQM in different organizational settings, e.g. project organizations. Further approaches and means to ensure management focus and support should be developed.

Main findings 3) It is a challenge to succeed with employee participation in TKA in the whole system, at the same time and on an ongoing basis:

- Aker Stord succeeded with many methods for direct and indirect employee participation in TKA, but not in the whole system, at the same time and on an ongoing basis.

Explanation:

Running a comprehensive participatory change process depends on management focus along the whole line of command in the whole company on a continuous basis, and requires a considerable amount of time. In the Aker Stord case employee participation in TKA very much depended on management focus. Whereas the manager of Prefabrication of Pipes promoted both direct and indirect employee participation in his department, this example was not followed by all department managers. Many line managers failed to invest the necessary amount of time and reported that they did not have the necessary competence to know what to do. In addition, Aker Stord became gradually more project organized. The initial TKA activities and the arrangements for employee participation were organized to fit in with a department structure.

Conclusion and way forward:

Management focus is important for employee participation in comprehensive change processes. Better understanding of what is needed to succeed and better approaches and tools to make employee participation happen in different organizational settings need to be developed, tested and evaluated.

Main findings 4) BPR and direct employee participation:

- Direct employee participation in TKA was reduced when BPR was introduced at Aker Stord.

Explanation:

Direct employee participation in TKA focused on continuous improvement at the shop floor operational level. Larger and more comprehensive projects came in focus when BPR was introduced. Fewer employees were invited to participate in these than in the incremental ones. The speed of change increased, and it was more difficult for employees to understand what was going on and to keep up with the development.

Conclusion and way forward:

Existing BPR approaches and methods do not ensure direct employee participation in process design and redesign. Approaches and tools to ensure employee participation in these areas need to be developed, tested and evaluated.

Main findings 5) BPR and indirect employee participation:

- Indirect employee participation in TKA was reduced when BPR was introduced at Aker Stord.

Explanation:

The introduction of BPR in the Aker Group reduced the position of Aker Stord as an independent company. More decisions concerning Aker Stord were taken at Aker Group level. The speed of change increased and it was difficult for shop stewards to stay ahead of the development. Full time shop stewards felt that information came late and that there was little time for reflection and discussion in e.g. co-operating bodies like the Works Council. Local shop stewards reported that they were scarcely able to keep informed about what was happening.

Conclusion and way forward:

The Aker Stord way of practicing Norwegian arrangements and traditions for two-party cooperation is not adequately suited to ensure indirect employee participation in BPR. Better approaches, structures, organization and tools to ensure that need to be developed, tested and evaluated.

Main finding 6) It is a challenge to succeed with TQM and BPR simultaneously in the whole system, at the same time and on an ongoing basis:

- Aker Stord did not succeed in running TQM and BPR in the whole system at the same time and on an ongoing basis. A predominantly antithetical approach to combining exploitation and exploration was practiced.
- The department Prefabrication of Pipes managed to run TQM and BPR, supported by direct and indirect employee participation, at the same time with encouraging results. A mostly dual/synergistic approach to combining exploitation and exploration was practiced.

Explanation*:*

There was a change of top manager at Aker Stord. The one who started TKA advocated a continuous incremental improvement approach. The successor advocated BPR and had little focus on continuous incremental improvement, even if TKA was meant to include both. Other managers along the line of command did likewise, and many stopped asking for results from continuous improvement activities. The comprehensive BPR projects at Aker Group level and at Aker Stord were in focus. Some work processes, e.g. pipe handling, were planned to be radically changed and many saw no point in doing minor changes before the radical change had taken place.

Prefabrication of Pipes had an appropriate structure and practice for handling continuous incremental improvement that also could be used to support radical change. The department manager prioritized and succeeded to keep focus on both incremental and radical process change (BPR) on an ongoing basis. Management commitment to the synergistic combination of forms of exploitation and exploration seemed to have a decisive effect on this development.

Conclusion and way forward:

Top management focus and visibility are important for how line managers direct development processes. Better understanding of what is needed to succeed in order to combine TQM and

BPR and also better approaches and tools to make this happen in different organizational settings need to be developed, tested and evaluated. This includes the question of how the combination of direct and indirect arrangements for employee participation can increase the potentials of combining TQM and BPR synergistically in theory and practice. In addition to leadership focus, the Prefabrication of Pipes case shows that employee participation can be an important enabler for a successful synergistic combination of TQM and BPR characteristics, at least in a Norwegian setting. On the basis of this brief summary of the Aker Stord case and the preceding discussion, I will specify my summary of the shortcomings and way ahead for direct and indirect employee participation in some more detail before proceeding.

8.9.3.2 Some important shortcomings of today's participatory practice that need to be researched

On the background of my theoretical and empirical analysis, I will now highlight some important challenges and shortcomings of today's participatory practice that I find important to the synergistic combining of characteristics of TQM and BPR. Table 7 sums up and illustrates by shading where I find the most important shortcomings of existing practice, and were I see some opportunities that could be exploited, at least in a Norwegian industrial democracy context.

Direct employee participation is important for TQM, and many means to improve empowerment and to promote employee participation in idea generation at shop floor level – "lower level/socio-technical", such as suggestion schemes and brainstorming, have been developed (square 1).

BPR, however, has not developed sufficient tools and techniques for direct employee participation in work process redesign at this level. I therefore developed the Actor-Process Diagram. This tool is meant to promote direct employee participation in the development of the employees' "own" work processes. The Actor-Process Diagram should enable employees at the shop floor level to take part in radical innovation and process exploration. Radical innovation is very often composed of a series of incremental changes (Cole 2001). I use a tool from an explorative tradition (BPR) and include some techniques from the exploitative tradition in order to come up with a radical change accompanied by a series of incremental changes. The Actor-Process Diagram is a method that can ease the synergistic combining of forms of exploitation and process exploration and contribute to the evaluation of the appropriateness of the dual/synergistic approach. The Actor-Process Diagram will be included in my operational framework for employee participation in my approach to combining TQM and BPR synergistically in Chapter 9, and will be presented and discussed in detail in Section 10.2. The Actor-Process Diagram can be used to ensure both direct employee participation in radical work process redesign at the operational level (square 1) and indirect employee participation in Business Process Redesign at upper/strategic/political level (square 4). The tool also makes a synergistic combining of direct and indirect employee participation possible.

Neither TQM nor BPR have utilized the potentials of direct employee participation in company strategy development (square 2). This kind of participation is not an issue in either of the concepts. Consequently, neither of the concepts have any good recipes for how this can be carried out. I find the question whether employees should participate directly at company strategy level, and in that case how, challenging. Such a practice can increase employees' ownership, ability and motivation to contribute to company vision, goals, strategy and values. Many issues at this level are of a radical and explorative character

Participation form	**Decision levels**	
	Lower level/socio-technical	**Upper level/Strategic/political**
Direct	Empowerment (1) Suggestion schemes Brainstorming	Enterprise Development (2) Conference
	Work process redesign	
Indirect	Departmental Committee (3) Work Environment Council	Board representation (4) Works Council

Table 7 Employee participation arenas to exploit

The Aker Stord case showed that employees in the period of change had no clear picture of company strategy and how they could contribute. Based on that experience I therefore developed the Enterprise Development Conference to be used as a tool for that kind of direct employee participation in the other companies that we (the RF-researchers) cooperated with in ED 200 and VC 2010. The conference uses techniques from the continuous improvement tradition to come up with initiatives for both incremental and radical change. This way the conference also tests out the use of techniques from an exploitative tradition to enhance forms of exploration. Consequently the conference is also a method that can ease the synergistic combination of exploitation and exploration and contribute to the testing of the appropriateness of the synergistic approach. Like the Actor-Process Diagram, this conference is integrated in my operational framework for the further development of an approach to combining TQM and BPR synergistically that will be presented in Chapter 9. Its background, how it is carried out and experiences with the conference will be discussed in detail in Section 10.1.

Furthermore, neither TQM nor BPR focus indirect employee participation either at company upper/strategic/political level (square 4) or at the lower/socio-technical level (square 3).

Consequently no good tools or recipes have been developed. Aker Stord used existing tools, practices and arenas and experienced some successes and some failures. In addition to the above mentioned ED Conference and the Actor-Process Diagram I find that most existing arenas for indirect employee participation can be utilized in the same way as Aker Stord and many other Norwegian companies do. Important in square 3 is the Departmental Committee and the Work Environment Committee. In square 4 the Board of Directors and the Works Council are the most important.

9 Towards a synergistic combining of TQM and BPR in a Norwegian industrial setting

With reference to the Aker Stord case and my empirical analyses, I will now present a process model showing what I find to be important enablers for the synergistic combining of TQM and BPR and what can happen when you combine the two concepts in the same system. It is this model that guides my action research in ED 2000 and VC 2010. My findings so far support theory about leadership as a decisive factor for the success of change processes. This was an expected finding. More surprising was the finding that Industrial Democracy can have a great potential as an enabling factor for the synergistic combining of TQM and BPR characteristics. Consequently, my contribution to the further development of the synergistic combining of these two concepts is very much about involving people directly and indirectly in generating and following up incremental and radical ideas. I will therefore in Section 9.2 extend my former theoretical discussions with a section on Industrial Democracy and clarify important terminology. This is necessary for understanding the contextual setting of my research and important contextual factors for the outcome of the different change efforts. Section 9.3 is a presentation of an operational framework for employee participation in the synergistic combining of TQM and BPR. This operational framework guides my choice of change methods and tools.

9.1 Approach to the synergistic combining of TQM and BPR

The practical consequence of TQM is a process for continuous improvement, and the main outcome is minor improvements to the ways things are done on a continuous basis. Methods, means and approaches that increase a systems' exploitation capability are introduced. See the upper left hand side of Figure 19.

The practical consequence of implementing BPR is radical process innovations, and the main outcome is abrupt, intermittent and radical changes in business – and work processes. Methods, means and approaches that increase a systems process exploration capability are introduced. See the upper right hand side of Figure 19.

In an antithetical combining of TQM and BPR efforts to improve TQM will come at the expense of BPR and vice versa. At Aker Stord as a whole BPR was introduced at the expense of TQM. However, one of the departments managed to combine the two synergistically, thereby building exploitation and exploration capabilities reciprocally reinforcing. The two most important enablers making this possible were leadership focus and an ID context where direct and indirect employee participation was combined synergistically. The result was that incremental and radical innovations were developed synergistically in the same system, at the same time and on an ongoing basis. This result together with theory motivated for further research. In figure 19 the synergistic combining of TQM and BPR, of direct and indirect employee participation and of incremental and radical innovations is illustrated by the lower circle. It is this model that I want to try out through my action research.

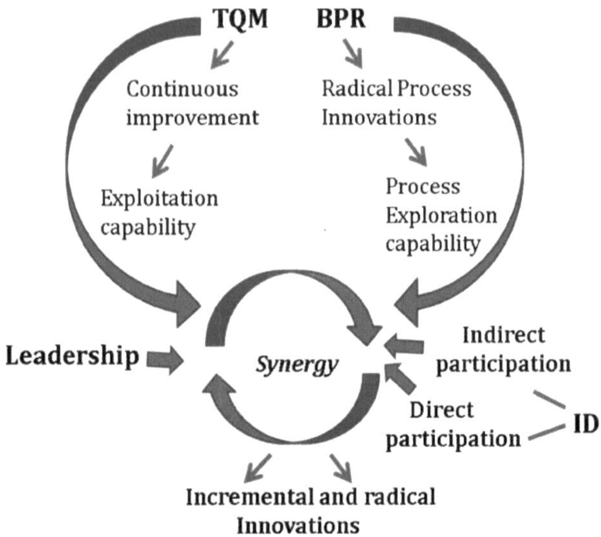

Figure 19. The process of combining TQM and BPR synergistically

TQM and BPR have some common characteristics, and each concept has unique characteristics. My synergistic approach builds on the strengths of both the TQM and the BPR ways of thinking, and at the same time, neutralizes some of the weaknesses associated with each concept. E.g., the approach builds on the TQM concept's thorough method for organizing a process for continuous improvement. On the other hand, the TQM concept has been criticized for focusing mainly on incremental changes of the way things are done today (Hammer & Champy 1993) and not on radical change, where BPR has its strengths.

9.2 Industrial democracy

In this section I will clarify what I mean by Industrial Democracy and Employee Participation in a Norwegian setting. This will be done by referring to some important historical lines and by relating these to industrial democracy theory. An important aspect of the Norwegian tradition has been the double ambition of both improving industrial democracy and the well being of the workers while improving productivity and competitiveness through the same development process at the same time (Gustavsen 1992).

I will now first discuss and define what I mean by employee participation and industrial democracy. Then, I will describe the practical approach that Norwegian industry has followed.

9.2.1 Employee participation defined

Participation is a widely used term, which, however, has been defined very imprecisely by several writers. In ordinary speech, the term is used in a wide sense to cover almost any situation where some form of worker-management interaction takes place (Pateman 1970). I need a precise definition.

French et al. (1960:3) say that participation in the industrial sphere refers to a *"process in which two or more parties influence each other in making plans, policies or decisions. It is restricted to decisions that have future effects on all those making the decisions and on those represented by them*". According to these authors, this definition excludes the situations; *"where an individual, A, merely takes part in a group activity; where A is merely given information on a decision affecting him before it is executed; where A is present at a meeting, but has no influence"* (French et al. 1960: 3). According to this definition participation refers to decision-making. A situation where employees are persuaded to accept decisions made by management beforehand is not participation, but *"pseudo participation"* (Verba 1961).

In the definition of participation above made by French et al. (1960) it is stated that participation is a situation where two or more parties influence each other in making decisions. However, according to Pateman (1970), referring to theories of participatory democracy (Cole 1920, Mill 1965, Rousseau 1968), political equality refers to equal power in determining the outcome of decisions. If you are in a position where you can take part in determining the outcome of decisions, you are in a position of power. This is *"full participation"* (Pateman 1970:70).

Influence in itself is not synonymous with power. To be in a position to have influence on a decision is not the same as having the power to decide on the outcome of the decision. In Norwegian industry, management has the final power to decide. Employees are in a subordinate position where they only have influence on decision-making. This is *"partial participation"* (Pateman 1970:70).

However, participating in decision-making is not sufficient in itself. All parties involved need relevant information to make adequate decisions (Pateman 1970). According to Gustavsen (1990) employees' possibilities to participate in the dialogue and exchange of meanings connected to decision making is of equal importance and closely linked to the decision making itself. The information obtained needs to be evaluated and discussed. Without this possibility, participation in decision-making may be more of a ritual kind of participation than a real expression of meaning.

Full participation is here defined as:

> *A process that gives all participants in a decision-making forum the same power to decide the outcome.*

Partial participation refers to:

> *Processes that put employees in a position to influence the outcome of the decision making process.*

That means being heard, but others are in the position of making the final decision.

As employee participation may be carried out by 1) direct participation or 2) indirect participation (participation through elected representatives), participation can be summarized in four fundamentally different ways – see Figure 20.

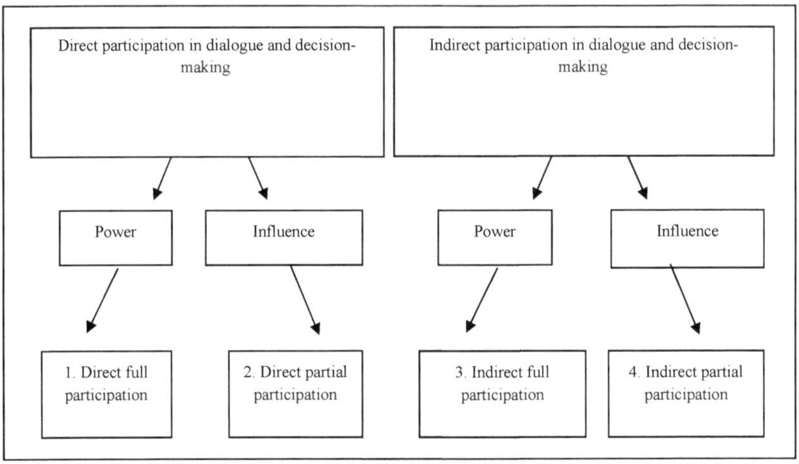

Figure 20 Four fundamental ways of employee participation in decision-making

Figure 20 departs from the two basic channels through which employees may participate in decision-making: the direct and indirect form. Together with the extent to which employees have the power of equal opportunity to decide the outcome of the decision or are only in a position to influence, this decides the kind of participation: 1) direct full participation, 2) direct partial participation, 3) indirect full participation or 4) indirect partial participation. An example of a situation of direct full participation is an employee-owned workshop where all employees have the same power to decide the outcome of decisions. An example of direct partial participation is a Search Conference (Emery & Purser 1995), a Dialogue Conference (Gustavsen 1992) or an Enterprise Development Conference (Tønnessen 2000) where all employees are invited to participate in decision-making, but where management makes the final decisions. An example of a situation with full indirect participation is employee representation at the board of directors, where the elected representatives have the same power to decide the outcome of the decisions as the other members. An example of indirect partial participation is employee representation in Works Councils through elected represent-tatives/shop stewards.

Nyhlen (1994) use the term "co-determination" and "participation" to distinguish between the participation form based on power and the one based on influence only. To simplify, I will follow Nyhlen (1994) and use the word "participation" in the sense of "partial participation" (Pateman 1970). Full participation will be used to describe situations of co-determination where the influence is based on power.

9.2.2 Industrial democracy, definition and limitations

The kind of industrial democracy that I will be referring to in this book is not democracy in its strictest political sense (Pateman 1970, Dahl 1992). There is no equal power to decide the outcome of decisions, but it is possible for employees to have influence on them. Referring to Figure 20, and according to Pateman's definitions, it is direct and indirect partial participation that will be the focus of my industrial democracy discussion. Participation (in the sense of "partial participation") does not require the democratization of authority structures (Pateman 1970). You can have "partial participation" at both the lower operational/socio-technical level and at the upper/strategic/political level (ref. Table 7) in an enterprise without having a democratic structure in its strictest sense. You can also have full participation at lower level without a democratization of authority structures. So the words participation and democracy cannot be used interchangeably (Pateman 1970).

The terms participation and democracy are used rather loosely in much of the literature (Thorsrud & Emery 1964, Pateman 1970). The two words are also frequently used interchangeably (Pateman 1970). Democracy is often used to describe a general climate that exists in an enterprise rather than merely a formal structure. It may be used to describe a situation of pseudo participation (Verba 1961) or even merely to indicate that a friendly atmosphere exists (Pateman 1970).

Pateman (1970) found this incorrect. If the authority structure in industry is to be parallel to that of a political democracy, then the "government" of the enterprise must be elected and removable by the whole body of employees in the enterprise. Alternatively, the whole body of employees must take the management decisions. A system of industrial democracy implies full upper and lower level participation by employees. Industrial democracy as a structure in its strictest sense can be defined as "a structure that allows for full employee participation at upper and lower enterprise level".

In democracy theory the interrelation and reciprocal dependency of participating at the two company levels in Table 7 and also the relation to democracy at societal level has been discussed by academics like Cole (1920), Mill (1965) and Pateman (1970). They all build on Rousseau's theory of participatory democracy (Pateman 1970) and they lift Rousseau's theories out of a societal perspective and into an economic life/industrial setting. In short, this theory states that participation at lower company levels motivates and trains employees for participating at higher company levels, and then further on for participation at society level (Pateman 1970). In the same way that participation in local government is a necessary condition for participation at national level because of its learning and motivating effects

(Rousseau 1968), so could "government" in the work place, and also other lower level authority structures, have the same effect (Mill 1965, Cole 1920). Pateman (1970) finds evidence that supports the theory that people learn to participate by participating. But the theory that lower level participation increases the motivation for and ability to participate at higher company levels still remains a hypothesis that needs to be tested.

"We cannot hope to answer the important question how far rank and file workers are likely to be interested in and take up such participation opportunities until we have information on the effect of a system that combines both lower and higher level participation" (Pateman 1970:83). Now in 2011, there is little empirical evidence to be found in the literature to claim that this connection exists.

Dahl (1973) discusses these kinds of issues and the limits of participation in decision-making processes. His point of departure is that power can only be legitimate if it is founded in a democratic process, and he proposes three main criteria for judging whether decision making processes are legitimate, democratic and fair: 1) personal choice, 2) competence and 3) economizing. As to personal choice Dahl find that, as social human beings interested in living peacefully together with other humans, we cannot as individuals achieve our own goals in a satisfactory manner if we do not allow others to pursue their goals on an equal basis. It is our choice, in general, to accept decisions made by a majority vote. Regarding competence, we accept that decisions are made by people who are better qualified and informed than us. And finally, as regards economizing, time and resources are valuable for us. We prioritize the decision making that we want to participate in. Dahl claims that the ideal democratic state will never be achieved because the costs of participating in all kinds of decision-making are too high. For these reasons Dahl (1973) claimed that the optimal democratic system will always be different from the ideal one.

Dahl (1992) lists the following five criteria for a democratic process: 1) equal voting rights, 2) effective participation, 3) information, 4) people's final control of the agenda, and 5) inclusion. Thus, it clearly takes a lot for processes to be considered fully democratic. Nor is it necessarily desirable for everyone to take part in all decision-making, and there may be good reasons why employees of an enterprise or members of a community accept that others make decisions on their behalf in matters that concern them. I will return to Dahl's criteria when I discuss my framework for employee participation and in my analysis of the action research. I will now show how industrial democracy and employee participation has been approached in Norway.

9.2.3 Industrial democracy in Norway – a practical approach

Norway has a longstanding tradition of promoting employee participation and work place democracy in industry (Thorsrud & Emery 1964, 1969, Gustavsen & Hunnius 1981, Gustavsen 1992, Karlsen & Munkeby 1998). This tradition is reflected both in laws and regulations and in collective agreements between employers' and employees' associations. Industrial democracy has been an important element in reform strategies and research on new kinds of work organization (Berg 1983, Gustavsen 1990, Gustavsen 1992, Gustavsen 1993, Gustavsen et al. 1998, Olberg 1995, Falkum et al. 1999, Pålshaugen & Quale 2000).

Many organizations, among others The Confederation of Norwegian Businesses and Industry (NHO) and the Norwegian Federation of Trade Unions (LO), strongly believe that industrial democracy will increase both productivity and the wellbeing of employees (LO & NHO 2005). Some refer to this tradition as a national competitive advantage (Karlsen & Munkeby 1998, Finne & Nilsen 1998). This tradition is important to take into consideration when management and business development strategies are considered.

Industrial democracy in a Norwegian industrial setting is not industrial democracy in its strictest meaning. In the early reform work that was carried out by Einar Thorsrud and Fred Emery in the 1960s, the aim was to pursue some basic and fundamental conditions for real democracy, and to adapt these conditions to a Norwegian setting in line with the prevailing understanding of the term among politicians and in industry (Thorsrud & Emery 1964).

Even if it is possible to agree on some basic conditions and a theoretical framework for industrial democracy, Thorsrud & Emery (1964) departed from the view that this kind of agreement disappears the moment somebody tries to make this term meaningful in a given social setting. Goals and criteria for evaluating practical results differ depending on the social setting. Accordingly they started out by mapping prevailing forms and goals for industrial democracy in a Norwegian setting. This mapping (Thorsrud & Emery 1964:25-26) can be summarized as follows (my translation and abbreviation):

- It is acknowledged that industrial democracy must be considered in relation to a Norwegian work life setting, with the existence of a considerable number of laws, regulations and agreements between the labor market parties
- There is a general agreement on taking steps towards an increased democratization of Norwegian work life, to a) direct the conditions in Norwegian work life in a direction more in line with general democratic principles in society, and b) improve the opportunities for increased employee involvement, which in turn will probably give better productivity and efficiency
- It is a common attitude that industrial democracy in Norway must be more than just formal arrangements with elections, representation, committee decisions, etc.
- This "more than" refers to a situation where people's democratic conditions in their daily work situation are in line with some general conditions that have to be met to develop a real democracy: a) All humans must be considered equal. b) All humans must be given enough freedom of movement in their daily life to be able to participate in the development of society. c) Leadership depends on the trust of the many, and must be responsible towards the many (my summing up of a summary of Mannheim (1956:177-79) in Thorsrud & Emery 1964:14)

Against this background, Thorsrud & Emery adopted a very practical and general understanding of industrial democracy in Norway as *"a distribution of social influence among all those that participate in work life as opposed to a concentration of influence in a few hands"* (Thorsrud & Emery 1964:13, my translation).

This early reform work on industrial democracy in Norway was done together with researchers using action research. In short, the so-called "Samarbeidsforsøkene" (The Industrial Democracy Program) in Norway, tested two forms of participation: 1) employee representation at board level and 2) direct participation at shop floor level. It started out with a phase A) where employee representation at board level was tested and evaluated. In other words, and with reference to Figure 20, "indirect full participation" at board level (Thorsrud & Emery 1964).

These experiments were not very successful for a number of reasons. There was a very vague understanding of industrial democracy and consequently a vague understanding of what the goals of these experiments should be. There were methodological problems related to the measurement of the results. It was difficult to see what consequences this kind of employee representation had for the company, the board and the employees. The researchers concluded: *"Our data indicate that employee representation at board level is not effective as a means to increase industrial democracy"* (Thorsrud & Emery 1964:11, my translation).

The researchers found nothing to indicate that employee representation at board level increased employees' influence on decision-making in the companies. They found that it was difficult to start the transfer of some of the decision making to employees by starting at the top level in the company. They concluded that in order to realize participatory democracy, they had to start at a level employees both had the competence for and where they were motivated for increased participation: at the shop floor level (Thorsrud & Emery 1964).

This reasoning introduced phase B) of the Industrial Democracy Program in Norway. This phase focused on direct participation at shop floor level (in Pateman's (1970) terms: "partial participation"). Direct employee participation would reduce employees' alienation at work, develop participatory competence and increase motivation for participation at higher company levels at later stages in the industrial democracy development process. Participatory democracy at shop floor level would prepare for employee participation at higher company levels, e.g. the board of directors (Thorsrud & Emery 1964). This reasoning is very much in line with the above-mentioned thinking of Rousseau (1968), Cole (1920), Mill (1965) and Pateman (1970), and also in line with research results and studies both in Norway (Holter 1964) and in the United States (Trist 1963) in the same time period.

Phase B) was run in the second half of the 1960s through a series of field experiments in four factories. The central philosophy behind these experiments can be summarized in three terms: 1) Socio-technical systems, 2) Autonomous work groups and 3) Psycho-social job demands (Gustavsen 1990).

Socio-technical systems focus on the optimal integration of two subsystems: the technical and the social. These systems should work complementarily, and the idea was to increase both productivity and the work environment at the same time in the same development process (Gustavsen 1990).

The second term, **Autonomous work groups**, directs the development process towards alternative forms of work organization. The basic idea was that groups of employees should:

- Have the right to make decisions on their own in certain questions related to how work is organized and carried out.
- Have the competence needed to take the kind of decisions mentioned above.
- Have a medium through which the group members can interact, e.g. dialogue.
- Perform tasks that are different, but that are reciprocally dependent and constitute a whole altogether (Gustavsen 1990).

The third term, **Psycho-social job demands**, takes its point of departure in an understanding of the human nature, a hypothesis of the kind of needs humans have in relation to work, and a hypothesis about how work should be organized in order to make increased democratization of the work place possible. These were expressed both at individual level and at group level. At individual level these needs are:

- Variation and tasks that demand more than endurance only.
- Possibilities for learning.
- Being able to make decisions.
- To be respected.
- The possibility to see how one's own work contributes in a wider context.
- The possibility to see the job in relation to a foreseeable future.

At group level we find the following:

- Tasks that are interrelated, job rotation and physical closeness between jobs that depend on each other.
- Jobs that are connected, common main tasks, common goals and marginal tasks.

The results from the industrial democracy program varied between the four factories that participated. Some improvements both in productivity and work environment were documented, but not in line with the ambitions and goals for the program. Possible "Hawthorne effects" were not mentioned in the evaluation of the results (Claussen 2001).

The program had an ambition to spread the ideas and good work practices to other companies, but little happened (Kelly 1982, Gustavsen 1990, Claussen 2001). However, the point of departure and way of thinking from this industrial democracy program have been followed up in later programs (Claussen 2001), influenced the prevailing way of thinking among the labor market parties (Hovedorganisasjonenes Fellestiltak 2005), is found in the Norwegian work environment legislation (Friberg 1970), and also in research programs like ED 2000 and VC 2010 (Gustavsen 2002). A considerable amount of the work done in the last 40 years has built on this ancestry from the 1960s.

9.2.3.1 Democratic dialogue

Participation in decision-making is important. Of equal importance is the ability to participate in discussions where own and others' viewpoints, knowledge and experiences can be shared, tested, refined and where learning can take place. This seems to be an important condition for rational choice (Gustavsen 1992).

Gustavsen (1992) has developed the theory of "Democratic Dialogue" as an approach to this kind of dialogue. This theory is an important basis for my action research. The theory of democratic dialogue (Gustavsen 1992) is based on the following criteria:

- *"The dialogue is a process of exchange: Ideas and arguments move to and from between the participants*
- *It must be possible for all concerned to participate*
- *This possibility for participation is, however, not enough. Everybody should also be active. Consequently each participant has an obligation not only to put forth his or her own ideas but also to help others to contribute with their idea.*
- *All participants are equal*
- *Work experience is the basis for participation. This is the only type of experience, which, by definition, all participants have*
- *At least some of the experience each participant has when entering the dialogue must be considered legitimate*
- *It must be possible for everybody to develop an understanding of the issues at stake*
- *All arguments that pertain to the issues under discussion are legitimate. No argument should be rejected on the ground that it emerges from an illegitimate source*
- *The points, arguments, etc. which are to enter the dialogue must be made by a participating actor. Nobody can participate "on paper" only*
- *Each participant must accept that other participants can have better arguments*
- *The work role, authority, etc. of all the participants can be made subject to discussion – no participant is exempt in this respect*
- *The participants should be able to tolerate an increasing degree of difference of opinion*

- *The dialogue must continuously produce agreements that can provide platforms for practical action. Note that there is no contradiction between this criterion and the previous one. The major strength of a democratic system compared to all other ones is that it has the benefit of drawing upon a broad range of opinions and ideas which inform practice, while at the same time being able to make decisions which can gain the support of all participants"* (Gustavsen 1992: 3-4)

9.2.3.2 Characteristics of Industrial Democracy in Norway and my approach summarized

With reference to my theoretical discussion above, the characteristics of the Norwegian Industrial Democracy context can be summarized as follows:

- Direct and indirect employee participation in decision making is anchored in Norwegian law, regulations and frame agreements between the labor market parties.
- Employee participation in decision-making can best be characterized as an opportunity to influence decision-making.
- In general there is a positive attitude to employee participation in decision-making in the Norwegian society.
- Employee participation is primarily directed towards decision-making at lower socio-technical level in the companies. Employees are represented at board level and at two party cooperative arrangements such as the Works Council, and can thereby influence strategic decisions.
- Employee participation includes participation in the dialogue leading up to decisions, as well as the decision making process itself and following up.

In my research the term industrial democracy is not used to describe a situation of full participation at upper and lower enterprise level. It is used to describe:

> A process where employees are given increasingly more opportunities to participate in dialogue and decision making, both at upper and lower company levels, and are trained and motivated to do so.

Given opportunities involves providing information and the possibility to exchange views and opinions. I will now shortly discuss the potentials of an ID context as an enabler for TQM and BPR conceptually. This is a basis for ID as an enabler for the synergistic combining of TQM and BPR.

9.2.4 ID as an enabler for TQM and BPR

In the TQM tradition direct employee participation is considered a means to business excellence. The TQM tradition has developed quality circles and similar tools for shop floor level/socio-technical participation. In several Japanese companies quality circles have proven to be a success (Imai 1986). However, the outcomes of this kind of participation have never been as successful in the West as in Japan (Hill 1991, Mak 1999). Norwegian industry has found other ways of using and strengthening already existing ways of direct employee participation, such as empowerment/self determination, suggestion schemes and departmental brainstorming (Tønnessen 2002). Apparently, the TQM way of thinking of direct employee participation at shop floor level (square 1 in Table 7) can easily be combined with Norwegian industrial traditions. In addition, basic elements in the TQM concept are very much in accordance with important Norwegian work environment regulations (Tønnessen 1992).

Direct employee participation at company strategy upper level (square 2 in Table 7) is not an issue in the prevailing TQM literature. Adequate tools are lacking. Indirect participation through elected representatives and company union cooperation are not topics in TQM neither at shop floor level nor at strategy/political level (square 3 and 4 in Table 7). TQM is founded on consensus in organizations and is not very occupied with the fact that one usually finds conflicting parties in companies.

So, TQM is absent from squares 2, 3 and 4 of Table 7. However, direct employee participation at shop floor level (square 1) is given a decisive position in TQM philosophy: One does not practice TQM without direct employee participation.

The basic elements in BPR were developed through the research program "Management in the nineties" at MIT between 1984 and1989 (Scott Morton 1991). Like TQM, many variations of the concept have been developed. As my point of departure I have used Michael Hammer and James Champy's bestseller "Reengineering the Corporation" (Hammer & Champy 1993). The ambitions are high: *"Fundamentally reengineering is reversing the industrial revolution"* (Hammer & Champy 1993:49), and further down the same page: *"Reengineering is the search for new models for organizing work. Tradition counts for nothing. Reengineering is a new beginning"*. Important is: *"identifying and abandoning the outdated rules and fundamental assumptions that underlie current business operations"* (Hammer & Champy 1993:3). Reengineering is the answer to *"the crisis that will not go away"* and a business life characterized by *"inflexibility, unresponsiveness, the absence of customer focus, and obsession with activity rather than result, bureaucratic paralysis, lack of innovation, high overhead ...."* (Hammer & Champy 1993: 30).

In addition to this, Hammer & Champy had the ambition to create a "*new world of work*". In the chapter about this new world of work, we find the following objectives bullet pointed: *"Work units change – from functional departments to process teams"," jobs change – from simple tasks to multidimensional work"," people's roles change – from controlled to empowered", "job preparations change – from training to education", "managers change – from supervisors to coaches", "organizational structure changes – from hierarchical to flat"," executives change – from score keepers to leaders"* (Hammer & Champy 1993: 65-82).

From a work environment perspective it all seems fine and in line with Norwegian legislation and organizational work environment ambitions. What about employee participation? According to Hammer & Champy (1993:103): *"The Reengineering leader makes Reengineering happen. He or she is a senior executive with enough clout to cause an organization to turn itself inside out and upside down and to persuade people to accept the radical disruption that Reengineering brings. Usually a senior executive is assigned the job of leader. It's a self-nominated and self-appointed role. Someone with the clout to carry it off becomes the leader of Reengineering when he or she is seized with a passion to reinvent the company, to make the organization the best in business, finally to get it completely right".*

The same top down approach is found in the recipe for organizing Reengineering: *"The leader appoints the process owner, who convenes a Reengineering team to reengineer the process, with assistance from the czar and under the auspices of the steering committee"* (Hammer & Champy 1993: 103). The steering committee *is "a collection of senior managers"* (Hammer & Champy 1993: 114).

This concept is very much top down and in conflict with Norwegian traditions of employee participation and union-company cooperation. In the Nordic translation "BPR Norden" (Ernst & Young and Willoch 1996) trade unions are for example not a subject at all. With reference to the Aker Stord case, the company needed to do some adjustments to the concept before they could use it. In Denmark research on BPR has shown that it is the "American way" of running BPR processes that is the dominant practice and not a participatory approach (Koch et al. 1999). I find this aspect the most challenging part of my attempt to integrate important characteristics of TQM with characteristics of BPR in a Norwegian industrial democracy context in a synergistic way.

Furthermore, most BPR methodologies do not focus on soft elements such as human and organizational factors (Ruessmann et al. 1994, Vakola 2000). Some see the inadequate treatment of the human aspect as the main reason for the failure of so many BPR efforts. (Bashein et al. 1994, Cooper & Markus 1995, Deakons & Makgill 1997). The experiences from three large European construction companies using the eight-phase CONDOR BPR methodology showed that the incorporation of human and organizational issues was a critical success factor (Vakola 2000).

Davenport (1993) considers human and organizational development approaches such as greater employee empowerment, reliance on autonomous teams and flattening organizational structures as important enablers of process innovation. He also claims that another important enabler for process innovation, namely IT, rarely is effective without simultaneous human innovations.

My study of international literature, company reports and reports from international conferences shows that the Reengineering movement has failed to utilize the potentials of company–union cooperation both at shop floor level and at company strategy level. With few exceptions, it is hardly a topic at all. Little company and union time seems to be spent on cooperation to achieve business excellence.

Against the background of some of these early experiences and the above-mentioned critique of the concept, it was with great interest that I looked into Hammer's next book (1996) "Beyond Reengineering" when it was presented three years after "Reengineering the Corporation" (Hammer & Champy 1993). The main message here is "process centering", which in short simply *"means that all people recognize and focus on their processes"* (Hammer 1996: 9). *"One of the first steps to process centering is that everyone in the company is becoming aware of their processes and their importance to the company"* (Hammer 1993:14). The term worker does not fit in to the process centered organization. Instead the terms *"process performers"* and *"process professionals"* (Hammer 1996:69) are introduced as terms for people who understand that in doing their work they are contributing to the performance of a process rather than simple isolated tasks. According to Hammer (1996) process centering replaces simple jobs and complex processes with simple processes and complex jobs. Process centering is professionalizing work. A professional is *"someone who is responsible for achieving a result rather than for performing a task"* (Hammer 1996:44).

Who then should perform change? According to Hammer (1996) everybody should participate, and an organization needs two systems: the "surface system" and the "deep system". The surface system performs the daily operations and creates customer value while the deep system accomplishes change. The primary processes of the deep system are: learning, redesign and transition. Process professionals should take part in all three processes. In addition to process owners and process professionals, you need some change experts – "reengineers". The reengineers are employees with special skills in bringing about change and work primarily in the deep system facilitating and leading change processes. A specific position in the organization should be appointed as owner of the three deep system processes: *"the Chief Transformation Officer"* (Hammer 1996: 221).

Like TQM, the reengineering philosophy advocates company-wide direct employee participation in change processes. However, the BPR change recipe is very much top down, which is also reflected in prevailing reengineering practice. The term process professional is first of all meant to highlight the dependence on professional skills in process-centered organizations. An important question is to what extent employees need to be familiar with some change methodology in order to participate in change processes. This is not a big issue in Hammer's book, though.

So, just like TQM, BPR is absent in the three squares 2, 3 and 4 of Table 7. It is only in square 1, direct employee participation at shop floor level that we find that the BPR philosophy gives some room for employee participation. Everybody should participate. However, unlike TQM where everyone is expected to participate in the development process itself and is considered an expert on the basis of his or her own work experience and knowledge, we see that the conduct of BPR processes depends very much on professional expertise. What this means to direct employee participation at shop floor level is a question yet to be answered.

Like TQM it seems reasonable conceptually to have both direct and indirect employee participation in BPR processes in a Norwegian setting. It is also in accordance with important stipulations in Norwegian laws, regulations and agreements between the labor market parties. However, BPR processes depend very much on management involvement, process consultants and specialists. The goal is to obtain dramatic results in a short period of time. Participation is dependent on support from change experts and on employees becoming process professionals. The need for union representation in BPR is not recognized in the literature. Company-wide employee participation and union representation takes time. Is there in practice room and time for the Norwegian tradition of employee participation? Is it inefficient? The Aker Stord case has shown that it is challenging. Should unions be involved in the practical change processes at all, or should they take a negotiating role outside the processes?

9.3 A framework for the organizing of employee participation in the further development of an approach to the synergistic combining of forms of TQM and BPR

This section focuses on the practical organization of employee participation in the synergistic combining of TQM and BPR. I consequently do not differ between ID arrangements for TQM and for BPR separately. Neither will I differ between innovations coming out of TQM efforts and BPR efforts, nor between incremental and radical innovations. I focus on organizing for developing incremental and radical innovations synergistically.

I will first outline what I find to be the most basic conditions for employee participation in innovation. Based on that I will show how direct and indirect employee participation can be organized at different company levels and discuss the interrelationship and mutual dependency of direct and indirect employee participation in innovation at and between these levels. For each company level I will give examples of different tools and practical methods that can be used to carry out direct and indirect employee participation. The selection of tools and methods is guided by and based on the list of basic conditions for employee participation.

9.3.1 Basic conditions for employee participation in a synergistic combining of TQM and BPR – motivation, ability and opportunity

Based on the theoretical ID discussion above, my experience from the Aker Stord case and also the twelve years with ED 2000 and VC 2010 I find that three basic conditions need to be present in order to ensure broad employee participation in change. Employees must have 1) the motivation, 2) the ability and 3) the opportunity for such participation. This means that employees should be motivated to do something about their workplace and be able to see the connection between their contribution to change and their own interests as employees. It means that training is important, both in job-related skills and innovation methods. And finally, arrangements must be made to make it possible for individuals to participate.

In the following I will list a few key criteria for each of these fundamental conditions. This list has not been tested scientifically. At this stage it is a practical tool in a development process. It is based on experience and empirical studies, and needs to be verified scientifically. An early version of the list was developed together with experienced Aker Stord personnel as a guideline for making the interview guides for the three audits. The list was modified as we gained experience and increased knowledge through the three audits. This list will be used as a guideline and basis for a) my recommendations for the organization of employee participation, b) for the development of change techniques, c) for analysis purposes, d) for guiding my action research and e) for choosing innovation methodology.

9.3.1.1 Motivation to participate

Creativity and innovation seem contingent on motivation and the desire to contribute to bringing about change. Organizational psychology has many ideas on how best to motivate employees (Howell & Dipboye 1986). The set of criteria listed below is based on my own experience with improvement processes in Aker Stord and in the other networking companies (Tønnessen 2000), and is supported by motivation theory (Howell & Dipboye 1986) and innovation theory (Deming 1986, Juran 1989, Hammer & Champy 1993). I have chosen to gather the criteria in two (interrelated) groups:

- Management commitment and involvement
- Self interest in innovation processes

Management commitment and involvement

Although it is not the managers who are in focus in my research, their role in facilitating and motivating for participation is crucial. Therefore, management commitment and involvement seem to be important conditions for ensuring that employees experience self-interest in taking part. Both my own experience with innovation processes and that of others have shown that management commitment and involvement throughout the line is one of the most important conditions for kindling the interest of employees and for succeeding with such processes (Deming 1986, Juran 1989, McDonald 1992). There are several ways in which managers can involve themselves in innovation processes and show commitment. I find the following to be most important:

Managers need to:

- Formulate vision, values, objectives and strategy, communicate them to all employees and translate them into practical action.

- Communicate through words and action that all job-related experience is of equal value as a basis for participation in the change process.
- Communicate the importance of the change process by taking the initiative to participate in innovation activities.
- Demonstrate a willingness to make decisions and to follow up the innovation activities with specific measures.
- Have the ability to delegate and demonstrate trust in the employees.
- Ask for results and give recognition.
- Have the ability to make arrangements to ensure that employees and groups succeed.
- Have the ability to coordinate activities and make prioritizations in working towards common goals.
- Have the ability to set aside sufficient resources for the innovation activities, among others in the form of time and facilities.

These criteria apply to managers at all levels, and they all seem important to succeed. This was clearly demonstrated at Aker Stord. To recall, the audits showed that some middle and first-line managers had too little information and training about TKA to facilitate others. They did not know how to translate all those fancy TKA words into practical action, and not much happened in their areas of responsibility. In the department Prefabrication of Pipes the situation was opposite. Managers throughout the line of command demonstrated commitment, and the department achieved outstanding results. The Aker Stord experiences show that middle and first-line managers should be given a defined place in change processes and required to produce results.

Self-interest in innovation

It is a great challenge to get to the stage where individual employees experience that the innovation process attends to their own interests as employees and as co-workers in the organization. The experience from Aker Stord was that many lacked interest in what was happening in the TKA process and some experienced the process as disruptive additional work. Many, particularly some old employees, did not consider it their duty to make suggestions for improvements. They considered this to be the management's task. This is in line with what we experienced in some of the other network companies in ED 2000 (Tønnessen 2000). On the basis of the Aker Stord case and the other ED 2000 experiences (Tønnessen 2000), the following criteria seem important if employees are to experience self-interest in the process:

- The change process should to be perceived as a means to increase job security through improving the company's competitiveness. Employees need to be assured that their contribution to increased competitiveness will not make them redundant in the next round.

- The change process should provide the opportunity to promote employees' own points of view, and to realize their suggestions.
- Improved opportunities should be given to participate in designing employees' own work situation.
- The change process should provide the opportunity for learning and self-development.
- Individual employees' competence and insight should be sought after.
- The change process activities should provide better qualitative job content and variation.
- The process should contribute to giving employees increased recognition among managers and colleagues.
- The activities should be arranged so that they do not lead to an increased level of conflict in the organization.
- Feedback and any rewards, for instance in the form of group or company bonuses, should be related to visible achieved results.
- Discomfort from participating in the process should be avoided.

The awareness of the importance of both national and international competitiveness among employees at Aker Stord was visible all the time, and the company had long lasting traditions of both direct and indirect employee participation in business development. I did not uncover any heavy resistance to change in general, except for some disagreements on prioritizations and some critical questioning mostly due to a shortage of knowledge of what was going on. However, the contribution to innovation varied considerably among employees. Employees at Prefabrication of Pipes were given good information. They were given the opportunity to present and discuss own viewpoints. The department was well-known for their innovation activity, employees took pride in what happened and got both monetary and non-monetary recognition for their contribution. The arrangements for employee participation were also arenas for learning, and many of the changes that were realized were both productivity and work environment improvements. All these factors motivated for more participation and gave visible results.

Regarding avoiding conflicts, the audits uncovered some disagreements about the suggestion box system. Some claimed that people were stealing ideas from colleagues and delivered them as suggestions to get monetary rewards. Others claimed that the monetary rewards varied considerably without people understanding why. This was reported as demotivating and resulted in some resistance to the system. This did not happen at Prefabrication of Pipes were the interviewed reported that there was a high degree of information-sharing, discussions and openness.

9.3.1.2 Employees' ability to take part in innovation processes

Participation requires more than motivation. The conditions must also be created to develop the ability of employees to participate, which in turn can help to strengthen motivation. Based on the Aker Stord case and other ED 2000 cases (Tønnessen 2000) I have differentiated between two main groups of criteria here as well:

- Job and company-related knowledge.
- Skills in and knowledge of innovation methodology.

Job and company-related knowledge

The following criteria seem to be most important:

- Theoretical and practical knowledge of the employee's own job.
- Theoretical and practical knowledge of related disciplines, suppliers, the company's products and manufacturing processes and customer requirements.
- Knowledge about the company's vision, values, objectives and strategy.

The importance of these criteria can be seen by referring to the Aker Stord case. Employee participation in cross-functional BPR projects relies on competence on own job, related disciplines, company strategy and the value chain. Customer-supplier interviews and the development of actions related to the interviews both increase and depend on knowledge of supplier – customer relations and demands. The importance of this kind of job – and company related knowledge is also supported by theory (Deming 1986, Juran 1989, Hammer & Champy 1993, Hammer 1996).

Skills and knowledge of innovation methodology

- The ability to identify problems and challenges, to analyze them, and to come up with proposals.
- The ability to identify, delimit and describe work processes through appropriate tools and aids.
- The ability to assess work processes and identify improvement measures through appropriate tools.
- The ability to implement and follow up improvement measures using appropriate tools.
- Cooperation skills and reliability.

One of the experiences from the initial phase of the Aker Stord TKA process was that fuzzy words and high ambitions alone did not make any change. The first training sessions were very theoretical, and some first-line managers reported that they experienced a situation where they gave power-point presentations with a lot of nice words that they did not understand the meaning of and did not know how to follow up in practice. It was first when people were introduced to tools like brainstorming, customer-supplier interviews and BPR techniques that change happened. Also these criteria are supported by theory (Deming 1986, Juran 1989, Hammer & Champy 1993, Hammer 1996).

9.3.1.3 Employees' possibilities to take part in innovation processes

The following sub-criteria apply to both direct and indirect participation.

- Internal routines, roles and responsibility for the innovation process must have been established.
- Suitable meeting arenas at all enterprise levels.
- Time allocated for innovation activities.
- Suitable premises and accessible materials.
- Change techniques, schematics and available IT support.
- Specialist support for running a synergistic approach to incremental and radical change available.
- Specialist support for important change techniques and methods available.
- Resources to realize prioritized change efforts available.
- Appropriate routines for coordination of the direct and indirect participation.

Prefabrication of Pipes had a well established, visible and well-functioning system for handling, storing and follow up of continuous improvement activities, with the Department Council (DC) as the co-ordination body. This structure proved useful also for handling the more radical process innovations connected to the BPR process. The manager also invited and used internal change specialists regularly to provide tools and knowledge. Departments that did not have any equivalent system or cooperated with internal change specialists reported little activity. The importance of having well functioning internal routines, roles and responsibility for the innovation process is well documented (Deming 1986, Juran 1989, Hammer & Champy 1993, Hammer 1996).

Allocation of time is also critical. Some employees reported that TKA was the last point on the agenda in their team meetings, and that this agenda point was usually skipped because of lack of time. Prefabrication of Pipes allocated at least half an hour and usually more to TKA in their regular team meetings. This department produced a considerable amount of proposals. Prefabrication of Pipes also experienced the importance of a good co-ordination between direct and indirect employee participation. All the innovation proposals from individual employees were discussed, prioritized and acted upon in the DC, which employees considered a legitimate and effective practice.

9.3.2 Coordination between direct and indirect participation in innovation

Not everyone is able or willing to participate in all types of decisions about innovations at all levels. There may be valid reasons why employees in an enterprise or members of a society will accept that decisions are made on their behalf in matters that concern them. Recalling from the theory discussion, Dahl (1973) listed three important criteria for accepting other people's decisions, "personal choice", "competence" and "economics". Therefore, even if for example Aker Stord had been a worker-managed enterprise, which it is not, it would nonetheless probably not have been desirable for everyone to take part in all kinds of decisions. Aker Stord is a limited company, where the owner's right to manage applies. At Aker Stord TKA was rooted in existing laws, regulations and agreements between the labor market parties, without challenging any fundamental elements in the owners' right to manage. Therefore, it is within this set of agreements that participation takes place. But even when it comes to decisions within this framework it is not desirable that everyone participates in all kinds of decisions. And this is where the representatives/local union branch comes into the picture.

There is no hard and quick answer as to how such participation should be effectuated in practice, as it needs to be adapted to the size and type of organization as well as to local conditions. As a starting point, and on the background of Dahl's theory (1973), I will suggest the following basic principle for work sharing between these two participation channels.

- In decisions that affect individuals and their immediate job situation and day-to-day tasks direct participation should be the dominant element in matters that are dealt with at employee and group level.
- In matters that affect several groups, are of a complex nature or require extensive changes, the representative channel ought to dominate.
- There should however be opportunities for both direct and indirect participation at all levels from employee level to corporate level.
- At corporate level, and especially in matters of a longer-term and more strategic nature, it is the indirect participation channel that is the most relevant one.

In the presentation below, which guides much of my action research, methodology and analyses, I will first address employees' direct participation, and thereafter the indirect channel.

9.3.2.1 Direct employee participation in innovation

Based on the Aker Stord case and the ED 2000 experiences (Tønnessen 2000) I find that direct in-house participation in innovation and learning should take place at the six basic organizational levels presented in figure 21.

Direct employee participation can take place at all six levels. Arenas at the two levels at the bottom of the figure are the traditional ways for employees to influence changes. Employees should have the opportunity for some self-determination (level 1). This is a legal right founded in the Norwegian Work Environment Legislation (Friberg 1977). There should also be opportunities for employees to propose changes through the line of command (level 2). Improvement activities at the participatory levels that I describe below should have a continuous focus on further developing these two forms of participation through organizationnal changes, job design, redesign of business processes, technological development, management training etc. At Aker Stord many did indeed report increased autonomy as a consequence of the BPR process.

Figure 21 Organizational levels where employee participation in innovation can take place.

These first two arenas for participation focus on the individual employee. A main focus of my action research concerns the individual as a member of a community of employees. My ambition is to create conditions that can improve the synergistic combining of TQM and BPR through increasing employees' opportunities to participate in various kinds of groups/projects where specific challenges and problems are identified, discussed and solved in dialogue and collaboration with fellow workers, managers and staff. These kinds of dialogue based innovation arenas are described below (levels 3-6).

The third level (3) in Figure 21 consists of the people who are going to participate or are involved in a specific job, and focuses on the planning of this specific job. "Safe-job analysis", "before-job discussions," "during job discussions" and "after job discussions" are examples of arenas that have been introduced for employee participation in innovation at this level. The aim of these kinds of dialogue-based arenas is to make better use of the employees' own experiences to come up with innovative ideas that can improve HSE, reduce breakdowns and increase productivity. These kinds of arenas are referred to as "task/ order" in Figure 21.

At the level above (4) we find existing work groups/departments. The focus at this level is mainly within the actual unit and the work performed at this level. Quality circles, department meetings with a development agenda and departmental brainstorming are examples of this kind of participation. Very often project groups are established to follow up and carry through ideas and proposals from these kinds of meetings. Representatives from the employees who will be affected by changes should be invited to participate in these project groups.

The next level (5) deals with a company's business – and work-processes. The company should have a continuous focus on process innovation and establish innovation teams for each project that has been decided on. These teams may be composed of employees from within one group or department, but the potential for radical changes and dramatic improvements are more often a result of changes across groups, departments and companies along the value chain (Juran 1989, Hammer & Champy 1993). It is often useful to strengthen the group with external expertise in order to get ideas from outside the company. At this level, innovation teams should be composed in accordance with the value chain and the groups, departments and companies involved (Ernst & Young and Willoch 1996).

On top (6) we find the participation arena that may involve all the employees in the company. The focus at this level may be on all kinds of innovation areas connected to the company's activities. This may encompass everything from challenges of a superior/long-term nature related to strategy and down to practical actions that can be implemented immediately.

In ED 2000 I developed the Enterprise Development Conference (The ED Conference – see Section 10.1 for details) for participation of all the employees in the generation of innovative ideas at enterprise level and for participation in business' strategy development. At the ED Conference each individual employee is invited to present suggestions on areas in the enterprise that should be given priority with regard to further development, and to suggest specific development projects within these areas along with practical initiatives. The suggestions are discussed and prioritized in workgroups. The follow-up and realization of the conference's results are planned through a joint committee.

These six levels for direct in-house participation in the generation of innovative ideas may be considered as a point of departure. In real business life arenas on these levels will take a variety of different forms, and various combinations are possible. It is important that the initiatives and activities at the different levels are coordinated. E.g. at the company level the results from a Dialogue Conference or an ED Conference will be a large number of incremental and some radical innovation proposals and ideas. These have to be transferred to the correct organizational level to be followed up. If a company does not develop good routines for handling proposals between different company levels, efforts like an ED conferences can be a waste of time. Furthermore, innovation activities at each level should be related to and carry out the company's superior vision, goals and strategy, and be in accordance with the values of the company.

9.3.2.2 Indirect participation in innovation

The Aker Stord case and other empirical studies that I performed (Tønnessen 1997, 2000, 2002, 2005) suggest that the planning, carrying out and supervision of the different forms of direct participation in innovation described above should be coordinated with and supported by the employees' chosen representatives through relevant channels for indirect employee participation at different company levels. The direct and indirect channels for employee participation in innovation should be run in a way that make them reciprocally reinforcing, while recognizing Norwegian rules, regulations and agreements in force between the employers and the employees' associations (LO & NHO 2006) and also Norwegian legislation (Friberg 1977).

The potentials and strengths of direct and indirect participation at the different company levels may vary. In many cases also arenas above company level such as concern, branch etc. as in the Aker Stord case may be of great importance, but will not be dealt with here. I will now show how the indirect channels for participation in innovation can be combined with and support the arenas for direct employee participation and vice versa. I will start at the company level and proceed downwards in Figure 21. The numbering below follows the numbering in the figure, and starts with 6 at top.

At the company level (6) the Norwegian Limited Companies Act (Justis og Politidepartementet 2011) arranges for employee representation at the board of directors. The employees' chosen representatives may use this arena to promote and support innovation proposals and ideas that are raised on arenas on lower company levels. Vice versa, important questions and innovation ideas from the board may be transferred to lower company levels and arenas for employee participation and discussion.

Furthermore The National Wage Agreement between The Norwegian Federation of Trade Unions and The Confederation of Norwegian Business and Industry (LO & NHO 2006) has a part B, which is an agreement between the parties on how to co-operate in order to promote business development in general. This part B prepares for the establishment of Works Councils or equivalent arrangements where both parties are represented. The Works Councils can initiate and supervise the overall planning and follow-up of dialogue based conferences like the above mentioned ED Conference. Vice versa, some of the ideas and innovation proposals that are the result of these kinds of conferences should be transferred to the Works Councils for further discussions and decisions. Shop stewards participating in this arena should promote and support proposals that are given a high priority. In addition, arenas at this level should play an active role in the planning and following up of company efforts like the introduction of a company-wide process for continuous improvement or a business process redesign effort.

Processes (5): A central element within BPR is to change organizations from being primarily function oriented to become more process oriented. ED conferences and similar arenas for employee participation can result in a number of suggestions for incremental and radical process innovations (Tønnessen 2000). Many of these will cross departmental boarders. Based on my case study experiences I recommend that the management group deals with the large and company-wide projects and takes responsibility for the overall planning and follow-up of such projects. However, Works Councils or similar cooperation bodies should play an active role in planning and following up such projects in addition to being kept informed about the projects. The Works Council may e.g. be given the responsibility for the overall planning and follow-up of some projects, or take the position of a steering committee, depending on the complexity and importance of the project.

In minor projects that include one or two departments, I recommend that the overall planning and follow-up of particularly important/complicated projects are done by the relevant Departmental Committees (DC). The National Wage Agreement, referred to above, recommends e.g. that DCs should be established in companies with 200 employees or more (LO & NHO 2006). DCs are arenas for company-union cooperation and established to increase productivity and to improve the work environment. For larger projects steering committees and/or reference groups may be established by and report to the DC. It is important that shop stewards take part in such a way that the dialogue with the DC is maintained.

The composition of project groups in important process innovation projects should cross departmental boarders in line with the TQM and BPR ways of thinking (Juran 1989, Hammer & Champy 1993). It is important that shop stewards and safety delegates are invited to take part, in addition to ordinary employees. They represent operational knowledge and through their participation they will obtain insight into the way these kinds of process changes are carried out. This will help them develop the skills needed to act as facilitators in smaller redesign processes at their own operational/departmental level.

The next level (4) refers to innovation activities within the department or group. If the company has a Departmental Committee (DC) or equivalent, it is recommended that this body plans, co-ordinates and follows up the local improvement work which is initiated at departmental level (e.g. ideas coming from department meetings, brainstorming sessions or quality circles).

At group level, formal co-operating bodies are rarely established. However shop stewards and safety delegates may be appointed. In this case the manager in cooperation with the shop steward or the safety delegate, can follow-up the group's innovation activities. Such a small local "committee" may also follow up innovation proposals that have been transferred to the group as a result of an improvement process or brainstorming session at a higher company level. This kind of two party cooperation has some advantages: It gives the process legitimacy, there will be a dialogue about the accomplishment of the initiatives in question and it also reduces the likelihood that the supervisor is looking after his/her own interests or is accused of doing so.

Employees should be given the opportunity to participate in the project groups/task forces that will be the result of all the above mentioned efforts in all innovation areas. This way experiences and insights from the shop floor are used to ensure satisfactory solutions. In addition, employees can be more motivated to carry through the innovations decided on if they have been part of the decision making process. There can be less resistance to change.

At the levels "Task/order", "Line of command" and "Self determination" (3 – 1) the channels for indirect employee participation in innovation have no formal organization. However, shop stewards and safety delegates can take part in arrangements like "safe-job analysis", "before-job meetings", etc. This way, challenges connected to the way work is carried out may be picked up and discussed in union-management cooperation bodies at higher company levels. In the lower level innovation arenas, shop stewards act as traditional employee representatives and follow up issues that cannot be solved through the line of command. This kind of participation is also an opportunity for shop stewards to obtain experience and knowledge about arrangements like “before-job meetings” in order to participate in the planning, coordination and improvement of the total set of participatory tools an enterprise uses.

My theoretical and empirical analysis, my cooperation with industry and study of company reports show that excellent ways of work, tools and techniques for most of the arrangements for employee participation in innovation mentioned in this chapter are available in literature, handbooks, manuals and in company procedures. My analyses also showed that TQM and BPR have not exploited some important potentials regarding employee participation in business development. I found that some new tools needed to be developed. I will now present the two tools that I found necessary to develop.

10 New tools to support a synergistic combining of TQM and BPR

My theoretical discussions and the Prefabrication of Pipes experiences indicate that an ID context can be favorable for developing a more synergistic approach to the combining of TQM and BPR. In order to make my approach to employee participation in the synergistic combining of these two forms of exploitation and process exploration more complete, I developed two new tools.

The first tool – the "Enterprise Development (ED) Conference" is a tool for employee participation in developing ideas in relation to companies' strategy and development processes. The second is a tool for employee participation in process exploration – the Actor-Process Diagram. Both tools are designed to enhance learning and increase the generation of ideas for incremental and radical innovations synergistically at all levels in an organization. I developed and tried out both tools through action research in industry, and will now present these studies and show how the tools were applied in two company cases, Titania and Serigstad.

10.1 Tool 1. The "Enterprise Development Conference"

I will first present the background for the ED Conference. Then follows a company case where the Conference (Titania) was applied. I will then give a recipe for the ED Conference as a tool in its basic form. This recipe builds on the experiences with the Titania case as well as the ten preceding conferences that I arranged. The ED Conference follows some basic principles. However, the practical arrangements vary and are adapted to the actual company situation. After the presentation of the case and the recipe for the ED Conference, the ideas the conference builds on and different aspects of the conference will be discussed both from a theoretical perspective and with reference to the case experiences from this last company case and from my other action research projects in ED 2000 (Buytendorp & Hansen 1997, Eidstuen 1997a, Eidstuen 1997b, Børve 1998, Eidstuen 1998, Tønnessen 2000, Bjørnevik et al. 2005). I will discuss the ED Conference as an enabler for the synergistic development of incremental and radical innovations that in a wider context is crucial for the synergistic combining of TQM and BPR.

The case follows the direction of the synergistic development logics, but focuses on one specific tool for the further development of ID as an enabler for the synergistic combining of TQM and BPR. So, it is not the integration of TQM and BPR concepts that is being tested in this case study.

10.1.1 Background – The ED Conference as a tool for planning incremental and radical innovation synergistically

Direct employee participation at strategy/political level and exploration is not an issue in the prevailing TQM and BPR literature, and adequate tools are lacking. In order to exploit the potential of this kind of participation, I developed "The Enterprise Development (ED) Conference". Initially, it was used to identify and create consensus about important improvement areas for companies that started up their TQM processes in Enterprise Development 2000. This included both radical and incremental innovation, in accordance with the EFQM model (EFQM 2003). The context and the rationale behind the initial development of the ED Conference are described in Tønnessen (2000).

TQM and "Continuous Improvement" have mainly been directed towards direct participation in operational/lower level decision-making and incremental improvements in the employees' immediate work situation. The focus has been on improvements that belong to the exploitation paradigm. Indirect employee participation is of little importance to TQM. Direct or indirect employee participation in exploration is not a focus in TQM.

In BPR, company-wide employee participation in whatever form does not usually play a role at all. This seems to apply both in Norway and internationally. Employee participation in the synergistic combination of forms of exploitation and exploration is the challenge that the ED Conference is exploring.

This challenge raises questions such as:

- How is it possible to expand direct employee participation in exploitation at shop floor level to also comprise forms of exploration and higher level decisions in the enterprise (related to company vision, values, objectives, strategy and overall planning)?
- If it is possible, what can be the consequences for a company's innovation capability, for employee participation in innovation and for learning how to participate in innovation?
- Can the ED Conference contribute to the completeness of an employee participation framework that promotes the synergistic combining of TQM and BPR? If so, how?

Direct employee participation at higher company levels and overall decision making is not new, either in Norway or internationally. In Norway, researchers at the Work Research Institute (WRI) have developed the "Dialogue Conference" (Gustavsen 1990, Ebeltoft 1991, Pålshaugen 1991, Gustavsen 1992, Naschold et al. 1993). Internationally as well as in Norway, this form of direct participation through conferences has also been carried out through a similar arrangement called "Search Conferences" (Cabana, Emery & Emery 1995, Cabana & Fiero 1995, Emery & Purser 1996, Purser & Cabana 1997). These two kinds of conferences have a number of common features, but have been organized in many different ways and for various purposes. The research institutes that have promoted them have given this form of conference different theoretical and ideological anchoring. E.g. the Search Conference (Cabana, Emery & Emery 1995) seems to focus more on the Conference's contribution to business development than the Dialogue Conference does, while the latter emphasizes democratic dialogue (Gustavsen 1992). The ED Conference shall promote both business development (in this case the synergistic combination of forms of exploration and exploitation) and democratic dialogue in a way that make these two goals reciprocally supportive.

The ED Conference is mainly based on a further development and appropriate adaptation of methods and approaches that have been developed within the following three traditions.

1) The ED Conference is anchored in the values and Norwegian tradition of close co-operation between the labor market parties on enterprise development. This tradition is especially represented by the cooperative NHO and LO body "HF" (Hoved-organisasjonenes Fellestiltak 2005).

2) It is partly based on the action research tradition that goes back to Kurt Lewin in the 1940s (Clausen & Lorentzen 1992). The Dialogue Conference (Gustavsen 1992) that has been developed by employees at the Work Research Institute in Norway can be seen as part of this tradition.

3) Finally, it is partly based on the quality tradition. This tradition has developed a number of techniques in order to ensure employee participation in improvement activities, e.g. "brainstorming" and the "affinity diagram" (Bergman & Klefsjø 1994).

The practical form that these conferences take differ in several ways, but it is the desire to uncover everyone's point of view, to promote democracy and equality in the choice of solutions, and to create broad acceptance of solutions that forms the basis for actions.

Accordingly, the following propositions were developed for the eleventh conference, which will be presented below, the **Titania** case:

- The ED Conference, including follow-up, generates a considerable number of incremental and radical innovation initiatives synergistically.

- The ED Conference, including follow-up, improves employees' direct and indirect participation in innovation synergistically.

10.1.2 The Titania case – identifying future challenges

10.1.2.1 Background and planning

Titania A/S was established in 1902 and is the largest company in the community of Sokndal situated at the south-west coast of Norway. Titania is a mining company that in 2004 accomplished an annual extraction of rock material from the open cast mine at Tellnes in Sokndal of nearly 8 million tons. From this, about 650,000 tons of fully processed ilmenite concentrate was produced. Ilmenite concentrate is the company's main product. Titania's raw material base is considered to be the world's largest deposit of ilmenite, and the ore reserves are considered to be sufficient for several generations to come.

Most of the company's 250 employees are trade union members. The company reports that they have good work environment conditions, a high degree of job satisfaction and had a sickness absence rate of less than 3.5% in 2004.

In 2004, a large number of jobs were under threat of redundancy and termination due to a drop in orders signaled by a major customer – Tinfos Titan & Iron A/S. Titania decided to start a comprehensive organization development (OD) project aimed at creating an even more robust company, improving competitiveness, creating better, more secure and more interesting workplaces and avoid redundancies. In this setting OD is referring to the development of the company in a wide sense and not delimited to organizational and people issues. Based on discussions in the combined Works Council and Work Environment Committee (WEC/WC), it was decided that the OD project at Titania should be carried out in accordance with important fundamental principles laid down in the Main Agreement between the Confederation of Norwegian Business and Industry (NHO) and the Norwegian Federation of Trade Unions (LO) (NHO & LO 2006) and in the Norwegian work environment legislation concerning all-party cooperation, co-determination and broad employee participation.

It was decided that the combined Work Environment Committee and Works Council (WEC/WC) should be the central organ to drive the process and promote employee participation. The practical work of planning and implementing the OD process was to be facilitated by a project group consisting of the head trade union representative, another representative from the employees appointed by both the management and the union and the Human Resources (HR) manager. The project group was to report to the combined WEC/WC, and should stay in close dialogue with the CEO, who would be responsible for the project. The HR manager was appointed project manager.

The CEO was familiar with the VC 2010 program and contacted LO's representative in the program to get support. The LO representative arranged a meeting between the company and IRIS, which had previous experience with this type of development processes. IRIS agreed to assist the company with this effort. The project was to be linked to VC 2010, and could therefore make a contribution to the development and research work that was ongoing in the Hordaland and Rogaland Development Coalition. From the very start, the Titania representatives made it clear that their objective was to ensure development and increase value creation through broad employee participation. Titania wanted to further develop the company on the basis of mutual trust and a good cooperation between the management, union representatives and employees.

On IRIS' advice, Titania started the OD project with a start-up conference to which all employees were invited to give their input to and prioritizations for improvement initiatives (both radical and incremental) for the company. The aim of the start-up conference was "... *to ensure participation and input from all employees with the planning and start-up of a new*

OD project, which is intended to build an organization for the future. This future organization will be based on broad employee participation. Participation will be important in the endeavor to realize Titania's annual production target of 1 million tons of ilmenite concentrate" (my translation from the conference program).

The WEC/WC decided that the OD project and the start-up conference were to be part of the start-up of a process for continuous improvement within the company. To the company, the expression Continuous Improvement included both incremental and radical change in line with the EFQM model (EFQM 2006). Arrangements were to be made to ensure that the points of view and suggested improvements of all employees should be presented, discussed and prioritized. The conference was intended to create broad acceptance for solutions, which in turn should create a basis for action. The start-up conference should create a real opportunity for employees to participate in the planning of the OD process. The WEC/WC decided that the start-up conference should be conducted according to the pattern from the ED conference that was developed by IRIS in ED 2000 to ensure participation of all employees in the further development of the company and in the strategy processes.

The start-up conference was arranged so that it could be conducted in the course of one day and be planned, carried out and followed up by the WEC/WC. Titania's existing, broadly employee composed cooperative bodies should do the practical follow-up of the results of the conference. Action researchers from IRIS should act as advisors during planning, implementation, summing up and follow-up of the conference. Four action researchers from IRIS were engaged in the planning, carrying out and in the reporting of the results from the conference. I was also invited to the first two WEC/WC meetings after the conference to give advice on the follow-up process with a special focus on how to organize the development process, methods and techniques. The cooperation with the researchers from IRIS started in spring 2005 and lasted until late autumn the same year. The researchers did not participate in the further follow-up process after that, but received some information on the phone and on e-mails.

I will now describe how the conference was organized and pay particular attention to the way activities were carried out in relation to how direct and indirect employee participation were combined to promote incremental and radical innovations synergistically. Furthermore, I will show how the theory of Democratic Dialogue (Gustavsen 1992) directed our way of planning dialogue, learning and decision making, and how techniques from the quality tradition were used to promote radical as well as incremental innovation. These issues and the Titania experiences will be analyzed later in this section and related to the experiences with the preceding 10 ED conferences that I had facilitated (Tønnessen 2000).

10.1.2.2 The Titania conference

The project manager for the OD project started the conference welcoming the participants and encouraging the employees to use the start-up conference to present ideas they had been thinking about, perhaps for years. He informed the participants about the work that had been carried out to prepare the conference, developments in the company over the past few years, and the objective of the OD project. He stressed that it was now important that everyone contributed with suggestions for:

- Initiatives that could give an early cost effect for the company and/or
- Technological and organizational initiatives with a longer time horizon.

Essentially, the project manager asked employees to come up with both incremental and radical innovations.

The Regional Director from NHO was invited to talk about NHO's attitudes related to improvement processes. Titania was described as a cornerstone company. He emphasized the importance of the company for the local community and said it was difficult to imagine that community without Titania. He also pointed out how important it was to engage employees in improvement processes. *"Growth is not generated from the top. It is mobilized from the bottom. Developments take place in individual companies, which are constantly working to discover how to do things better"* (my translation). He praised management for taking the initiative for this work.

The Head Union Representative said that it will be a hectic and different sort of day and asked: *"Can we turn negative thoughts into something positive? Can we make our everyday lives more secure for the next century?* (my translation).

He pointed out that it is the employees who are best placed to suggest improvements that will make Titania a better and more secure place to work. Furthermore, *"We can have some good ideas emerging at this conference. After that, it will be a challenge to do something about them, but now at least we are going to write them down and prioritize them"* (my translation).

The head of the local union thereby asked all employees to contribute to the further development of the company.

The Managing Director followed up by saying: *"Friday 13th May is an important day for Titania. Titania is going to advance and secure its future. All of us employees are an incredible resource, and the responsibility is ours. It is us employees who will be doing the job. The reason why we will be doing it is that it is our company. Some people think this company is owned by Americans. I don't think so. Those who own shares do not feel ownership in the same way as we who work here. It is we who own the company.*

The company is doing well now, but nothing is more dangerous than things going so well that we just sit on our hands and do nothing. There are many examples of companies that were doing too well to discover the dangers that were lurking – now they are history. We must grab our chance and face the challenges.

Right now we are on schedule and reaping excellent results. So far this year (April 2005) we have achieved pre-tax profits of 71 million NOK. Last year profits were a total of 94 million

NOK. The key to developing Titania is that things are going well for Titania. The present focus is to increase the production volume towards 1 million tons and to achieve permanently lower and falling costs per ton in the future. In the WEC/WC we have decided to:

- *Work towards zero injuries in a good work environment.*
- *Conduct an active and visible environmental protection policy.*
- *Develop employees and the organization so that individual employees feel stronger ownership of their own workplace and results.*
- *Secure the supply of ore to cover the long term aspect and needs of the market.*
- *Increase the strategic investments that support our focusing.*
- *Increase our efforts to develop and market the use of ilmenite as weight material in drilling mud offshore based on the pilot study by 2006.*
- *Find new markets for ilmenite"* (my translation).

After these introductions, two hours were spent on group tasks in which all employees participated. The program for the Enterprise Development Conference at Titania, 13th May 2005 is presented in Table 8.

Time	Activity
10:30	Lunch
11:00	Welcome and start-up
11:15	Perspectives on the OD work from the head union representative and the managing director
12:00	Brainstorming and group work
14:30	Break with food and refreshments
14:50	Brainstorming and group work
15:30	Plenary presentations and discussion
16:45	Rounding off with union representative and management
17:00	Conclusion of conference and follow up activities
17:30	Evening event and dinner
21:00	Close of conference

Table 8. The programme for the ED Conference at Titania 13th May 2005

A total of 197 employees were present during the start-up conference. The participants were divided into 22 groups, and distributed between various meeting rooms throughout the building. Each group was required to concentrate on certain production areas. They were handed a task and asked to prioritize 1-4 proposals for presentation at the plenary session. The organization and practical implementation of the group work at the conference is presented below.

The organization of the group work

Organization

- The group coordinator is elected in advance, and will function as discussion leader.
- Take five minutes in silence to read the task.
- Discuss the task briefly to ensure everyone understands what to do.
- A spokesman for the group is elected to present the result in the plenary meeting.
- The coordinator and spokesman agree on who will do the writing, and how to organize the making of the joint sheet of paper and the transparencies for presentation.

Brainstorming and discussions

Individually

Each person writes down their suggestions for improvements in big, clear letters on post-its or on paper. Use a broad felt pen. Make one note for each suggestion. This is done in silence. Allow 15 minutes.

Presentation in the group

Each participant reads his suggestions out loud, then explains and gives the reasons for his suggestions to the group while handing the yellow post-its to the coordinator. The coordinator/spokesperson sticks the post-its to a large joint sheet of paper. It is not permitted to criticize or discuss the suggestions at this stage. Clarifying questions are OK. When everyone has presented their suggestions, the discussion leader asks if anyone has come up with any more. These are added to the joint large sheet of paper.

The group coordinator arranges the proposals together with the group

The suggestions are sorted. Identical and similar suggestions are considered jointly. No discussion yet. Each idea is given a number and arranged in sequence.

Individual prioritization

The ideas are now prioritized. Each person notes the number of the three ideas they feel are most important.

The group's prioritization

The participants then state which numbers they have noted down and the coordinator places a cross for each idea that is read out. The coordinator counts the number of crosses for each suggestion and marks the prioritized sequence with 1 as the highest priority. The 2-4 ideas with the highest priority are discussed. If there are more ideas than this that stand out, it is also OK to discuss them if time allows. Now, at last, participants are allowed to voice any objections.

Preparations for the plenary presentation and handing in of the group's ideas

The plenary presentation is prepared by writing down keywords representing the most important arguments for the 2-4 prioritized suggestions and any objections that have been raised on a joint sheet of paper. The 2-4 suggestions are also written on a transparency for display to all the participants. Each group writes the group number and name of its group leader on the joint sheet. The coordinator checks that everything can be understood by those who will be summarizing the conference afterwards. It is particularly important that the suggestions are readable, adequately described and that the prioritized sequence is visible. The sheets are left on the table, and collected later. All the suggestions are kept and taken further after the conference.

The pictures below show the groups at work and how the suggestions were gradually arranged on taped flip-over sheets and visible to everyone.

Starting up the group work

Preparing a flip-over sheet

Yellow post-its

10.1.2.3 Results and following up

A total of 672 improvement proposals were presented by the employees. Most of these can be considered incremental. However, there were also some radical ones among the 627 proposals, such as the proposal for a new power plant, a new organization, a new way of organizing maintenance work, the development of a new mining field, "Tellnes East" and a new way of crushing raw material/stone. The proposed improvements also cover several innovation areas such as new products, new processes, new markets and organizational changes. Thus, the method promoted developments in several innovation areas.

Afterwards the 672 proposals from the group work were collected and reproduced in a conference report with assistance from action researchers at IRIS (Bjørnevik et al. 2005), and then followed up through the company's further goal, planning and budgeting processes, and through immediate initiatives and innovation projects. This follow-up constitutes the main content of Titania's improvement process in the immediate future. It is a demanding project and a comprehensive change process that has now been initiated.

The WEC/WC decided that the practical follow-up of the conference's results shall be achieved by using and further developing Titania's own existing broadly employee composed bodies for employee – employer cooperation and its practice of direct and indirect participation in enterprise development. This way direct and indirect employee participation can work mutually reinforcing to improve the synergistic combination of forms of exploitation and exploration. Large projects encompassing the entire company are planned and followed up by WEC/WC. In practice, project groups are established with their own project leader for each project that reports to WEC/WC. The status in spring 2007 was that the WEC/WC among others had organized a considerable amount of the 627 proposals in seven company-wide projects. These are:

- Work environment
- Competence development

- Future organization
- Infrastructure
- Improvement concept and approach
- Internal transportation
- Maintenance organization

By spring 2007 the "Maintenance organization" project had been completed. The building of a new power plant started in 2008. This is a 40 million NOK investment, and will produce 110 GWh. This power plant will make possible another of the radical innovation proposals, namely the new mining field, Tellnes East. The development of the new mining field will double the company's total mining-areal. In addition to the large company wide projects, the WEC/WC also prioritized and followed up some minor projects like e.g. "Improved communication in the laboratory and between the laboratory and other departments". The company representatives do not differ between radical innovations and incremental innovations. The development work is to promote both and build up knowledge, tools, procedures and systems that support both.

At Titania, improvement projects are also used as a means for the employees to learn a systematic way of working through the different phases of problem-solving, from the current status situation to identification of problem areas, causal analysis, consideration of solutions, starting up actions and follow-up. The intention is for employees to learn the most fundamental principles with regard to project management, such as being able to describe the current situation, the desired situation and the target for the project. They learn how to make a progress plan with activities, deadlines, persons in charge and status reporting. A good systematic method of implementing the improvement projects is applied to strengthening and further developing the company's structure, system and procedures for development work. It is worthwhile mentioning that part of the training of shop stewards, managers and facilitators were carried out by two shop stewards from a customer company (Tinfos Titan & Iron A/S) that also participated in VC 2010 and had several years of experience with this way of union – company co-operation from ED 2000.

The bottom-line effects of these kinds of change processes are difficult to measure. The projects are intertwined with each other and also with other business processes. In addition, Titania's results are very much affected by fluctuations in the market price for their products. However, compared to the years before the start up of the improvement project both the returns and the investments increased considerably the following years (Brønnøysund Registeret 2009). This is no proof of the success of the project, the ED Conference and the follow-up activities. It is only a good indication.

The ED Conference is a highly flexible tool that can be adapted to different needs and businesses. In my action research it has been used in different settings and for various purposes (Tønnessen 2000). Based on the Titania experiences and the experiences with the 10 preceding conferences, I will present some basic principles and methodology for the ED Conference next.

10.1.3 The ED Conference – basic principles and methodology

The ED Conference has been developed to enable individual enterprises to carry out the conference without disturbing company production too much, and in a way that enables all employees to participate. The ED Conference can be arranged in the course of a working day or an afternoon/evening after work. This is a short time frame and requires some work to be carried out in advance. Furthermore, it requires that something happens afterwards and that the conference is included as one of several elements in a strategy and development process. The practice described here assumes that the enterprise also practices several forms of direct and indirect participation in enterprise development to support the conference.

The ED Conference should be adapted to each individual enterprise. It is particularly important to take into account the skills and experience of individual employees in part-taking in enterprise development and improvement activities. The status of the enterprise's strategy work and its ability to conduct strategic work must also be assessed along with the co-operation climate between management and employee representatives.

The event should also be adapted to the requirements of shift work, hiring out activities, etc. In some cases it will not be possible to bring together all the employees at one time. For instance, the ED 2000 enterprises Westcon and Stord Offshore, that at the time of holding the conference had extensive hiring out activities, conducted the conference with a selection of participants consisting of representatives from the management, staff and the unions from the various parts of the organization (Tønnessen 2000). However, this practice should not be taken too far if direct, rather than indirect, employee participation is the goal. An alternative is to do like one of the other ED 2000 companies, Kværner Hetland. They did not have the opportunity to close down the factory in order to give all employees the possibility to participate. Instead the company divided the organization into two and arranged two conferences. This way all employees could be invited. The results of the conferences were integrated and united into one conference report including a plan for following up (Tønnessen 2000). The conference has mainly been developed for SMEs, but can also be adapted to large enterprises. In the latter case it can be carried out in individual departments separately, followed up by summing up the results from the different departments in a joint plan (Tønnessen 2000).

The ED Conferences can be divided into three phases: 1) Preparation, 2) Holding the Conference and 3) Follow-up.

Preparations for the ED conference

Based on my action research experiences (Buytendorp & Hansen 1997, Eidstuen 1997a, Eidstuen 1997b, Børve 1998, Eidstuen 1998, Tønnessen 2000, Bjørnevik 2005) and ID considerations I recommend that the conference is planned in a co-operation between the top management and the employee representatives. For companies with Works Councils (WC) or similar committees for employee-employer cooperation, these may be suitable planning arenas. Some of the enterprises we collaborated with in ED 2000 and VC 2010 had no strong tradition of co-operation between the parties. In these instances, the body, with representatives from both parties, which as part of ED 2000 had been established to manage the improvement process, coordinated the overall planning of the conference. This was done in the ED 2000 company Kværner Hetland (Tønnessen 2000). In the ED 2000 company NOHA Brannteknikk, a steering committee with representation from all levels of the company and the unions, managed the overall strategy process (Tønnessen 2000). In most cases the planning of details was carried out as a co-operation between the general manager, the improvement process facilitator and action researchers. In some cases the corporate shop steward also took part in working out the details (Tønnessen 2000).

An invitation including the conference objective, agenda, expectations and any division of duties concerning individual topics should be sent to all participants well before the conference. This enables participants to prepare themselves, obtain information, discuss the matter with their colleagues and prepare for the task at hand. An ED Conference represents a completely new situation and way of working for many. Everyone is expected to be creative and to come up with proposals for improvements in important areas in a short period of time, and everyone is expected to prioritize between several initiatives. Even if these conferences are organized as brainstorming sessions with no prior preparations required, experience shows that it is easier to get off to a good start if some preparations have been made in advance.

Holding the conference

Holding the conference has four parts, a) Presentation of status and challenges facing the enterprise, b) Brainstorming in groups, c) Plenary presentation and d) Wrapping up.

Presentation of status of the present situation and important challenges facing the enterprise

The conference can start up by having a chosen moderator welcoming everyone and introducing the program for the conference. Then the general manager and an employee representative, for instance the corporate shop steward, open the conference by presenting a brief internal and external analysis and saying a few words about what they consider to be important challenges for the future. This contribution may be based on the strategy plan. In that event, the strategy plan is not presented in detail, but those areas are highlighted that the employees may have a special interest in and are able to discuss. This was done at both Kværner Hetland (Tønnessen 2000) and Titania (Bjørnevik et al 2005). If we make the strategic plan the point of departure, the presentation needs to be made in a language that is easy to understand. A strategy document may not always be well suited for reading out loud. Introductions of this nature are important. It is also essential for the follow-up of the conference's recommendations that the prioritized initiatives are formulated on the basis of a holistic perspective and in line with other important choices made by the enterprise.

Brainstorming in groups

To make sure that everyone is given the opportunity to speak, to engage with others in dialogue, to exchange ideas, to learn and to take part in prioritizing/selecting the initiatives of the conference, the next phase of the conference is group work with a maximum of 7-8 persons in each group.

If the topics to be discussed by the groups are determined beforehand, the groups should be composed with this in mind. Acknowledging that the greatest improvement potential is found across departmental boundaries (Juran 1989) and that the key value creating processes in an enterprise are found across disciplines and departmental boundaries (Kelada 1996), the groups should be made up of a wide cross-section of disciplines. The groups should also consist of both managers and employees, as they often represent different perspectives on the challenges facing the enterprise.

In order to avoid that a few people dominate and decide what is most important, the brainstorming starts with a quiet period, during which the group members write down their suggestions individually. Thus, the input from the conference to the company's business strategy starts at the level of the individual employee. The suggestions are then presented and justified to the other group members. The group secretary displays the suggestions on the wall to make them visible to all as soon as they are presented. When everyone has presented their suggestions, the group members are encouraged to present any new ideas that came to their minds while the others presented theirs. When everyone has presented their ideas, a discussion is initiated. In this way important principles of Democratic Dialogue (Gustavsen 1992) ensure learning among the participants.

It is hardly ever possible to implement all good suggestions at the same time. The next phase consequently consists of selecting a few ideas that the group will recommend the enterprise to prioritize. An easy to use voting technique can be used in this phase. The 2-5 highest prioritized suggestions will be discussed until final prioritization is reached.

The group work can be organized in many different ways, depending on time, the skills of the participants, level of ambition, etc. The activities mentioned above are the most basic ones, but can be combined with other exercises. Kværner Hetland (Tønnessen 2000), for instance, evaluated the prioritized initiatives using an Ishikawa diagram (Ishikawa 1985).

Group work is the most important element of the conference and an important arena for learning. It is the results achieved by the groups that are taken further.

Plenary presentation

When the group work is completed, everyone comes together in a plenary session for a presentation of the results arrived at by the different groups. This presentation is important, for four reasons in particular:

- It helps focus the participants on producing a result that can be presented to the others within the available time. Experience shows that there will always be a few groups that do not take the task all that seriously if they know the end product will not be presented at a plenary session (Tønnessen 2000).
- It is important for learning reasons to show all participants the diversity of ideas that emerge in these meetings, that people view things differently, that there are different proposals for solutions to the same problems and to show how others plan to solve challenges,
- It is necessary to show the participants that they may prioritize things differently, and that it is impossible in practice to do everything at once. Without such understanding, the follow-up of the conference can be a great disappointment for most people when they discover that their own suggestions are not immediately followed up.
- We get immediate reactions from others in the assembly in the form of objections, support, further development of the idea, new ideas, etc.

Wrapping up the conference

The general manager wraps up the conference with a preliminary summary of the incremental and radical innovation proposals that have been presented and with a binding statement as to how the conference will be followed up. The latter is very important, and the follow-up procedure must be planned before the conference is held. Unless the employees are informed of the further process, they may have unrealistic expectations regarding the realization of the proposals, followed by disappointment or indifference. Many enterprises have experienced that similar kind of sessions have not been followed up and have come to nothing (Quale 1995). The ED Conference could easily become yet another example of this.

Follow-up of the conference

The conference is neither a decision-making nor a concluding body. Above all else, it is a brainstorming session where the participants are also given a lot of information about what is happening in the enterprise, where they learn through dialogue with colleagues and are given the opportunity to discuss issues of concern with colleagues, shop stewards and managers. If such a conference is prepared and organized well, there is no difficulty in acquiring a large number of good ideas from the employees. It is the follow-up that decides whether the ED Conference is a good investment or a waste of time and source of frustration. It is essential that good, clear plans for follow-up are drawn up before launching a conference of this nature. With regard to the follow-up, different activities are recommended, as in the Titania case. Based on my action research with 11 conferences (Buytendorp & Hansen 1997, Eidstuen 1997a, Eidstuen 1997b, Børve 1998, Eidstuen 1998, Tønnessen 2000, Bjørnevik et al. 2005) the following seven are to be considered a minimum of measures that should be initiated:

1. Organization and division of responsibilities for the further follow-up

2. Idea bank
3. Minutes and action plans
4. Integration of conference results in the company strategy, planning and budgeting processes
5. Immediate measures
6. Innovation projects
7. Continuous distribution of information on status to all

Organization and division of responsibilities for the further follow-up

Most enterprises have an established co-operative body, Works Council (WC), Liaison Committees (LC), or a steering group composed of various parties, which may also have taken part in planning the conference. It is recommended to use such a body for further management and co-ordination of the subsequent development work. In several of our partner companies many reported that existing bodies for cooperation in general did not function too well. A task of this nature could help revitalize such co-operative bodies. Depending on how many representatives there are in such a body, it will often be appropriate to establish a small working committee including representatives of all parties of the actual co-operative body. This working committee is given responsibility for the conference report, for preparing matters related to the follow-up, for presentation to the relevant co-operative bodies and for following up the decisions between meetings in the actual co-operative body.

In cases where there is no broadly composed body suitable for this task, a separate committee should be appointed. The fact that these broadly composed bodies are made responsible for controlling and coordinating the follow-up, does not mean that this is where the specific improvement activities are to be conducted. As soon as practically possible these should be delegated to the line and into the employer-employee co-operative arenas where they organizationally belong. The innovation activities shall not be conducted outside the organization but be integrated into the other operational activities.

Idea bank

Many large enterprises have established idea banks where suggestions from employees are registered. These might be ideas stemming from suggestion boxes, brainstorming sessions, reports following undesired events or non-conformance reports. Enterprises without such an idea bank should establish one to keep management, employees, and union representatives updated on the follow-up of ideas, both collectively and individually. It is recommended to make updates electronically. In addition, updated lists should be made available to all employees, for example on notice boards.

Minutes and action plans

Minutes from the conference should be prepared as soon as possible to present all the groups' ideas and prioritizations e.g. in the form of a conference report. Ideally, the group leaders should summarize the results from their own groups in writing, which will then be included in the minutes/conference report. In most cases action researchers or staff personnel have simply collected the flip-over sheets and transparencies prepared by the groups and written their content into the conference report (Tønnessen 2000). It is this report that constitutes the written basis for the follow-up of the conference.

The entire company will have ownership of the ED Conference's results. Therefore, the follow-up must clearly be the result of a structured prioritization process, where different groups feel that their interests are attended to. An action plan for the follow-up should also be prepared and may be attached to the conference report. This action plan need not be detailed, but it should describe what happens next and who is responsible for the various actions. If such a document is not provided, the entire follow-up process may appear a little haphazard.

Integration in the targeting, planning and budgeting processes

One should avoid letting the follow-up process live its own life apart from the enterprise's other planning processes. The prioritized ideas should therefore be integrated in the enterprise's vision, values, objectives, strategy, budgeting processes and other action plans. Some initiatives involving the entire enterprise may be rather costly and may need to be included in one of the following years' budget. Other initiatives may be delegated down to the correct organizational level for further assessment and then included in the current or next year's operating or maintenance budget. An important principle is to delegate as much as possible of the practical follow-up to where it belongs. This will ensure better ownership of the chosen solutions, ensure learning at the operational level and will integrate larger parts of the organization into the development process.

Immediate measures

Immediate measures can have an immediate effect on motivation (Juran 1989). Many of the ideas that emerge in the brainstorming sessions are suitable for immediate implementation without significant expense or major changes. If some immediate actions are taken, it becomes apparent to the organization that things are happening after the conference and that management will follow up. What can be initiated immediately should be decided in the first meeting held by the body responsible for following up the conference. The rest of the follow-up process needs to be conducted over time, and many of the initiatives will normally need further study and planning.

Innovation projects

A number of the suggestions that are tabled during an ED Conference will need further study/ investigation/analysis before they can be implemented. Experience shows that complicated issues such as communication, responsibility, co-operation between departments and material and production control routines are given high priority in the conferences (Tønnessen 2000). These may be issues that have been burdensome for quite some time, that everyone knows about, but are difficult to solve.

Starting actual innovation projects (radical and incremental), whether local, cross-sectional and/or company-wide, is something I gave high priority when introducing the various development processes in the companies I have co-operated with in ED 2000 and VC 2010. In order to initiate a development process that is rooted in broad participation, emphasis has been placed on teaching all employees some easy to use innovation techniques. I have applied practical innovation projects prioritized by the employees as the most important tool for teaching employees a systematic way of working their way through the different phases of an innovation process; proceeding from the present situation, identification of problem areas/ ideas/possibilities, analysis, development of new solutions, planning for implementation, piloting/initiating, implementation and follow-up by evaluating and improving. By doing that employees have learnt and become acquainted with some basic project management skills, such as describing the present situation, the future desired situation and the purpose of the project, and making a progress plan with activities, milestones, division of labor, responsibilities and status reporting.

Continuous information about status

An ED Conference creates expectations for action. However, action is necessary but not sufficient. It is essential to provide continuous information regarding the follow-up of the conference. A common mistake is that the management and the body charged with following up of the conference expect all employees to see as clearly as they do that the actions taken are connected to the conference and that employees are observing everything that happens. I have often experienced frustration spreading across the shop floor because "nothing is happening". After a while, people are not able to connect initiatives that are implemented in the company to the conference, nor do they observe everything that takes place. Without proper information it is difficult to understand that it will take time to assess some of the suggested initiatives. Employees may also find that nothing happens with their own suggestions (Tønnessen 2000).

The idea bank should be made available to everyone. The ED 2000 company Kværner Hetland, for example, posted monthly bulletins on the notice board about the suggestions, what was being done about them, as well as their status regarding decision-making and the time for planned implementation. This included information about who was responsible, which suggestions were under consideration, which needed to wait for the time being, which had been completed, and which would not be implemented (Tønnessen 2000).

For the last-mentioned category, an explanation should be presented either in the form of minutes of a meeting of the follow-up body, a departmental meeting, etc., depending on where in the organization the suggestion ended up. Employees who submit suggestions that are subsequently rejected or given low priority must be given a reason for this. This form of feedback is crucial for motivational and learning reasons.

The status of the follow-up process should also be presented in staff meetings, departmental meetings and in company newsletters. If the company has internal newsletters these should be used to describe some of the activities.

10.1.4 Developing incremental and radical innovations synergistically through the ED Conference – empirical analysis

In this section, I will sum up the experiences with the 11 ED Conferences. I will relate the use of the conference to different stages in a company's strategy process, discuss how the conference fits in with my framework for employee participation in innovation, sum up important criteria for succeeding with employee participation in innovation and discuss how the conference promotes the synergistic combining of direct and indirect employee participation in developing incremental and radical innovations synergistically.

10.1.4.1 The ED Conference at different stages in a strategy process

At Titania, which had a well communicated business strategy, the ED Conference was used as tool for identifying radical and incremental innovations and further directions for the company. It was used as a tool to ensure company-wide employee participation in innovation and to give content to an enterprise development process. Based on the experiences with the 10 proceeding conferences (Tønnessen 2000) and the Titania case (Bjørnevik et al. 2005), I find that the ED Conference could be used to handle different company demands and can be used to support different stages in a company's business strategy development process. This is described in Tønnessen (2000). In short, the conference can be used: 1) As part of the process of mapping facts, challenges and innovation needs related to the enterprise's present situation. 2) In connection with the enterprise's formulation of goals, strategies and innovation direction. 3) In the interpretation and concretization stage of elements in a company's goal and strategy process where innovation initiatives are planned. 4) At the stage where the experiences with the current strategy and innovation initiatives are evaluated and revised.

10.1.4.2 Identifying innovation initiatives for the development and the concretization of the company's strategy

The Titania conference and the 10 preceding conferences have without exception given valuable contributions from the employees to the formation and concretization of parts of the companies' strategy. Important challenges facing the company were communicated and discussed. Important objectives and strategies were identified, developed and directed towards specific innovation initiatives. All the companies followed up the results of the conference, and many of the suggested innovation initiatives have been implemented. For the IfS ED 2000 partner companies this is documented in half-yearly status summaries from individual companies (Buytendorp & Hansen 1997, Eidstuen 1997a, Eidstuen 1997b, Børve 1998, Eidstuen 1998).

The ED 2000 companies gathered the suggestions for improvement measures that were the result of the ED Conferences into prioritized improvement areas. For example, Sunnhordland Mekaniske Verksted prioritized the following three areas: 1) Management of and responsibility for tool storage facilities, 2) Material and production management and control, and 3) Communication/responsibility. Stord Offshore prioritized the areas: 1) Training/skills, 2) Information/communication, 3) Welfare, and 4) Production flow and tools. The other ED 2000 companies prioritized corresponding areas and all of them have run a number of incremental and radical innovation projects (Tønnessen 2000).

During its first conference in January 1998, the ED 2000 company Kværner Hetland, with a headcount of 250 employees, generated 125 suggestions related to several of the main elements in its strategy. Nine months later, one third of these suggestions had either been realized or started. All the time, Kværner Hetland had two major company-wide innovation projects going and one waiting/being planned. Minor local projects were handled in the departments/groups. Annual savings after nine months resulting from the realized proposals are difficult to measure, but were estimated by the company at NOK 3-6 million. And then there were the gains achieved through increased opportunities from freeing up time and resources. The total equivalent of invested time and resources after nine months was estimated at some NOK 1 million. The ED Conference and follow-up also contributed significantly to the development of a culture for continuous improvement (including both incremental and radical innovations in line with EFQM) with broad participation within the company (Tønnessen 2000).

Experience in other ED 2000 companies confirms this picture and shows that many of the specific initiatives suggested by employees during the ED Conferences were realized (Tønnessen 2000). The ED Conference promotes employee participation in strategy development and implementation. What is more important, though, is to see whether the ED Conference and follow-up lead to permanent changes in the company's ability to develop strategies and continuously make improvements through employee participation. If the ED Conference was an isolated event, the activities would probably fade away gradually. If it influenced other more permanent arrangements and practices for employee participation in innovation, the effects would probably be longer-lasting. I will therefore use some space to assess the experience gained in light of the question of possible long term effects.

10.1.4.3 Improvements in the companies' ability to continuously innovate through employee participation

The ED Conference has been developed as one of several means to enhance enterprises' ability to exploit the potentials of employee participation in innovation. On the basis of my experience from the Aker Stord case and thirteen years with ED 2000 and VC 2010, I described three basic conditions that need to be present in order to get a process for employee participation in innovation moving in Chapter 9. Employees must have 1) the motivation, 2) ability and 3) opportunity for such participation.

In the following I will discuss the ED Conference's contribution to each of these fundamental conditions and the criteria for employee participation listed in Chapter 9.

The ED Conference's contribution to increasing employees' motivation to participate in innovation and the importance of leadership

Literature (Deming 1986, Juran 1989, Hammer & Champy 1993, Hammer 1996) and the Aker Stord case show that management focus has a decisive effect on employees' motivation to participate in innovation processes. In the ED conference vision, objectives and strategy are communicated by the top manager. The top manager asks employees for help with inputs to the strategy process, thereby emphasizing that their job-related experience and insights are valuable. The importance of participation is underlined by setting aside considerable time and resources for planning, implementing and follow-up of the conference. Great emphasis is placed on broad participation in the implementation of innovation initiatives and projects in the follow-up of the conference, and those making proposals are as far as practicable possible involved in formulating the measures. Many found that not only did the initiated innovation projects improve effectiveness in the affected area, but they also solved work environment problems. And, not least, many expressed a positive feeling of having been listened to. They had received a lot more information and had learnt (Tønnessen 2000).

However, middle managers and first-line managers have no special roles in the implementation of the conference and participate on equal terms with the other employees. This is done to underscore equality. My experience so far in ED 2000 and in VC 2010 is that companies tend to spend insufficient time on training and motivating middle managers and first-line managers who must be given a defined place in the process, and are expected to deliver results. Many at this level have insufficient knowledge of processes of this nature. Some of them also perceive increased employee participation as a threat to their own position in the company. After all, employees are being asked to become involved and find solutions to challenges that many managers consider to be part of their job. In several cases, managers have also been pressured into implementing ideas suggested by employees in their own department that these managers were not particularly interested in (Tønnessen 2000).

On the other hand, several middle and first-line managers were selected as group leaders during the group work or as project managers for implementing incremental or radical projects. This is a method of ensuring management involvement. Middle managers and first-line managers generally represent a challenge in this type of development processes (Cole et al. 1995). However, as previously mentioned, success seems to depend on the participation of management, which illustrates the need to develop new approaches and mechanisms. In some of the departments/groups at Titania and in the companies that participated in ED 2000 (Tønnessen 2000) there has been little activity, which seems partly due to a lack of interest and commitment on the part of middle and first-line managers.

Employees' ability to take part in innovation processes

In order to participate it seems not enough simply to be motivated. Employees need some knowledge and skills to be able to participate. In turn participation seems to strengthen motivation (Tønnessen 2000). In Chapter 9 I differentiated between two main groups of criteria for employees' ability to take part in innovation activities:

- Job and company-related knowledge.
- Skills in and knowledge of innovation methodology

I will comment on these two groups of criteria below.

The ED Conference's contribution to increase job related knowledge

In addition to giving insight into vision, values, goals and strategy, the conference also facilitates increased insight into colleagues' problems, challenges and points of view. Otherwise the conference contributes little to skill development in relation to an employee's own job and the jobs of other people or about customers and suppliers (Tønnessen 2000).

The ED Conference's contribution to increase skills in and knowledge of innovation methodology

With regard to knowledge of change methods, each conference allows for training (by doing) in the use of various techniques. Through group work and plenary discussions, attempts are made to create an understanding for when certain techniques should be used and why. My experience so far, however, is that, once back at the workplace, the different improvement techniques that have been introduced at the ED Conferences have not been used very often. Some enterprises have utilized some of them. For example, the IfS-company Pyrox Produksjon can document significant gains in efficiency by applying some of the techniques when they did changes to layout and work routines in the production department (Tønnessen 2000). Others have only applied the techniques in joint meetings with researchers present. This may be due to some techniques being too complicated to use, and they may need to be developed further. I have emphasized that the techniques should not be used for their own sake, but only when this type of aid is really called for. In my experience, it is the very simplest techniques, such as brainstorming and the use of post-its, that have been most used afterwards.

The opportunity of employees to participate in innovation

In Chapter 9 I identified six basic levels for direct and indirect employee participation. The following discussion will be structured according to these levels. To make it easier to follow the discussion, the levels are repeated below using the same numbers as in Chapter 9:

6. Company

5. Business processes

4. Group/department

3. Task/order

2. Line of command

1. Self-determination

With reference to Chapter 9, increased self-determination at work (level 1) is a key development goal. The same applies for good co-operation between managers in the line of command and employees (level 2). Challenges should be solved at the lowest possible level and with the least possible use of resources. However, there will always be a great number of issues that affect several people or large parts of the business, and have to be assessed in relation to the enterprise's vision, values, goals and strategy. It is primarily the potentials of the latter that the ED Conference explores. In the following I will limit the discussion to the significance of the ED Conference for participation in innovation to levels 3-6. Increased self determination in one's own job and a change in attitude in relation to facilitating the commitment of all employees on a daily basis to big and small improvements may very well be a result of the specific projects that arise from an ED Conference. However, it is the lasting changes in the way employees participate in innovation in co-operation with others that is my focus. I will first comment on direct participation, and then on indirect participation.

Direct employee participation in innovation in dialogue with others in different kinds of groups/arenas

The importance of the ED Conference for stimulating direct participation at the different levels seems to increase the higher up on the six levels one goes (Tønnessen 2000). Participation in the development of the company's strategy (level 6) was discussed in Sections 10.1.4.1 and 10.1.4.2. The next level is business processes (level 5). Improving business processes was a new practice for most people in the ED 2000 and the VC 2010 enterprises. The ED Conferences at Titania and at the other ED 2000 companies generated a number of process innovation projects across departments (Tønnessen 2000).

One of the objectives of ED 2000 was that our partner-enterprises should at all time have a certain minimum number of parallel innovation projects going on. When one was finished, they just moved further down the priority list and started up the next. Many of these were process innovation projects. All our ED 2000 and VC 2010 partner enterprises introduced such a practice. Thus the ED Conference became very important in ensuring that innovations were initiated and implemented across departmental boundaries (Tønnessen 2000).

The next level (4) refers to innovations within the department or group. Many of the initiatives that are aired at an ED Conference can be handled internally in individual departments and are sent there. For learning, motivation and efficiency reasons my recommendations to the participating companies was that the further prioritization after the conference should be done at the lowest possible company level and within a body that represents the parties. Also at the department/group level the ED Conference has achieved great importance for the instigation of innovation projects and for a certain dissemination of innovation methods. The main weight of the department-wide proposals arises on a continuous basis through internal brainstorming sessions, gatherings or development meetings. These should be treated in the same system as the results of the ED Conference.

Basically, the ED Conference does not address level 3 (tasks/jobs). However, several of the ED Conferences have revealed important weaknesses in the way the work is planned and have started processes aimed at improving work processes. E.g. in the ED 2000 company Westcon the ED Conference resulted in the introduction of "Before and After Job discussions" to which all who will be participating in the job concerned are invited to take part in discussing the job (Tønnessen 2000).

Experience gained so far from 11 conferences has shown that the ED Conference can be a useful tool for initiating innovations of importance to the enterprises' development and concretizing of the company's strategy, and for introducing a change process with broad participation. This has been achieved by stimulating increased direct employee participation in decision-making, especially at levels 6-4, and by introducing methods to ensure that this participation is goal-oriented, appropriate and real.

The ED Conference's significance for employees' indirect participation in innovation

The highest level (6) for indirect participation in enterprise development at company level is representation at the board of directors. I do not have data indicating that the ED Conference has had any significance for the role and importance of union representatives here. But it is reasonable to assume that the mobilization of all employees in the company's strategy process could provide employee representatives with a better foundation for their contribution to the board's strategy discussions and prioritizations.

In my framework I have recommended that the ED Conference is planned and followed up by a body broadly composed of both parties. In some cases a separate committee has been established for this purpose. In other cases an already existing body has been allocated this task. Planning need not necessarily involve as many persons as follow-up. If the company has a Works Council (WC) or similar arrangement, this kind of cooperating body can be suitable. Titania used the combined WEC/WC. In some enterprises, the ED Conference has helped to revitalize/establish this type of co-operative bodies because it has been necessary to have some kind of cooperative body to follow up the results of the conference and manage activities that are of great importance to the company (Tønnessen 2000).

At department/group level the follow-up of the results of the conference and the union representatives' participation in innovation projects have given the union representatives a more active role in enterprise development activities (Tønnessen 2000). In addition to shop steward involvement, broadly composed committees (or just the head of department and the department union representative, if the department is small) undertake the local follow-up and prioritization of the improvement measures proposed at the conference (Tønnessen 2000). A suitable body for this purpose is the Departmental Committee (DC).

Union representatives as well as managers at Titania and from our ED 2000 partner enterprises report that a change of attitude took place in both parties in relation to direct and indirect participation in innovation activities following the ED Conferences and the other participation-oriented initiatives the companies introduced in connection with ED 2000 and VC 2010. The parties seem to have gained a better understanding of each other's positions. The awareness from both parties of the advantages related to co-operation between the parties regarding enterprise development increased. The climate for cooperation improved. Employees and union representatives are to a greater degree regarded as assets rather than opponents. There is also greater awareness of the connection between operational issues and work environment. For instance, some managers expressed a great deal of skepticism prior to the first ED Conferences, as they feared a great deal of expressions of dissatisfaction and complaints. But much of what the employees were dissatisfied about concerned production control, planning, diffuse lines of responsibility and inefficiency. Weaknesses of this nature were reported to create a bad work environment (Tønnessen 2000).

10.1.4.4 Success criteria for achieving employee participation in innovation

The ED Conference was designed to promote contributions from employees in strategy and business development. My empirical research based on Titania and the preceding 10 conferences show that the following conditions seem to be critical for a successful outcome:

- The top management must lead the way, reserve resources, participate actively, ask for results and set demands.
- All company and work-related knowledge must be considered to be of equal value.
- Persons with experience and knowledge of innovation methodology must be available and facilitate the generation of ideas and prioritization.
- A plan for recording and follow-up of the conference's results must be ready for communication during the conference.
- A suitable body to follow up the results of the conference must be selected. This must be composed such that the parties and the various groups of employees feel represented and have trust in the body.
- Line managers must be given roles and responsibilities for following up local initiatives.
- All employees must be regularly informed about the status of the follow-up of the conference results.
- The ED Conference has to be developed as part of, and should be included in, a structured, holistic and continuous enterprise development process in which broad participation is a key enabling factor.

- Appropriate routines for coordination of direct and indirect channels for employee participation in innovation must be established.
- Mutual trust between employers and employees.

10.1.4.5 The combining of direct and indirect employee participation in innovation

The idea generation at the conference starts at the level of the individual employee, and it is the total number of the participants' ideas and how they are prioritized that constitutes the product of the conference. This represents direct employee participation in business strategy development and innovation. The employees' chosen representatives take part in the planning of, the introduction to and the summing up of the conference, and in the management and the follow-up of the conference's ideas and proposals. This represents indirect employee participation. From a central co-operative body the ideas are delegated for further planning and follow-up at the level in the organization where they belong. At this stage local co-operative bodies where employee representatives and management meet, take care of this. Then in the next phase where projects are started in order to realize the ideas, individual employees are invited to participate.

This way the ED Conference strengthens both channels for employee participation in innovation and makes them mutually supportive. Indirect employee participation prepares for employee participation in idea-generation, strengthens the follow-up process and contributes to the realization of ideas for incremental and radical innovations. Innovation ideas from employees' direct participation give the shop stewards valuable ideas and good arguments that shop stewards can use to strengthen their own position and importance in the development of the company. It is difficult to see how direct employee participation in business strategy development could be realized effectively without the use of an indirect channel for employee participation to follow up ideas. It is equally difficult to see how the results of an ED Conference could be followed up and managed without chosen representatives. To have all employees participate in the following up phases does not seem very realistic or cost efficient. To let management follow up alone could threaten legitimacy, would not be very democratic and could reduce employees' interest in and contribution to the process.

A good structure and practice for indirect employee participation seem to be necessary if the company wants to take advantage of the energy and creativity of all its employees in business strategy development. In Titania a structure of co-operative arrangements agreed between the labor market parties, such as the Works Council/the Work Environment Committee and Departmental Committees, were used for this purpose. By supporting direct employee participation, indirect employee participation thereby supports the synergistic combination of forms of exploitation and exploration.

10.1.4.6 Initiating incremental and radical innovations synergistically

The ED Conference increases employees' contributions to strategy development, strategy implementation and in the process of identifying further development needs, ideas and opportunities for the actual company. Much of this participation is being done through different arenas for dialogue. In the Titania case we recognize many of the central principles from the theory of Democratic Dialogue (Gustavsen 1992). It was possible for all concerned to participate. Ideas and arguments floated freely between the participants. Everybody was active, wrote down, presented and argued for their ideas. All participants were considered equal. Work experience was the basis for participation and all arguments were considered legitimate. Efforts were made to make sure that everybody developed an understanding of the issues at stake. The dialogue produced agreements that provided platforms for practical action, and everybody had an equal vote when ideas were prioritized. The setting created a platform for learning and awareness of important challenges for the company, other departments' needs and of different colleagues' positions and for the exchange of information, viewpoints and knowledge. Both incremental and radical ideas were produced. The conference promoted both incremental and radical innovations by using tools from TQM, e.g. brainstorming, changing between silent individual reflection periods, followed by group and plenary discussions and simple prioritizing techniques that ensured equal voting.

The use of principles of Democratic Dialogue (Gustavsen 1992) in these arenas ensures that all employees learn to use methods that promote both incremental and radical innovation, which is fundamental for the synergistic combining of forms of TQM and BPR. The conference itself is an arena for democratic dialogue and some of the arenas used for follow-up and implementation of ideas follow some of these principles. In all the 11 conferences employees were trained to use participatory methods for identifying and prioritizing ideas, and the same methods were used to promote both incremental and radical change. Employees and shop stewards that were asked to take part in follow-up activities and innovation projects that the conference produced needed to learn more innovation methodology than others. In sum a larger proportion of employees came into situations where they learnt how to contribute to incremental and radical innovations in the company than before. Consequently, in this case the ED Conference promotes learning for the synergistic combination of these two development strategies.

The ED Conference increases the number of ideas from employees that can improve quality, efficiency and operational effectiveness and it increases the number of ideas for radical change that can give improved strategic flexibility. In addition the conference promotes the interlinking between the company's strategy processes and participatory idea-generation and follow-up. In other words, it promotes both forms of exploitation and exploration. In the Titania case, a sense of urgency was created by the managing director who was very clear in his message to all employees. Both radical changes for the future survival of the company and short term continuous improvement were needed.

Various means are used to promote the synergistic combination of the two. As mentioned above, tools from TQM are used to come up with both incremental and radical changes. It is a goal to teach employees some easy methods that can be used for participation in both generating incremental and radical innovations. Much of the structure and procedures for identifying, prioritizing, evaluating, planning, implementing and following up of ideas are the same for incremental and radical change. Some are different. E.g. radical projects usually need more project management tools and competence than minor improvements. However, both types of changes should go through the same phases from needs identification/idea-generation to evaluation/follow up. When people are trained to think and act in according to these development phases, they are trained to handle some of the methods and logics that support both radical and incremental change. Consequently, when employees take part in incremental improvements they also get experience, competence and motivation for participation in more challenging and radical projects. The experience from the 11 ED Conferences show that forms of exploration can support forms of exploration in line with the synergistic approach to the balancing of exploitation and exploration expressed by, among others, Sutcliffe et al (2000) and Boer et al (2006).

To conclude, I find that the ED Conference can contribute to the synergistic combining of TQM and BPR by contributing to the completeness of an ID framework that promotes the synergistic generation of ideas for incremental and radical innovations. Also, the conference, including planning and follow up, creates arenas for learning how to combine the development of ideas for incremental and radical innovation synergistically.

I will now proceed with the second tool that I find important to make my contribution to the further development of a participatory approach to the synergistic combining of TQM and BPR more complete, the Actor-Process Diagram. Whereas I developed the ED Conference to promote employee participation in business strategy development and innovation, the Actor-Process Diagram was developed to promote employee participation in radical process redesign, including both business processes and work processes.

10.2 Tool 2. The Actor-Process Diagram and the combining of TQM and BPR characteristics synergistically

Since the mid-nineties an increasing number of organizations have applied a process redesign approach (Hammer & Champy 1993, Davenport 1993, Moltu 2000) as a means to improve performance. Process redesign efforts are mainly carried out by management, consultants and specialists. Employee participation in these kinds of change processes is rather limited (Koch et al. 1999). For many redesign efforts one of the goals is "process centered organizations" (Hammer 1996). However, in order to form a "process centered" organization, employees should become "process professionals" who understand that their job involves contributing to the performance of a process rather than performing simple isolated tasks (Hammer 1996). If it is a goal to develop process centered organizations and process professionals, is it then wise, as many do, to run the redesign processes in a way that does not involve, treat and train people as "process professionals"? Is it wise to let others design the work processes that these future professionals shall perform? Or should employees and their chosen representatives be more involved, and should this kind of development work in Norwegian industry be carried out more in accordance with Norwegian Industrial Democracy traditions?

10.2.1 Business process redesign and the combining of TQM and BPR synergistically at Serigstad

The Norwegian company Serigstad thought alternatively to the prevailing reengineering practice referred to above and chose to involve employees and their representatives systematically when they needed a redesign to improve business performance. I will now present a redesign approach and a tool that facilitates training and involvement of people more systematically in designing their future work: the Actor-Process Diagram. The approach and the practical use of the diagram will be shown by referring to the Norwegian iron foundry Serigstad. In a wider context, the company developed a combined TQM and BPR approach to business development. So the case will be evaluated in relation to my research question of combining TQM and BPR synergistically.

Based on the Aker Stord experiences and with reference to my research objective and research question, the following specific questions were developed for the Serigstad case study:

- Will a participatory approach to redesign improve the company's capability to do incremental and radical innovations synergistically?
- How far and wide should this participation be taken?
- Who should participate: The employees directly involved, representatives from other departments, shop stewards or all?
- Is it necessary to both involve employees directly and the employees' chosen representatives (shop stewards, safety delegates)?
- Can TQM and BPR be combined synergistically?

Based on these questions the following propositions were developed for the Serigstad case:

- The Actor-Process Diagram generates a considerable number of incremental and radical innovation initiatives synergistically.
- The Actor-Process Diagram increases direct and indirect employee participation in innovation synergistically.
- The Actor-Process Diagram promotes the synergistic combining of TQM and BPR

I will first say a few words about the company and its reasons for starting up a redesign process. Next, I will present the project organization that was set up for the redesign project and show how this was composed to ensure both direct employee participation in the change process and indirect participation through chosen representatives/shop stewards. Subsequently

I will show how the workforce was informed and mobilized for the change effort and how they were involved. I will then present the Actor-Process Diagram and describe how this tool was used to involve employees in the actual change process. I will describe how the project was expanded to cover the development of the whole plant and also widened to become a combined TQM and BPR effort. Next, the results of the change process will be summed up, together with a few comments on what seemed to cause these results. I will discuss the Actor-Process Diagram as a as a tool for involving employees directly and indirectly in the process of initiating incremental and radical innovations, as an arena for learning how to combine incremental and radical innovation synergistically, and as a means to the synergistic combining of TQM and BPR. As to the latter and in a wider context, I will discuss the Serigstad case in relation to my research question of combining TQM and BPR synergistically.

10.2.1.1 Background: Productivity challenges at Serigstad Iron Foundry

After three years with low productivity, too many rejected products and poor economic results, Serigstad Iron Foundry in Norway invested in a new production line in 1998 in order to increase productivity. One year later, in 1999, there were no signs of improvements. On the contrary, the investment created new problems and increased maintenance costs. The operation time for the main production line was 58.8 %. Something had to be done. Serigstad contacted IRIS, which was known for doing action research on business development together with industry in the region. The company's initial idea was to follow up the technical innovation (the new production line) with continuous improvement in order to realize the potentials of this new investment in technology.

I informed about the research program "Enterprise Development 2000" (ED 2000). After two meetings with IRIS, Serigstad decided to join ED 2000 and go for a redesign process in accordance with my approach to combining TQM and BPR methods synergistically. At that time the market was considered promising, as long as the plant managed to increase productivity and reduce prices.

Serigstad Iron Foundry produced cast iron products, had about 60 employees at that time and was part of the Serigstad Concern. Serigstad A/S was established in 1864. It is situated at Jæren in the south-west of Norway and in the middle of one of Norway's largest agricultural districts. When the researchers from IRIS started cooperating with the concern, the main products were steel and iron products for agriculture, offshore and industry. At that time the concern was divided into four. 1) A machine producing plant and administration at Tu, 2) the iron foundry at Bryne, 3) a steel foundry at Klepp and 4) a daughter company producing bolts – Bolt A/S, also situated at Bryne. Both the concern and the iron foundry had long lasting traditions for close union-company co-operation, and reported a high degree of mutual trust between the parties. At the very beginning it was decided to cooperate closely with the union both in the planning, organization and follow-up of the change process. This differs significantly from prevailing BPR practice and recommendations from the BPR literature

(Hammer & Champy 1993, Davenport 1993, Coulsen-Thomas 1995, Zairi & Sinclair 1995, Hammer 1996, Koch et al. 1999, den Hertog et al. 2004).

The Serigstad case will give the reader some ideas about how company-wide employee participation in redesign may be carried out in practice. Innovation at Serigstad refers to the ability to come up with new work processes, new organizational solutions, new technical solutions, more suitable raw materials, new or improved products and new markets. As to the latter, the company's sales department was not part of the iron foundry, it was geographically situated another place in the concern and not part of the redesign effort. This situation will be commented on later.

10.2.1.2 Organizing the redesign project at Serigstad

The company's decision to run this redesign effort as part of the action research program Enterprise Development 2000 was taken late autumn 1999. It was organized as a project, and the Concern President took ownership of the project. The plant manager was appointed project manager, and the joint union-management Departmental Committee (DC) should act as a reference group. In this respect, the Iron Foundry was considered a department in the concern. The DC was considered important to ensure employee-management co-operation and was given a central role in the redesign process. The project, scope of work, strategy and the co-operation with IRIS were discussed and decided on in the DC. This committee also appointed the project group.

The project group was composed of one employee from each department in the iron foundry, one of the department managers, the company's productivity shop steward, the Concern Shop Steward (who had his daily work at the plant), the project manager and two action researchers from IRIS. In accordance with international experience, and to get a holistic redesign process (Koala 2000), the project group was multidisciplinary and represented the whole line of production. In order to involve as many employees as possible, it was decided that task forces composed of representatives from the employees who would be most affected by the actual redesign should be established continuously when required in order to carry out the specific development tasks. The composition of the project group, the organization of the project and important links to other parts of the Serigstad organization are presented in Figure 22.

The project organization is presented on the left-hand side of Figure 22. The most important links to the standing organization and meeting arenas are presented on the right-hand side. Close communication and co-operation between the project, the board of directors, the concern management and the plant organization were considered very important. The President reported plans and progress to the board and to the concern management group, both of which were involved and made some important strategic decisions during the change process. The redesign project was on the agenda at all plant management and department meetings, and the changes decided on in the project were carried out through the plant's ordinary line of command. The shop stewards reported and discussed the process in union meetings, and it was an important issue in Work Environment Committee meetings.

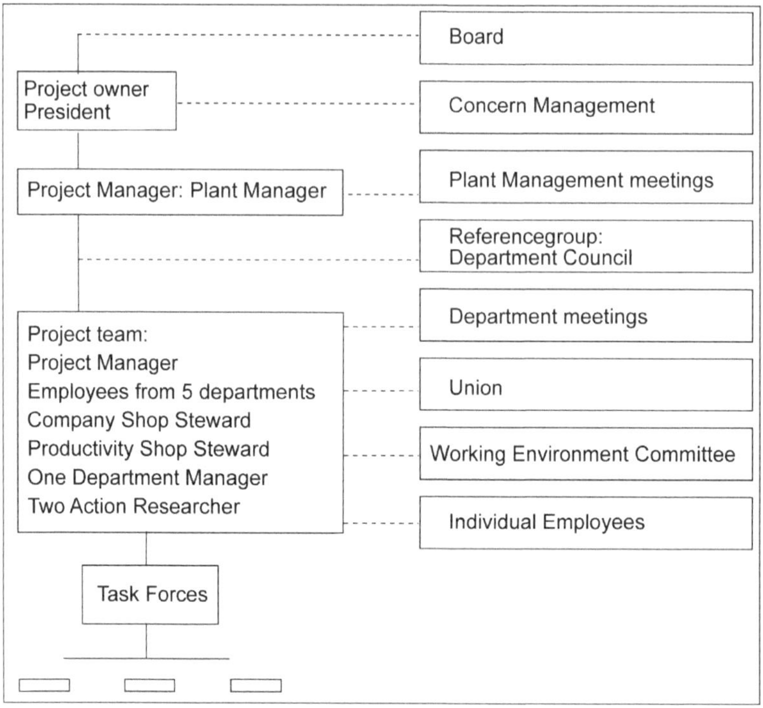

Figure 22 The project organization at Serigstad and important links (dotted lines) to the formal organization

Thus, as a point of departure, the redesign project was based on the company's tradition of close management-union co-operation and indirect employee participation. The company wanted broad direct participation of the employees who would be affected by the actual changes, believing this strategy would create better solutions and stimulate the willingness and necessary motivation for change and productivity improvements. This viewpoint is in line with international research (West 1998, den Hertog et al 2004).

As described earlier, union involvement has hardly been a subject at all in either BPR or TQM literature. However, research results from Aker Stord as well as other companies that participated in "Enterprise Development 2000" showed that in a Norwegian industrial setting direct participation and indirect participation worked reciprocally supportive to companies 'implementation of quality management' (Hansen 1997, Tønnessen 2002). These results strengthened the action researchers' belief in the positive effects of direct and indirect participation in the project.

10.2.1.3 Mobilizing for the redesign process

On the background of a discussion of the factory's situation, and having acknowledged the low operation time on the main production line as a main challenge, the Departmental Committee formulated the main goal for the project as follows:

Increase the running time of the main production line to 90% (my translation)

After having decided on the project organization and the goal in the DC, the project manager arranged a meeting where all employees were informed about the project and the serious economic situation that made the project necessary. At the meeting both the union and the management stressed the importance of employee participation and support. All employees were invited to participate in the change process. The researchers were introduced to the employees, became acquainted to the plant and spent some time talking with the employees in their daily natural work environment.

The first meeting in the project group was arranged the first Tuesday after the decision to start up together with IRIS had been taken in the DC. The DC had proposed that the 90% running time goal should be achieved within a year, without an increase in staff while the work environment should be improved at the same time. The company and the employees should work smarter, but not harder. About 90% of the company's production passed through the main production line. With the 1999 running time of 58.8% as a point of departure, the 90% running time goal seemed very ambitious. The representative of the supplier that had delivered the company's main production line considered it possible, but hard.

It is important to keep in mind that the stock holders were very impatient after three years with poor economic results. Time was running out for the factory. In fact, this might be their last chance. The goals had to be ambitious.

At this first meeting in the project group the factory's situation and the project goal were discussed and agreed on. Most of the members of the project group had never participated in planning a change process, and I advised the group to formulate and agree on some "rules of the game". Eleven rules that regulated the group members' behavior both inside and outside the project group were formulated. These were not based on theory but rather on the employees' practical experience and what they found appropriate. However, the researchers were familiar with the principles of "Democratic Dialogue" (Gustavsen 1992). We made sure that these kinds of principles were discussed, but did not interfere when the rules were chosen. It was "their" rules. These rules became important and created a mutual trust and common understanding among the group members from the start. The rules are listed below (my translation):

- All employees are considered equal
- All proposals from employees are of equal importance and value
- All employees shall be talked about in respectful terms
- The discussions in the project group shall be open

- No expressions of opinions shall have any negative consequences for the persons expressing these opinions
- All proposals shall be treated in a positive and constructive manner
- Project members shall not express negative attitudes outside the project group. The discussion of controversial matters shall first be discussed inside the project group
- All kinds of opinions are welcome in the project group
- Matters concerning individuals shall be discussed with a positive attitude and in a positive atmosphere
- Nobody is allowed to talk about other members of the project group or their proposals and opinions in negative terms outside the project group
- All employees shall be informed continuously about the process and be stimulated to come up with proposals and comments. All employees shall be included in the change process. Nobody shall be shut out from process activities or information

During our one year action research project there were no indications or examples of anybody violating these rules.

After having agreed on the "rules of the game", the group decided to have two-hour meetings at least once a week. This arrangement ensured continuity in the change process, and made it possible for the group members to perform their weekly ordinary work during the whole process. A specific room was dedicated to the project.

The second meeting in the project group focused on breaking down the main goal into a hierarchy of operational sub-goals at two levels, where work environment improvement was an important part. The overall goal "to increase the running time of the main production line up to 90%" was first broken down into four sub-goals, which in turn where broken down into a set of subordinate goals. The four sub-goals were:

- Increased productivity: Reduced downtime on the main production line
- Better service: Customer delivery by time and quality
- Sick leave: Avoid sick leave due to lack of motivation and work factors
- Work environment: Further develop the good social work environment in the plant

During the second meeting in the project group, "increased learning through the job" (my translation) came up as a new and parallel goal. Learning was considered important both for the further development of the company and also as a means to reach production goals. This additional learning goal was broken down into three sub-goals, each with a set of subordinate goals. The sub goals were:

- Learn from mistakes
- Employees shall have full insight in relevant production demands
- All employees shall know the whole production process (my translation)

These introductory phases and the time spent on the development of a goal hierarchy gave the project group a good start and a good common identification with some common goals. Thus the start up, the anchoring and the mobilizing activities before starting the redesign process, presented in the sequence in which they were performed, were:

- Management obtaining acceptance for a redesign project in the DC Agreement on hiring an external expert as facilitator for the change process
- Two meetings between management, union and action researchers, including overall planning of the company project as part of the national action research program Enterprise Development 2000
- Anchoring the project and the cooperation with action researchers in the DC
- The establishment of the DC as a reference group to the project, appointment of project owner, project manager and project group. Formulation of overall goal
- Meeting between researchers and top management in the concern. Gaining acceptance for the change approach and the action research design
- General meeting where all employees were informed about the project and invited to ask questions and give comments
- Researchers introduced to the plant. Meeting and talking with operators in their natural work environment
- First two meetings in the project group to decide on meeting schedules, overall plan, "rules of the game" and to break the overall project goal into operational sub-goals

Now, the redesign process in the company could start.

10.2.1.4 Starting up the redesign process at Serigstad

Being part of the project group, the action researchers were given the role as external process redesign facilitators and instructors. A "learning by doing" way of teaching was chosen. Several techniques and methods, mostly from the quality management tradition, like brainstorming, Pareto analysis (Juran 1989), the Ishikawa diagram (Ishikawa 1985) and other tools were introduced and used during the change process together with the Actor-Process Diagram. I will present the Actor-Process Diagram below. This tool had a very central place in the Serigstad case and was developed as part of the toolbox in ED 2000. The presentation of the tool below and the way it was used, will therefore also a represent a short version of the Serigstad case as a whole.

In the third project group meeting, my approach to combining TQM and BPR methods, to process redesign and the Actor-Process Diagram were introduced and accepted by the group. Before that, the approach and the tool had been presented, discussed and accepted in the initial meeting between the company, the union and the researchers. After acceptance, the group started out by identifying the company's main business processes in a traditional reengineering way (Coulsen-Thomas 1995) and chose to focus on the production process. The other main processes, e.g. marketing, were excluded from the reengineering effort.

Next, the group defined the production process, its goal, where it starts, its main inputs, main throughputs, where it ends, main outputs, its customers and main deliverables, also in a traditional reengineering way (Galloway 1994). After having defined and delimited the production process in relation to the other main process and support processes, the group could start mapping the production process in detail. This was done on large brown papers on the walls and by means of the Actor-Process Diagram methodology. Before describing the Serigstad case any further, I will now shortly comment on the reasons for developing the Actor-Process Diagram and then present its basic principles.

10.2.1.5 The Actor-Process Diagram – background

The Actor-Process Diagram builds on existing re-engineering tools, but I found it necessary to do some further development among others in order to ensure employee participation in redesign processes as well as good work environment conditions through redesign efforts. Employee participation and work environment improvement are not usually among the design criteria when redesign efforts are planned and carried out (Koch et al 1999). Experience and research show that Hammer & Champy's (1993) ambition of creating "a new world of work" has not been realized in real business life. Their claim about increasing employee empowerment through reengineering has even been called fanciful by critics (den Hertog 2004).

One of the reasons for the lack of development in a work environment perspective in reengineering processes might be the lack of employee involvement in change processes. Many writers on management have held that quality of working life unproblematically falls into place as a more or less certain outcome from their recommended top-down ideal approaches to improved working conditions (den Hertog et al 2004). Den Hertog finds that this is not necessarily true.

Furthermore, a lot of companies have increased productivity and accomplished cost reductions through re-engineering, but most of the re-engineering efforts have failed to realize the expected favorable outcomes (Knights & Willmott 2000).

Finally, the need for a better combining of continuous incremental improvement and radical innovation has been expressed by many scholars (Davenport 1993, Kaleda 1994, Cole 2002). BPR alone is not enough. A view also expressed by Hammer (1996) in his writings a few years after the 1993 Hammer & Champy classic, where Reengineering was introduced as a better alternative compared to TQM: *"TQM and Reengineering fit together over time in the life of a process"* (Hammer 1996:82). However, as referred to in the theoretical discussion, Hammer's understanding of TQM is rather limited. When comparing it to Reengineering he claims that *"TQM assumes that the design of the process is sound and that all it needs is some minor adjustments"* (Hammer 1996:82).

Thus, on this background my ambition has been to develop a tool for re-engineering of business processes, which not only helps pursuing traditional productivity and customer satisfaction goals, but also:

- Ensures that a larger proportion of the redesign efforts succeed
- Involves people more systematically in the redesign processes
- Increases learning about business process – and redesign-methodology
- Improves job content
- Is easier to use on the shop floor operational level
- Depends less on external reengineering experts
- Can be combined with Continuous Improvement in developing an approach for a more synergistic combining of TQM and BPR

10.2.1.6 The Actor-Process Diagram for mapping, evaluating and redesigning work processes – basic principles.

The Actor-Process Diagram is based mainly on experiences with the use of flow charts as a tool for business process redesign (Lowenthal 1994) and experiences with the STEP (Sequential Time Events Plotting) methodology for accident investigation (Hendrick & Benner 1987). Compared to mainstream techniques for process redesign, the STEP methodology adds one important advantage to redesign methodology; it is actor oriented. During the mapping and the analysis of the focal process, all relevant actors/group of actors are placed on the left-hand side of the diagram and the events/elements in the process are plotted to the right in line with the corresponding actor in the correct sequence of time, as in Figure 23. This makes the method actor oriented and ready for the development of meaningful and challenging jobs with a learning content. The most important elements in the Actor-Process Diagram are summarized and presented below.

When a process is being analyzed, the project group discusses problems and challenges connected to all the different sub-processes and activities in the process. This is shown in Figure 23 by "Act" written in the squares. Of special interest are the changeover phases, which often has the character of internal customer-supplier relations. All problems or challenges that are identified are given a number and plotted on the correct place in the Actor-Process Diagram in diamonds in Figure 23. All challenges are discussed in the project team in order to come up with some kind of action or further analysis. Each challenge and the corresponding action the redesign team has decided on, is described underneath and given the same number, as demonstrated in Figure 23. Sub-processes of special interest are broken down into activities and analyzed accordingly.

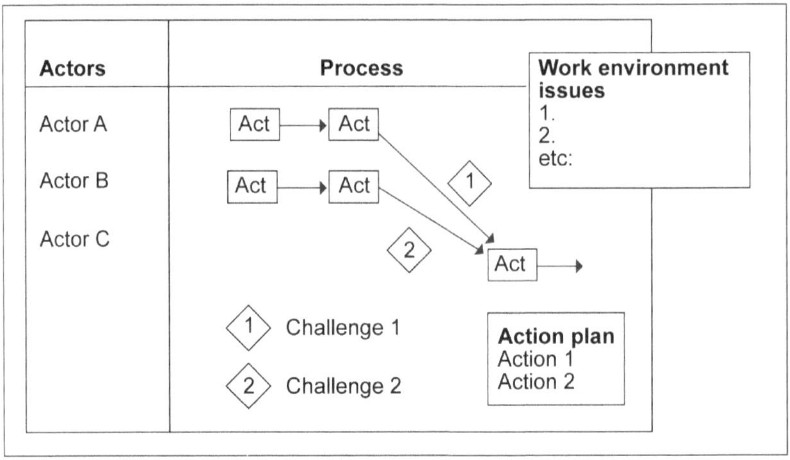

Figure 23 The Actor-Process Diagram

In addition, redesign groups are advised to develop a special work environment card where all the most important work environment aspects (including job content and employee participation) are listed. This is shown in the upper right-hand corner of Figure 23. The list is used as a checklist during the whole process of analyzing and redesigning the focal process. The list contains the work environment issues the employees find most important and not the viewpoints of HSE experts. It is "their" list rather than a "perfect" list made by an outsider. However, the making of the list can be facilitated by personnel familiar with HSE issues.

As mentioned above, for both participatory and practical reasons, the analysis of the focal process and the redesign are done on large brown papers taped to the wall. The activities are written down with a big black felt pen on post-it stickers. The brown papers function as posters, and are visible to all participants all the time and also to colleagues who are not part of the redesign team in between meeting sessions or when results are being presented. The results from the group meetings are stored electronically after the meetings.

10.2.1.7 Following up the results from the Actor-Process Diagram analysis of the existing production process at Serigstad

When the project group at Serigstad had finished mapping the existing production process according to the Actor-Process Diagram methodology, the group started acting on the list of challenges and actions that had been identified and developed as part of the mapping. It was considered necessary to commence working with these improvement proposals for actions at once instead of waiting until the whole redesign plan had been completed. There were different reasons for that decision. The company needed improvements at once for productivity reasons. There was no point in waiting. It was also necessary for the project to show the workforce some "quick wins" for motivational reasons. In addition the process was designed as a combined BPR and TQM effort. Continuous improvement was an important element in the further development of the company's change capability. However, all the time, the project group was well aware of the fact that comprehensive and radical changes were necessary. Thus they avoided minor improvements that might come into conflict with later changes or could turn out unnecessary in the long run.

The methodology used to come up with improvement proposals was a mixture of quality management tools and re-engineering tools. Along with the Actor-Process Diagram, methods from the quality management tradition such as brainstorming, the affinity diagram and the Ishikawa diagram were used. More than 60 proposals were developed. Three of these innovation proposals were characterized as radical and comprehensive: a new organization, a new Production and Material Planning System and the use of pig iron as raw material instead of scrap iron. These three radical innovations would demand comprehensive changes in the company and new ways of working, if decided. Some proposals were easy to implement, e.g. a new system for labeling production forms, a move of the spectrograph to a better position, a new telephone in the control room, better routines for cleaning up before leaving for the day, training more people to run the main production line, etc. Some needed to be treated by a separate task force, e.g. a new procedure for handling orders, a new routine for unloading, transporting and storing raw materials for the iron furnace and a new shift-working arrangement.

Some proposals needed more planning, e.g. to stop producing products that were produced manually and outside the main production line. These products were time consuming and produced with loss. Other proposals had to wait for the more comprehensive redesign to be planned, e.g. changing over to alternative raw materials and a proposal for a new way of co-operating and communicating with customers. Minor changes were delegated to department managers. It was a general principle that the employees who would be most affected by the changes should participate in the practical development and realization of the improvements. External and internal experts supported the project groups when necessary. E.g. a systems supplier was hired for planning and implementing a new Production and Material Planning System.

10.2.1.8 The redesign process and widening the scope

The change project started out with a focus on production and the main production line. However, after a few weeks it became obvious that the whole plant needed a radical technological and organizational redesign with a change in culture and new knowledge and skills as part of this. However, the production process was the main focus all the time. After having mapped the existing production process, identified challenges and actions, and having agreed on the status of the present situation, the project group could start preparing the design of a radically new production process by using the same Actor-Process Diagram methodology and some of the additional quality management tools that were used to analyze the old process.

According to the Cobra methodology (Coulsen-Thomas 1995; see Chapter 4), the analysis of the old processes can be skipped if the intended changes are radical and very different from the existing processes. However, and in line with international experience (Kaola 2000, den Hertog 2004), the project group initially found it necessary to analyze the existing processes before designing new ones for a number of different reasons. Among others, tools and techniques had to be learnt. This was the first time the members of the project group and sub-process redesign groups participated in this kind of change process. Furthermore, it was assumed that recognizing existing problems would help ensure that they were not repeated in the new process. It would also increase the participants' understanding of existing processes. But, most importantly, it would create a common understanding of the present situation, a common background for the design of the new process, facilitate communication among the participants in the design process, give them a common understanding of and a motivation for the job at hand and for the overall business goals.

After having analyzed the present main production process, the wall papers with the map of the existing production process could be taken down from the walls, and fresh wall papers were hung up instead. The goal hierarchy and the list of work environment issues that the project group had prioritized were visible on the wall all the time.

The development of the new production process was based on the set of goals and sub-goals that the project group developed prior to the redesign process. This goal hierarchy became the design criteria of the new process. In order to develop a process and change oriented organization, the project group found it important to teach the employees a few improvement and change techniques that could be used for various purposes and in many different situations. Examples are brainstorming and cause-effect analysis. The design of the new production process was done in the same manner as the mapping of the old process, and important challenges and corresponding actions were written down.

The total redesign of the production process depended on good solutions for some critical sub-processes. The methodology for analyzing processes permits breaking down a company's business and support processes into a hierarchy of sub-processes down to smallest unit which is an "act" in the form of a verb, a subject and a descriptor (Hendrick & Benner 1987). This was done for the production process at Serigstad. Some of these sub-processes needed to be changed and sub-process redesign groups/task forces were established. These groups were managed by members of the project group who had their daily work with the actual sub-processes, and was facilitated by me in my role of action researcher. The task forces used the same methodology as the project group. For each sub-process a new process map was developed on a new wall paper in the same manner.

During the whole process, the design criteria that were developed initially, including learning and work environment issues, were visible on the wall and used as a guideline. As for the quick wins, sub-process changes were implemented as soon as they were decided on and parallel to the overall redesign process.

An example of a sub-process that was given a special treatment was the process of bringing raw materials (scrap iron) to the furnace. This sub-process was considered very dangerous, mostly because of explosion risks. But it was also very time consuming because of logistics problems and constant reloading.

Also support processes, based on the needs of the production process, were redesigned in the same manner. An example of a support process redesigned using the "Action-Process Diagram" was "handling of customers orders".

Thus, the methodological approach and all the change techniques that were used were introduced by the researchers. The goal was to spread this kind of process redesign approach and knowledge to as many employees as possible during the change process. To ensure the learning of process redesign skills and quality management methods, motivation and direct employee participation, the task forces/sub-process groups were composed of the employees that would be most affected by the planned changes. The company wanted to increase its ability to continuous improvement as well as radical redesign. The term "Process Professionals" (Hammer 1996), was not used, but that was the intended direction. Neither was the expression "process centered organization" used, but the company wanted to change from a primarily function-oriented to a primarily process-oriented organization. This is a radical change in itself.

The decision to give the project group its own project room, was important both for continuity, for information purposes and for the possibility to involve all employees in the change process. The brown papers on the walls gave the project group a common point of reference that was visible to all and always there.

In order to involve all employees, department meetings were arranged in the project room. People were informed about the change process, new improvement proposals were discussed, and employees had the opportunity to ask questions and present their own proposals. Employees were also given the opportunity to write down proposals on a specific brown wall paper whenever they wanted. The door to the project room was always open, and several employees used this opportunity. New proposals from employees outside the project group were discussed in the project group, and decided on continuously. In addition, the change process and the proposed solutions were continuously discussed with the trade unions, the DC, the supervisors not involved in the project group and presented at information meetings where all employees were invited to express viewpoints and to come up with proposals.

10.2.1.9 Important changes

The redesign process resulted in the introduction of a few radical innovations and more than one hundred incremental. The most important changes are summed up below:

- A new process-oriented organization, where department borders were broken down

- Fewer supervisors and a change in managers' roles towards facilitating, teaching and personnel development rather than supervising
- A new and more flexible shift work arrangement, flextime and a widening of production hours from 05.00 am to 06.00 p.m.
- The gradual introduction of a new material and production management system that affected the work processes in all parts of the factory
- Computer availability for employees at the shop floor
- Employees were empowered in different ways, e.g. they were given more information about production demands and results, more training and more responsibility for control and reporting. They reported their own time, their sick leave, they were allowed to stop production, and they could now report errors and problems. They also cooperated more across functions, and new job instructions and training programs were introduced
- Enriched job content, among others by job rotation, more complete tasks, individual fulfillment of whole tasks, increased autonomy, multi-skilling and increased integration of maintenance and production tasks
- Tasks that earlier were spread, were now gathered as part of the new process-oriented organization, e.g. procurement, documentation, wrapping, grinding and machining
- Logistics were improved by reducing the number of times goods had to be moved and reloaded in line with the new process-oriented production mode
- No large investments were done, only minor technical improvements
- The involvement of employees in the change process increased motivation, learning and the willingness to succeed. Employees reported a stronger sense of common destiny with the company

The new Production and Material Planning System transformed the factory's planning, analysis and control-functions from being predominantly manual and disintegrated to being computerized and integrated. Management and employees got access to more, earlier, updated and integrated information about production demands, results and status on the most important production variables and functions in the company. Examples are customer information, sales, prices, products, production, raw materials, quality figures, quality analyses, product-specifications, stock, spare parts, maintenance and personnel data. To a traditional iron foundry this was a radical change. With reference to Hammer & Champy (1993), this radical change had a direction in accordance with the authors' objectives for the "New World of Work". Several of these changes depended on and were introduced together with new information technology, both hardware and software. With reference to the theory discussions in Chapter 4, the utilization of ITC in reengineering processes is a very common and recommended change strategy (Hammer & Champy 1993, Willoch 1994, Moltu 2000, Koala 2000).

10.2.1.10 Results

The company measured and evaluated performance on several parameters from month to month. From 1999 to 2000 the company registered the following improvements:

- The running time for the main production line increased from 58.8% to 70%
- Productivity increased by 28%
- On-time delivery increased by 10%
- The percentage of produced goods that were rejected was reduced by 38%
- Sick leave was reduced from 7.4% to 5.6%

In addition, the employees reported improvements in the organizational work environment, including increased participation. This was reported in a work environment survey conducted by an external company providing medical and work environment services to the company: *"When we look at the comprehensive organizational changes that has been done in the period, we are impressed. Usually, employees receive too little information in a situation like this. It is often lack of or too little information employees experience in a situation like this. This results in a situation where the feeling of cooperation and well-being suffer. In this case, the management has done a good job. Less strengthening/repetitive work is another positive element in the work environment. To rotate between different tasks reduces strain and creates more interesting jobs"* (Skjærpe & Ewing 2000, my translation).

Even if the 90% running time goal on the production line had not been reached after a year, both management, the union and the employees considered the redesign process successful both in terms of productivity and work environmental aspects.

At the time when our action research project in the factory came to an end, a considerable number of incremental innovations had been started up and many implemented. Two radical innovation projects had been started up, namely the new process-centered organization and the new Production and Material Planning System. The third proposed radical innovation project, namely changing over to pig iron as raw material, would demand heavy investments and had to wait. In addition a large number of ideas for future changes had been identified, such as changes in logistics and storing. The redesign process in itself as a means to increased productivity was considered a success. However, the plant met other challenges meaning that, unfortunately, the story does not end here.

10.2.1.11 Marketing and sales problems

Even if productivity was considerably improved, the sales department did not succeed on the market. A new sales manager was employed during the action research period, and a task force was organized to come up with proposals that could increase sales. The company's number of customer visits increased, but sales did not increase correspondingly. It is also a fact that the competition from low cost countries from Eastern Europe increased in the period. The company focused on high quality and high tech iron products like engine parts, e.g. engine valves for Volvo. This product sold with profits. Volvo even decided to transfer 100 tons of the deliveries they received from other suppliers to Serigstad in that period. Although this only counted for 5% of the factory's production volume, it was received as good news and a sign of things were going in the right direction in a difficult period. Some of their small scale products were sold with loss in order to be competitive. After another year with red figures on the bottom line, the owners decided to sell the factory. The production equipment, products and methodology were bought by a competing company in Norway.

10.2.2 Empirical analysis: The Actor-Process Diagram and the combining of TQM and BPR characteristics synergistically

I will now discuss the appropriateness of employee participation in business process redesign, the Actor-Process Diagram as a means to this kind of participation and as a means to learning. With reference to my initial propositions about the Actor-Process Diagram as a tool for promoting incremental and radical innovations synergistically and also as a tool for combining direct and indirect employee participation in innovation synergistically, I will sum up the experiences and my analysis in relation to the research objective of combining TQM and BPR characteristics synergistically. Remember, the Serigstad case is neither a full TQM nor a full BPR implementation. It is the combining of approaches and techniques from both concepts that is being tried out.

10.2.2.1 Building a process oriented organization. How much employee participation?

My participatory approach to redesign, as demonstrated in the Serigstad case, differs from prevailing BPR practice and from the text book approach, where redesign is seen as a process managed and performed by managers, specialists and few individuals (Hammer & Champy 1993, Davenport 1993, Hammer 1996, Koch et al 1999, Moltu 2000, den Hertog 2004). The Serigstad and the Aker Stord cases demonstrated two ways of organizing employee participation in process redesign and of introducing redesign methodology to employees. However, internal and external experts were included in the project groups and in the development work in order to ensure the best available solutions and new ideas both at Aker Stord and at Serigstad. In addition, benchmarking with successful other companies was used as a means to increase the employees' ability to come up with new ideas. The project groups in both Aker Stord and Serigstad visited other companies to get ideas during their redesign efforts. To some extent benchmarking can compensate for lack of internal knowledge about new technology and good practice.

Before discussing my approach to employee participation in redesign any further, I will remind the reader about the industrial democracy setting that the cases and my research is part of. This background is important for the understanding of the cases and my theoretical and methodological approach to business development in general. The approach can be more difficult to use in other industrial settings.

In process-centered organizations focus moves from a task focus to a process focus. Workers become process performers and process professionals rather than individuals trained to carry out simple tasks according to detailed instructions from their supervisors. In organizations where the notion of the simple worker prevails, there is a complex structure of narrow tasks that have to be tightly overseen by supervisory management (Hammer 1996). In building a process-centered organization, work processes need to be developed that treat people like process professionals (Hammer 1996).

However, somebody has to design and develop these work processes. This discussion started out by questioning whether it was wise to leave this to the management, consultants or specialists or whether employees should be involved more directly in designing their future work processes. Is it reasonable to give employees the training and skills necessary to develop their own future work processes? Or should this be performed by specialists, among other reasons, in order to save time, which is important in redesign processes (Hammer & Champy 1993). To succeed, do you also need redesign processes that train and involve people in the process of continuously designing their process-oriented organization? Hammer (1996), for example, wants all employees to participate in redesign, but is very vague about the role of "ordinary" employees in how to get there. He is very clear about the roles of the process owners and some reengineering specialists, who are given different labels like the "the chief reengineering manager", "the reengineering officer", the "Tsar", the "reengineer", etc.

My experiences with Serigstad, Aker Stord and other companies that participated in ED 2000 (Tønnessen 2000) are that knowledge about and skills to work with innovation techniques and methods are scarce. In addition it takes a lot of effort to acquire these skills and it is easy to forget them. Employees do not use change methodology very often.

However, the experiences from my case studies show that training employees in some innovation methodology increases their motivation and the company's ability to change. E.g. in Serigstad some employee representatives in the project group were appointed project managers for redesigning local sub-processes together with their colleagues and with some assistance from me as action researcher. They performed according to expectations, seemed highly motivated, delivered good results and represented an efficient way of distributing change techniques and motivation in the organization. After having participated once, it is easier to contribute the second time.

The more people become familiar with change methodology, the easier it may be for local representatives to arrange change processes on the shop floor operational level. However, my experience from these case studies is that companies should pick out a few techniques that are easy to learn and use, and may be used for various purposes.

An additional effect of active participation in a change process is the learning effect. Employees at Serigstad reported a better understanding of the total production process after having participated in the redesign processes. This may make it easier for people to perform and deliver in a manner that eases the work of fellow colleagues along the production line, and also to see improvement potentials.

Participation in changing the production process also gives a better understanding of the working of the company as a whole, including internal and external demands. At Serigstad employees eventually became motivated for participating in the development of other company areas unfamiliar to most production workers. This is in line with theories of participatory democracy (Pateman 1970). In group meetings, people often expressed an eagerness to contribute to product evaluation, product refinement, the development of new products and customer contact.

Thus, based on my rather limited experiences from a Norwegian industrial setting, the Serigstad and the Aker Stord case, the answer to the initial question of whether employees should participate in the redesign of their future work processes is yes. Employees should participate, learn some easy techniques and become familiar with other techniques that are more challenging and that need facilitator competence. However, they should not become redesign specialists. In order to support the employees to learn redesign techniques, I find that some assistance from trained internal or external redesign specialists is necessary.

It is also important for the employees to get access to and use internal and external specialists depending on the different kinds of challenges they are facing. A serious pitfall can be to rely too much on the employees' own knowledge and not exploit the potentials of relevant new technology and knowledge developed elsewhere. Through my research I also find that the ability to find and exploit relevant external knowledge should be part of the training given both to internal redesign experts/facilitators and to some degree to employees (Gandrud et al. 2004).

10.2.2.2 The comprehensiveness of the participation

Looking back, a few lessons can be learnt about the organization of the Serigstad project and the lack of integration with sales and marketing. The Production Department and the factory itself had little contact with the market. The Sales Department was responsible for marketing and customer contacts. The Sales Department was not part of the factory. It was located at a different place in the concern and reported directly to the Concern President. The Sales Department never became part of the redesign process, even if the project group requested that kind of participation. The employees considered this a mistake. During the whole change process they continuously came up with new product proposals, proposals for a better use of the production facilities and identified products that were produced with loss. Only a few of these ideas reached the Sales Department.

In addition, the employees requested close product development cooperation between customers and production workers. The question was raised whether these two factors, a better integration of the Sales Department in the change process and a better contact between customers and production, might have changed the situation. Even if the redesign process in the Production Department itself was a success, the Serigstad experiences raise the question whether wider employee participation might have given a better outcome for the factory.

Anyway, the Serigstad case shows that company-wide employee participation in process redesign may be introduced as a new practice and with good results in an organization, at least in a relatively short period of time (in this case a year). The experiences also showed that participation, when first introduced and experienced, motivates for more participation and for participation on new areas, as e.g. in this case on areas where the employees had no experience, such as product development, sales and marketing. It also showed that participation from the production workers alone is not enough in a situation like this. In this case employees requested and saw the necessity of participation from the Sales Department in such a comprehensive redesign. Increasing productivity and quality and thereby competitiveness did not help, when sales failed.

10.2.2.3 A new world of work and organizational performance

Hammer & Champy (1993) had the ambition of creating a new world of work: *"Work units change – from functional departments to process teams, jobs change – from simple tasks to multidimensional work, people's roles change – from controlled to empowered, job preparations change – from training to education, managers change – from supervisors to coaches, organizational structure changes – from hierarchical to flat, executives change – from score keepers to leaders"* (Hammer & Champy 1993: 65-82). An important question is how realistic these ambitions are in practical re-engineering. The Serigstad case had a participatory approach to redesign, and also the Aker Stord case, albeit to a lesser degree. The reengineering efforts did in fact result in many organizational changes in line with most of these characteristics. E.g. new work organizations where department borders were broken down, a change in managers' roles towards facilitating and teaching rather than supervising, and employees empowered in different ways. E.g. they were given more information about production demands and results, they were given more training and more responsibility for control and reporting and job content was enriched, among others by increased autonomy Both Aker Stord and Serigstad transformed in the direction of the ideals of the "New World of Work" and the process centered organization.

Both the employees at Aker Stord and at Serigstad related the improvements in productivity to these kinds of organizational changes. The results are very much in line with European research. A growing number of studies indicate that organizational innovation and new forms of work organization have a positive impact on organizational performance (den Hertog et al. 2004). It is recognized that new approaches create the conditions for wider and more fundamental changes. A study in which ten leading European steel firms participated showed that radical reconstruction of the steel industry would not have been possible without multi-skilling, competence building, flattening of the organizational structure and removing functional boundaries (den Hertog & Mari 2000).

Furthermore, a survey from Denmark, France, Sweden and the Netherlands indicates that new forms of work organization enable firms to profit from the implementation of new technology. It also worked the other way around. New information and communication technology can strengthen the positive impact of new organizational forms (Banke & Nordskøv 2000, Huzzard 2000, Sacquepee & Dufau 2001, den Hertog & Verbruggen 2001). E.g. both in the Serigstad case and in the Aker Stord case increased computer availability at shop floor level made less supervision and improved operator autonomy possible.

Thus, both in the Serigstad and in the Aker Stord case I find that a participatory approach may contribute to the success of a reengineering project and also stimulate organizational changes in line with Hammer & Champy's (1993) ambitions of a "*New World of Work*".

10.2.2.4 Lessons learnt for ensuring employee participation in business process redesign and for the reciprocally reinforcing combining of direct and indirect employee participation

Based on the case presented here, on my analysis and theoretical discussions, I will present some practical and general recommendations that should be considered added to traditional re-engineering practice at least in a Norwegian ID setting in order to ensure employees' contributions in redesign processes. Some of these recommendations should also be considered in other cultures and companies having participatory ambitions.

In order to move in the direction of the process-centered organization, where employees become real "process performers" and "process professionals" (Hammer 1996) my experience from Aker Stord and Serigstad indicates that employees should participate more systematically and continuously in designing their own work processes. I find that a participatory approach is a learning approach. I conclude that the mainstream process redesign literature and methodology need a more participatory approach, and that the Actor-Process Diagram in combination with other practical quality tools can contribute to that.

The answer to the question of whether enterprises should give employees the training and skills necessary to participate in designing their future work processes is yes, but it is a matter of degree. You need internal and often external specialists to facilitate these kinds of change processes. One or a few persons in the enterprise should develop these kinds of skills internally while the rest of the employees should be trained in and become familiar with a few techniques that are easy to learn, easy to remember and can be used for many different kinds of change efforts.

Like the ED Conference, I see that direct and indirect employee participation should work reciprocally supportive in order to utilize the creativity and potentials of employee participation in business development. The local co-operative bodies, where employee representatives and management meet, took part in both the planning and follow up of the reengineering initiatives both at Aker Stord and at Serigstad. At Serigstad it was the Departmental Committee that decided on the overall redesign goal, they decided on the composition of the project group, they decided to use action researchers as partners in the program and decided on important issues along the road. All important matters were discussed with the DC. That way the DC represented all employees in the company and the decisions were considered legitimate. The alternative of involving all employees in these kinds of discussions is not very efficient.

Two shop stewards also represented a link between the project group at Serigstad and the DC. They forwarded ideas and view points from employees to the project group. They forwarded viewpoints from the members of the project group and the task forces to the DC. Shop stewards were represented both in the project group and in the DC. Not all employees were directly involved in the redesign process. But many used the opportunity to come up with ideas in department meetings, individually on the brown wall paper and through shop stewards. Most of the direct employee participation in the redesign process took place in the form of operators participating in the project group and in the task forces. This way the Actor-Process Diagram strengthens both channels for employee participation, makes them reciprocally supportive and thereby combines them synergistically.

Some of the lessons that I consider important for succeeding with utilizing the potential of employee participation in business process redesign can be summarized as follows:

- Employees should be represented both directly and indirectly in redesign teams. Shop stewards as well as employee representatives from the actual departments should be appointed
- The most relevant joint union-management committee should act as a steering committee or a reference group for redesign projects
- The redesign team should appoint special task forces for designing sub processes or local changes. Employees from the local departments should be represented in these task forces
- All employees in the departments affected by changes should get the opportunity to comment on the planned changes using, for example the department meetings for this purpose
- There should be a possibility for individuals to comment on any changes directly
- The follow-up of the results from the redesign processes should be done by utilizing and strengthening the company's ordinary arenas and routines for direct and indirect employee participation
- Redesign processes should be integrated with continuous incremental improvement and organized as part of ongoing business development processes

10.2.2.5 Towards the synergistic combining of TQM and BPR

Like the ED Conference, the Actor-Process Diagram increases employee participation in strategy development and innovation. As mentioned above this holds for both direct and indirect employee participation and for the synergistic combination of the two. In Serigstad both the project group and the task forces constitute arenas for dialogue where arguments and knowledge are exchanged between the participants. The use of the principles of Democratic Dialogue (Gustavsen 1992) in these arenas enhances learning, which is fundamental for the synergistic combination of forms of exploitation and exploration.

Furthermore, like the ED Conference, the Actor-Process Diagram is combined with tools from the quality management tradition, which can be seen also in other BPR processes (Coulsen-Thomas 1995). The way the diagram was used at Serigstad also contributed to and was combined with a process for continuous improvement. Like the ED Conference the Actor-Process Diagram at Serigstad used many approaches and techniques that belong to an exploitation oriented development direction, such as brainstorming, along with the radical BPR approach to promote both exploitation and exploration. The way the diagram was used at Serigstad promoted the interlinking between radical process redesign and participatory idea-generation and follow-up approaches well known from TQM. In other words, the diagram promotes both forms of exploitation and forms of exploration and the synergistic combination of the two. In Serigstad the use of the diagram increased the number of ideas from employees to improve quality, efficiency and operational effectiveness at the same time as it was the main tool for a radical change process.

Three radical innovations were identified during the research period at Serigstad, 1) a new organization, 2) a new Production and Material Planning System and 3) the use of pig iron instead of scrap iron. Only the first two were implemented. The third one demanded heavy investments and was never started. Along with this came about 100 incremental innovations that were accomplished at Serigstad during the research period by the use of some of the same methodology, the same system, the same change process and the same democratic structure for following-up. Many of these came as a result of and were related to the use of the radical change tool, the Actor-Process Diagram. These improvements were planned, evaluated, implemented and followed up by employees, shop stewards and first-line managers together. Shop stewards used the training they had received as trade union representatives when they co-operated with fellow employees in task-forces and they also used some of the time that was allocated to trade union work to participate in business development. In some cases external specialists were needed for limited periods, but these worked together with employees in minor task forces. This participation gave a lot of learning about the production process and demands and about methods for both radical and incremental innovation. Consequently, in this case the diagram promotes learning for the synergistic combination of TQM and BPR. This practice increased both single-loop and double-loop learning (Argyris 1996). Most the 100 proposals that came up had the single-loop character of “solving the problem at hand”. Others, such as a new organization and the development of a new way of communicating and co-operating with customers, was more of a double-loop character where “the way the business was done in general” was questioned.

Furthermore, if we analyze the radical innovations separately, actually a considerable number of incremental innovations appear. For example, the different elements that constitute the radical organizational change may not be classified as radical individually. For example, the number of departments was reduced from five to two. Department barriers were broken down and the production was more process directed. The foremen's role changed from supervising to facilitating and teaching. A new shift working arrangement was introduced. Job rotation was introduced. Production and maintenance tasks were integrated. Jobs were enriched and people became more autonomous, managing and reporting their own working hours. All these and some other changes may not be considered very radical individually, except perhaps for the persons most directly involved. But seen together, people experienced a radical change, and the organizational change as a whole can be considered radical. The factory transformed from a primarily functional oriented organization to a primarily process oriented organization.

The two radical innovations that were introduced did not reduce the efforts to improve quality and efficiency incrementally. The about 100 incremental changes that came along with and at the same time as the radical ones increased quality, efficiency and employees' well-being. In other words, in the Serigstad case exploration did not come at the cost of exploitation, rather the contrary. It was also the other way round. The radical changes depended on a large number of minor improvements. E.g. neither the new organization nor the new Production and Material Planning System would have been possible without moving some of the computers from the offices down to the production floor. Both innovations demanded easy operator access to computers. Moving the computers may be considered incremental in itself, but decisive to the success of the two radical innovations.

The Serigstad experiences also show that working on incremental changes can have radical ideas as a result. E.g. the radical proposal of using pig iron as raw material instead of scrap iron would change the production and logistics process up to metal-smelting quite radically. This radical idea came up as a result of an analysis of the sub-process iron-scrap transportation. Changes in this process were considered to be minor and incremental. However, the Actor-Process Diagram, an essentially radical change methodology, was used as the main tool for the analysis. During the analysis a considerable number of problems connected to this process were identified, including work environment, serious safety hazards and inefficiency. On this basis the employees started thinking radically new, and brainstorming was used to come up with new ideas. One of the ideas that came up and was prioritized was the use of pig iron instead of scrap-iron. In this case the Actor-Process Diagram was used to analyze a minor process where the idea was to come up with some incremental improvements. A quality management tool, brainstorming, was used to generate ideas, and a proposal for a radical change became the main outcome. This example shows how incremental and radical change can be intertwined, and that the same kind of methodology can be used to promote both radical and incremental change. In other words this is an example of a situation where the company builds forms of exploitation and exploration capability synergistically through employee participation. Employees learn to use some new methodology that supports the combining of TQM and BPR.

The Serigstad case can be described as an effort to start up a long term process and a way of work towards a synergistic combination of TQM and BPR. Thus, similar to the findings regarding the role of the ED Conference, the experiences with the Actor-Process Diagram are in this specific case in line with the synergistic position taken by among others Sutcliffe et al (2000) and Boer et al (2006) contrary to March's (1991) primarily trade-off position to the balancing between exploitation and exploration. However, my experiences do not cover the totality of the idea of exploitation and exploration.

The choice of a synergistic contrary to a trade-off position is important and can have serious consequences for the way learning is planned and carried out. If you take March's trade off position as a point of departure, it seems logical to focus on exploration capability and exploitation capability separately. If you take the synergistic point of departure from Sutcliffe et al (2000) and Boer et al (2006), learning should promote the synergistic combining of the two. One of the goals at Serigstad was to teach employees some easy methods that can be used for participation in developing both incremental and radical innovations. Much of the structure and procedures for identifying, prioritizing, evaluating, planning, implementing and following up of ideas are the same for incremental and radical change. However, some are different. E.g. radical change projects usually need more project management tools and competence than minor improvements. At Serigstad this was solved by establishing task forces to run the projects that also hired expertise from outside the company. Otherwise the logical sequences and many of the same tools can be used in the process from idea generation to the following up phase of the realized idea. Consequently the learning at Serigstad did not support either the one or the other forms of exploitation and exploration, but the synergistic combining of both development strategies.

Furthermore, like the manager at Titania and the head of "Prefabrication" of Pipes at Aker Stord the Production Manager of the iron foundry at Serigstad was very clear in his message to all employee that he wanted radical change to happen at the same time as he invited all employees to continuously come up with daily improvements that could improve efficiency, work environment and employee participation. The employees had to show monthly improvements at the same time as they needed to come up with a radical change. He also created a sense of urgency and crisis. His favorite expression was: *"If we do not manage to make money, the café is next"* (Production manager, my translation). In all three cases, leadership seems to be important for the possibility to move in the direction of a mostly dual/ synergistic approach to the balancing of forms of exploitation and exploration.

11 Discussion and conclusion

This chapter will discuss my findings and empirical analyses in relation to theory, and sum up my contribution. I will critically reflect on my methodological approach and sum up limitations. On this background I will suggest directions for further research and finally conclude.

11.1 Discussion

My theoretical discussions have focused on the capability of organizations to develop both incremental and radical change in the same system, at the same time and on an ongoing basis (Boer & Gertsen 2003, Boer et al. 2006). Many scholars have tried to find ways to combine the two change strategies (Kaleda 1996, Tushman et al. 1997, Cole 2001, Boer & Gertsen 2003, Gibson & Birkenshaw 2004, Petersen et al. 2004, Boer et al. 2006, Andriopoulus & Lewis 2009, Raisch et al. 2009, Fang et al. 2010). In order to meet customers' demand for a high variety of a fair priced, high quality, increasingly customized products delivered quickly and on time, Boer et al. (2006) argue that companies increasingly need to be more capable of combining operational effectiveness and strategic flexibility. An organization is operationally effective if it satisfies today's customers in terms of price, time, quantity and place (Boer 2001, Boer & Gertsen 2003, Boer et al. 2006). Operational effectiveness is based on exploitation capabilities. An organization is strategically flexible if it is able to develop new configureations of products, market approaches, processes, technologies, competencies, organization and management systems that enable the satisfaction of (the day after) tomorrow's customers. Strategic flexibility is based on exploration capabilities (March 1991, Boer 2001, Boer & Gertsen 2003, Petersen et al. 2004, Boer et al. 2006).

My overall research objective is to contribute to the further development of the theory of balancing exploitation and exploration (March 1991). I do this through analyzing and proposing theory on the combining of TQM and BPR, which I use to make March's (1991) theoretical constructs exploitation and exploration operational in an industrial setting. TQM and BPR do not cover the whole range of, respectively, exploitation and exploration characteristics, but represent forms of the two. My research question is: "Can TQM and BPR be combined synergistically, if so, how?" If they can be combined synergistically, how can this be accomplished, and what enabling factors seem to be most important.

In my theory section I described TQM as a full management concept including philosophy, strategy, systems and methods. Even if TQM in theory includes radical change (EFQM 2006), the concept has in practice primarily become a concept for continuous improvement through incremental changes. I find TQM to be an appropriate concept to depart from for the further development of an approach to the building of companies' exploitation capabilities.

In my theory section I further found that efforts to combine forms of exploitation and exploration synergistically are in need of good approaches both theoretically and methodologically. The potential for utilizing BPR philosophy, strategy, methods and tools in the further development of approaches to the synergistic combining of forms of exploitation and exploration have not been exploited. That is one of my contributions to the further development of theory and practice. BPR is primarily an approach to radical process innovation. I find BPR to be an appropriate concept to depart from for the further development of an approach to the building of process exploration capabilities. However, innovation in other areas like technology, raw materials, products, markets and organization is often part of process innovation too. Consequently BPR is a concept that can promote radical change in all innovation areas.

In order to strengthen enterprises' capability to combine TQM and BPR in a Norwegian industrial context my initial case study showed that two enabling factors seem important – leadership and ID. The importance of leadership was no surprise. More surprising were the importance of a better utilization and a better co-ordination of direct and indirect employee participation in innovation. Researching the importance of ID as an enabler for the synergistic combining of TQM and BPR consequently became another of my central contributions. To this discussion, it is particularly two of the theories of the Industrial democracy tradition that I find important, "Participatory Democracy Theory" (Pateman 1970) and the theory of "Democratic Dialogue" (Gustavsen 1992). Important in the first theory is the proposition that lower level/socio-technical employee participation motivates and trains employees for strategic/higher level participation. Innovation activities at lower level belong mostly to the exploitation paradigm, whereas ideas of an explorative character will mostly be found at strategic/higher company level. So, if a company succeeds with direct employee participation, where exploitation is prevailing, the theory proposes that this trains and motivates people for also participating in explorative innovation activities, where the indirect channel for employee participation is the most important. A well functioning indirect channel for employee participation is important for the following up of explorative ideas from the shop floor, as well as the exploitative, at higher decision levels in a company.

Important in the theory of "Democratic Dialogue" is the notion of equality and the importance of dialogue as a basis for sound decision making. The ability to participate in discussions where own and others' viewpoints, knowledge and experiences can be shared, tested, refined and where learning can take place is considered to be an important condition for rational choice and action (Gustavsen 1992). To see if, and then, how employee participation can contribute to a synergistic building of exploitation and exploration capability is important in my research.

I will now conclude the discussion of my research results in relation to the following questions:

1) What are the similarities between TQM and BPR? Where do they conflict and where can they reinforce each other?

2) Where do Norwegian traditions of industrial democracy conflict with TQM and BPR, and can an ID context enable the two concepts to reinforce each other and work synergistically?

3) If TQM and BPR can be combined synergistically, what kind of practical tools, methods and new approaches can strengthen employee participation as an enabling factor?

Finally, pulling together questions 1-3:

> So, can TQM and BPR, representing characteristics of, respectively exploitation and process exploration, be combined synergistically in a Norwegian industrial context, and if so, how?

11.1.1 Combining TQM and BPR

What are the similarities and differences between TQM and BPR? Where do they conflict and where can they reinforce each other?

Characteristics of TQM were described in Chapter 3 and summed up in Chapter 5. TQM has its main focus on continuous incremental change. Change is typically done in small steps, it comes gradually, details are important and everybody is involved. Management leads the change processes, but the change process and activities are very much bottom up. Groups and collectivism are important, conventional know-how and state of the art solutions are prevailing, a lot of effort is put into keeping the process going, but no large investments are made. Information is open and shared, and everybody takes part in a learning process.

Characteristics of BPR were described in Chapter 4 and summed up in Chapter 5. BPR has its focus on radical, dramatic and quick change. Changes typically come in big steps and can be abrupt and volatile. The change process is top down, few employees are involved, specialist competence is needed, technology – especially information and communication technology, often plays an important role, the change often requires large investments, the state of the art knowledge in the company and existing in house practice is often of little importance, and information sharing is scarce.

In spite of these differences between TQM and BPR, which can create conflicts of interest regarding focus, priorities and practical organizational development, the two concepts share some of the same language, the same focus on customer satisfaction and some improvement and change methodology and techniques. The most important common feature is that both concepts are process oriented. These factors and especially the last one can ease the combining of the two.

My theoretical analysis concluded that these two approaches to business development can be combined in the same development process and support each other in order to ensure business success over time. Most authors have advocated a binary (Sutcliffe et al. 2000) approach to the balancing of these concepts. In other words, they are both present in the same system, but not at the same time or the same place in the system and the interaction between them is not exploited. The Aker Stord case is an empirical example of this kind of balancing, even if Aker Stord had the ambitions to combine these concepts in a dual/synergistic (Sutcliffe et al. 2000) mode.

My conclusion is that balancing the two concepts in a binary manner works. This was no surprise and supports theory about the interdependence between radical and incremental change (Imai 1986, March 1991, Levinthal & March 1993). I also conclude that balancing characteristics of TQM and BPR in a synergistic manner is difficult and challenging. The companies involved in the two action research cases accomplished some degree of balancing the two during the research periods, thereby supporting theory about the dual/synergistic way of combining forms of exploitation and exploration (Sutcliffe et al. 2000, Boer & Gertsen 2003, Petersen et al. 2004, Boer et al. 2006). However, the potentials of this way of combining the two still needs to be further researched.

11.1.2 TQM, BPR and ID

Where do Norwegian traditions of industrial democracy conflict with TQM and BPR, and can an ID context enable the two concepts to reinforce each other and work synergistically?

The Norwegian industrial democracy tradition encompasses both direct participation of employees in business development and indirect participation. Direct employee participation is given a decisive position in the TQM philosophy and is considered a means to business excellence. Based on my theoretical discussion and empirical findings I conclude that the TQM way of thinking about direct employee participation at the shop floor/daily operational level can easily be combined with and supported by Norwegian industrial democracy traditions. These findings support TQM theory (Deming 1986, Juran 1989). However, direct employee participation at company strategy level is not an issue at all in the prevailing TQM literature, and adequate tools are lacking. The fact that indirect participation through elected representatives and company union cooperation also are no topics in TQM is not in line with Norwegian industrial democracy practice and Norwegian laws and regulations. I found this to be a challenge to Norwegian companies. However, my case studies and the national quality management survey showed that it is possible to practice quality management together with indirect employee participation. Furthermore, indirect employee participation can support the implementation of TQM. The Aker Stord case is an example of that.

The international BPR literature and practice do not focus company-wide direct or indirect employee participation. This is a challenge to Norwegian industrial democracy traditions. However, both the Aker Stord case and the Serigstad case show that direct and indirect employee participation can contribute to success of BPR efforts, provided that adequate tools and approaches are available and adapted to local conditions.

This relation between employee participation and business development is supported by research indicating that a purist development strategy where you either focus on employee participation and quality of working life perspectives (den Hertog 2004) or on serving capital without employee involvement and quality of working life aspects tends to fail (Buchanan & Badham 1999). A challenge is to develop a combined approach serving both objectives. My research and experiences from the three case studies lead me to believe that a combined and reciprocally supportive participatory and business oriented approach can give the best results in a Norwegian industrial context.

The TQM and BPR literature advise the establishment of specific arrangements such as steering committees, advisory groups, project groups, etc. for concept implementation. My research shows that the functions of these specific arrangements can be performed by existing cooperative bodies like works councils, department councils, work environment committees and other arrangements that involve shop stewards and safety delegates. My case studies show that they can ease the introduction and follow-up of the change processes, while at the same time strengthening the position of these co-operative bodies. Experiences from Aker Stord show that employee representatives can participate in promoting the introduction of new company development concepts, while at the same time strengthening their own position and the position of the union in business development processes and innovation. Among others, they can take important roles in following up and realizing innovation proposals from employees.

My case studies suggest that some conditions enhance company-wide employee participation as an enabler to innovation both directly and indirectly. Important is:

- Participatory leadership
- Mutual trust between employers and employees
- Trade unions that want to use trade union time to co-operate with management on promoting business development
- A practical framework, organization and arrangements for the co-ordination between direct and indirect employee participation
- Practical innovation methods and tools that ensure participation
- Available time

Thus, ID can support the implementation of TQM and BPR. So, can ID support a synergistic combining of these two forms of exploitation and exploration?

My theoretical discussion concluded that a Norwegian ID context can support the synergistic combining of TQM and BPR characteristics in the same development process in order to obtain long term business success. The Prefabrication of Pipes case and the Scrigstad case support this. Direct employee participation increases learning about change methodology, the company's value chain, work processes and how to improve or redesign work processes, and motivates for more participation in constantly new areas. It increases the employees' interest in, ability for and willingness to contribute to the further development of their own company. Indirect employee participation and formal arrangements for union – management co-operation can be used to ease the construction of a legitimate structure and practice for managing both incremental and radical innovation. Arrangements such as the Works Council and the Departmental Committee were used to plan, manage and follow-up both incremental and radical innovation activities synergistically.

My framework for the practical organization of employee participation arranged to support the synergistic combining of TQM and BPR, the practical adaptations of different change techniques, the development of the ED Conference and the Actor-Process Diagram all build on the guidelines from "Democratic Dialogue" (Gustavsen 1992). This kind of dialogue has the benefit of drawing upon a broad range of opinions and ideas, while at the same time being able to promote decisions supported by all participants. An important ambition of Democratic Dialogue is to produce agreements that provide platforms for practical action. Most of the techniques and ideas from the international management concepts TQM and BPR are developed to create practical action for change. The experiences from the case studies show that these techniques, if adapted to a Norwegian ID setting, increase practical employee participation in both incremental and radical change processes. Examples are brainstorming techniques, the Ishikawa diagram, the affinity diagram and process redesign tools. The case studies showed that both the ED Conference and the Actor-Process Diagram, which both included the use of these kinds of tools, promoted and anchored employees' participation in business strategy development in a way that included both incremental and radical innovations and their combination. Thus, ID can support the synergistic combining of TQM and BPR if arranged for.

11.1.3 The need for adaptations and further methodological development

If TQM and BPR can be combined synergistically, what kind of tools, methods and new approaches can improve employee participation as an important enabling factor?

To ensure fundamental conditions for employee participation such as motivation, ability and opportunity to participate in development work at different company levels, I found that some adaptations of existing tools and development of new practices and methodologies are needed.

First, to be motivated to contribute to the development of their workplace, employees should first of all be able to see the relation between their own contribution to innovations and their own interests as employees. Secondly, in order to build ability to participate, training is important, both in job-related skills and in change methodology. Thirdly, to give employees opportunities to participate at the different levels, changes in management attitudes and innovation practice are needed.

Regarding the first of these three fundamental conditions for employee participation – motivation, I find managers' role to be crucial. Top management focus is important. All three cases showed that. This supports both TQM theory (Deming 1986, Juran 1989) and BPR theory (Hammer & Champy 1993, Hammer 1996). Also a more systematic involvement of middle and first-line managers in innovation processes appears to be important.

Regarding the second fundamental condition for employee participation – ability, I find job-related knowledge to be important for both TQM and BPR. However, TQM and BPR alike need some methodological developments, among others on how to improve the employees' knowledge about related disciplines, suppliers, the company's products, manufacturing processes and customer requirements. Both concepts lack good approaches to ensuring employees' knowledge about company objectives and strategy, which is important for the ability to tune their own contributions to a larger setting. For the further development of a synergistic approach to combining forms of exploitation and exploration, I found that some development work is needed in this area.

A large arsenal of techniques exists for ensuring employee participation in continuous improvement. However, appropriate techniques for ensuring employee participation in radical change are lacking. For the further development of a more synergistic approach, a more systematic way of ensuring employees' participation in radical change is needed.

Also the third condition for employee participation – opportunity, needs some further development. In the literature and also in company examples I find many good approaches to ensuring direct and indirect employee participation at most company levels. However, neither in the literature nor in the company cases do I find examples of companies that have succeeded in ensuring employee participation at all these levels in the same system, at the same time and on an ongoing basis. Neither do I find any examples where participation at the different company levels is well coordinated. Some companies can document success with employee participation at some of the levels, other companies at other levels. I conclude that a comprehensive framework for the practical organization of employee participation in innovation at different company levels is needed, including some new tools to complement the practical realization of this framework.

There is no exact answer to the question of how much participation should take place and how the direct and indirect channels for participation should be coordinated. The framework I propose for ensuring employee participation in the more synergistic combining of incremental and radical innovation identifies six basic company levels where participation in decision making should take place:

- Self-determination in the immediate work situation
- By proposing changes through the line of command
- Through the planning and carrying out of specific tasks or jobs together with colleagues and managers

- Through participating in changes in own group or department
- Through participating in developing the company's business – and work processes
- Through participating in developing the company's vision, values, goals and strategy

In real business life arenas at these six levels will take a variety of different forms and combinations. There is a large arsenal of existing approaches, methodologies and techniques for participating at these levels. However, I did not find any good tool for company-wide employee participation in innovation activities related to the development of an organization's vision, values, goals and strategy. I therefore developed the Enterprise Development (ED) Conference. I did not find any good tool for employee participation in the development of an organization's work – and business processes either, and developed the Actor-Process Diagram. As to the existing tools designed for the other levels, few are ready to be implementted as they are. They have to be adapted to the specific company situation and in relation to the total approach and the techniques and procedures used at the different levels. Experiences from the case studies show that many of them are easily adapted.

The Enterprise Development Conference

Without exception, all the 11 ED Conferences that were run in ED 2000 and in VC 2010 have resulted in valuable innovation proposals from employees related to the formation and/or concretization of the companies' business strategy. Important challenges facing the companies involved were identified, communicated and discussed. Important objectives and strategies were identified, developed and directed towards specific innovation initiatives. All the companies followed up the results of the conferences, and many of the suggested initiatives have been implemented. The ED Conference, including the follow-up structure, is anchored in two-party co-operative bodies and activities, designed to handle both incremental and radical changes and also to support the synergistic combining of the two. The experiences show that the conference supports the combining of changes belonging both to the exploitation and exploration paradigm. E.g. in all 11 cases an adapted variant of brainstorming was used to generate both incremental and radical ideas. Furthermore an adapted variant of the affinity diagram was in some cases used to group many incremental improvement proposals into comprehensive ones that in total can be considered radical. In some cases the Ishikawa diagram was used to evaluate both incremental and radical proposals.

However, not all employees took part in the follow up activities, and the role of middle and first-line managers has to be further developed. In spite of these shortcomings, my research shows that the ED Conference can be used as a tool to promote innovation activities related to developing business strategy. Furthermore, the ED Conference fits in with and contributes to the further building of my operational framework for employee participation in innovation. The conference easily interlinks with, supports, but also depends on the other parts of the framework. It fills in one of the gaps needed to build a complete approach for employee participation in the synergistic combining of TQM and BPR. A process where incremental and radical innovations are developed synergistically also needs to be closely interlinked with overall business strategy. The ED Conference can contribute to this interlinking by generating important contributions to business strategy as part of such a development process. Finally, the ED Conference, with subsequent follow up activities, appears to improve the strategy development ability of SMEs in general.

The Actor-Process Diagram

For many redesign efforts one of the goals is to develop "process centered organizations" (Hammer 1996). However, in order to form a "process centered" organization, employees should become "process professionals" who understand that in doing their work they are contributing to the performance of a process and not simple isolated tasks (Hammer 1996). I find it reasonable to run the redesign processes in a way that involve, treat and train people as "process professionals, and developed the Actor-Process Diagram.

The Serigstad case gave promising results and suggests that using the tool leads to significant improvements in business performance, work environment, employees' understanding of the company's business processes and production demands, as well as insight into and practical understanding of relevant redesign and change methods and approaches. The Serigstad case showed how the tool can be used to promote incremental change, radical change and their synergistic combination.

In a wider context the Actor-Process Diagram proved useful as a tool that can support the synergistic combining of TQM and BPR. It contributes to the completion of my approach to the synergistic combining of direct and indirect employee participation. Employees learnt some simple, easy-to-use change methods in order to participate. Some of the more advanced techniques take some time to learn, and are easily forgotten when they are not used regularly. The Actor–Process Diagram is designed to be used at the shop floor level and for ensuring contributions from the people involved in the actual process being designed or redesigned. Little time and effort is needed to get employees involved and engaged in these kinds of development processes.

11.1.4 The synergistic approach

> **So, can TQM and BPR, representing respectively exploitation and process exploitation characteristics, be combined synergistically in a Norwegian industrial context, and if so, how?**

March (1991) holds that "maintaining an appropriate balance between exploration and exploitation is a primary factor in system survival and prosperity" (March 1991:71). Both my theoretical discussions and my empirical research support that.

Imai contrasts the Japanese way of running continuous improvement, called "*KAIZEN*" with (radical) innovation. He sees these two roads to progress as the "*gradualist approach*" and "*the great leap forward approach*" (Imai 1986:23), and as being interdependent over time for an organization's success. Imai represents a binary (Sutcliffe et al. 2000) approach to balancing exploitation and exploration. They alternate in time. This is was what happened in the Aker Stord case. The first phase of the TKA process focused on incremental continuous improvement/exploitation. The next phase focused on radical innovation/process exploration. Also the company's initial plan for Serigstad was to follow a pattern were exploration and exploitation alternated in time: a radical technological innovation had been implemented and was planned to be followed up by continuous improvement. However, a dual/synergistic approach that encompassed several innovation areas was agreed on with the researchers.

March (1991) further claims that balancing exploitation and exploration has the character of a trade-off between the two. Investing more in one of them goes at the expense of the other. Sutcliffe et al. (2000) have challenged this binary antithetical balancing of exploration and exploitation, and argue that there may be other, that is, dual/synergistic solutions. According to Boer et al. (2006), the need to develop such solutions is borne in the increasing need to combine operational effectiveness and strategic flexibility synergistically. So, theory suggests that operational effectiveness is based on exploitation capabilities. Strategic flexibility is based on exploration capabilities (March 1991, Boer & Gertsen 2003, Boer et al. 2006). Theory suggests a synergistic/dual approach is possible. What, however, about reality?

The challenge of combining TQM and BPR

Aker Stord started out on a TQM/exploitation journey. After a while they found that their approach needed to be strengthened by a process orientation. They introduced BPR and experienced that it was difficult to focus on both approaches in the same system at the same time. So, the TQM efforts declined and many activities stopped up. The Aker Stord case supports March's theory (1991) of the balancing between exploitation and exploration as a trade-off. More precisely, at Aker Stord process exploration capability was improved (building on exploitation capability – see below), but exploration activity went at the expense of exploitation activity. However, one of the departments, Prefabrication of Pipes, albeit not from the very start, eventually developed in a mostly synergistic direction – management commitment and employee participation seems to be the key factors explaining this. Based on these experiences and theory I developed a process model showing what I find to be the most important enabling factors and what can happen when you combine TQM and BPR synergistically. See Figure 19.

The Titania case and the Serigstad case support and add to the Prefabrication of Pipes experiences. In none of these cases was my approach fully implemented, but important tools and the direction of combining forms of exploitation and exploration synergistically were tested. Both cases support the position that it is possible to combine exploration and exploitation activities in the same system, at the same time and on an ongoing basis, though challenging. In addition to management commitment and involvement, the use of appropriate tools (such as the ED Conference and the Actor-Process Diagram, combined with well known TQM/CI and BPR tools) and a motivated, capable workforce that is in the position and feels the urgency to participate actively in idea generation and follow-up, either directly or indirectly, seem to explain these findings.

The case study results support and shed light on recent research and empirical studies (Sutcliffe et al 2000, Cole 2002, Boer & Gertsen 2003, Osri et al 2005, Boer et al. 2006, Greve 2007, Isobe et al 2007) that question whether March's theory of balancing exploitation and exploration mostly as a trade-off is always correct. Like the Prefabrication of Pipes case, the Serigstad case developed in a direction in line with my process model for combining TQM and BPR synergistically (Figure 19), thereby supporting an alternative to the antithetical approach. However, further methodological development, research and empirical evidence are needed, both for the forms of exploitation (TQM) and exploration (BPR) that I have focused on and on other forms.

Exploitation supporting exploration

All the three case studies showed that forms of exploitation provide a good basis for the development of forms of exploration. At Aker Stord:

- Fast learners of the existing TQM codes also became fast learners of the new BPR codes
- Employees developed a sense of competence in the predominantly continuous improvement phase of TKA that made employees able to respond effectively to the BPR phase. In Prefabrication of Pipes TQM and BPR approaches were used simultaneously and reciprocally supportive
- The BPR projects profited from a structure and practice for planning, implementing and following up changes that had been developed during the predominantly continuous improvement phase of TKA and the training that employees had got in using it
- The training and process awareness "embedded" in TQM through customer-supplier interviews, brainstorming and the Ishikawa-diagram, prepared employees for BPR
- Participation in continuous incremental improvement motivated and prepared employees for participation in constantly new areas and at higher company levels
- The BPR projects profited from being integrated in a more continuous change process where the radical changes were being further exploited and continuously improved

The Titania and Serigstad cases confirm this picture. At Titania, the ED Conference triggered both incremental and radical change simultaneously by using methods belonging mostly to the exploitation paradigm. At Serigstad the Actor-Process Diagram, a tool derived from the radical process change methodology BPR, was supported by tools belonging to TQM. Neither of the companies had problems whatsoever adopting and absorbing these tools and put them to work to generate, conduct and successfully carry out both incremental and radical changes. So, although radical changes may require additional capabilities in terms of, for example, project management and expert knowledge, the development of exploitation and exploration capability can be combined in a reciprocally supportive way.

Using exploration and exploitation synergistically

Having forms of exploitation and exploration in the same system all the time does not necessarily mean that they are or can always be deployed at the same time, as the Aker Stord case makes clear. How come?

First, the synergistic use of the two appears to depend a great deal on leadership. This supports theory about the decisive importance of leadership in these kinds of change processes (Deming 1986, Juran 1989, Yukl 2006). The top manager at Aker Stord who shifted the company's focus towards BPR got a predominantly antithetical approach to the balancing of forms of exploitation and exploration. The department manager at Prefabrication of Pipes and the top managers of the two other case companies focused on a predominantly synergistic approach and managed to develop the organizations in that direction.

Second, much of the organization structure, methodology, tools, routines, motivation, attitudes, competence etc. developed for exploitation can also be used to support process exploration (Prefabrication of Pipes). Similarly, tools that support exploration (e.g. the Actor-Process Diagram – Serigstad) can also be used to support exploitation. In other words, much of the structure and tools needed to support the two codes could be used in both development logics, but not necessarily at the same time and all the time.

Third, the human factor plays an important role. Sense of urgency (Serigstad) is a strong motivational factor, and is not always present. Training in the use of tools and actually using them in practice leads to capability. However, the motivation for using the tools may not be present all the time. Industrial democracy (all three cases) provides the opportunity to participate actively, either directly or indirectly. To fully exploit this opportunity is demanding.

Overall, in terms of the way that I operationalize my research question and under conditions outlined and discussed above, TQM and BPR can be used together in the further development of an approach to building forms of exploitation and process exploration synergistically in line with the process model presented in Figure 19 indeed.

In effect, March's (1991) theory on the antithetical nature of exploration and exploitation should be considered modified in favor of the suggestions put forward in more recent publications (e.g. Sutcliffe et al 2000, Cole 2002, Boer & Gertsen 2003, Osri et al 2005, Boer et al. 2006, Greve 2007, Isobe et al 2007). Under favorable conditions, no trade-off is needed: Forms of exploitation and exploration can be combined synergistically. Similarly, March's (1991) theory of "slow" and "fast learners' impact on exploration should be considered modified as well. This theory was developed through simulation and suggests that slow learners increase exploration while fast learners reduce exploration. My study shows that employees who know exploitation related codes can profit from that knowledge when they contribute to process exploration initiatives.

I will now conclude my theoretical discussion by presenting some propositions for further research.

11.1.5 Propositions

An antithetical (Sutcliffe et al. 2000) and a sequential approach to the balancing of forms of exploitation and exploration works, and is supported by theory and practice (Imai 1986, March 1991). From a theoretical position it seems reasonable to exploit the potentials of a synergistic approach to the balancing of forms of exploitation and exploration (Sutcliffe et al. 2000, Boer et al. 2006). My case studies show that it is possible, however challenging. Based on the analysis of my empirical research in a Norwegian industrial setting some propositions can be formulated for further research:

March's (1991) theory of "fast learners" should be considered modified. The Aker Stord case show that fast learning of the existing TQM codes supported fast learning of the new BPR codes. This gives the following proposition:

Fast learning of existing exploitation codes can support fast learning of new process exploration codes.

March's (1991) theory of balancing exploitation and exploration mostly as a trade-off is not always the best way and should be considered modified. Exploitation capability can improve forms of exploration capability and thereby support exploration. The Aker Stord case showed that TQM capability supported the building of BPR capability. Consequently, I will propose that:

Exploitation capability can support process exploration capability.

The Prefabrication of Pipes case and the Serigstad case combined TQM and BPR approaches reciprocally supportive and accomplished to develop incremental and radical innovations in the same system and at the same time. Many factors seem to contribute to that. I found leadership focus and employee participation to be the most important factors enabling this. I will therefore propose that:

The forms of exploitation and exploration that is represented by respectively TQM and BPR can work synergistically. Two enabling factors are important for this to be true: 1) A strong leadership-focus and direction. 2) Employee participation and involvement.

The importance of a strong leadership focus on development direction is in line with theory (Deming 1986, Juran 1989, Yukl 2006). Employee participation as an enabling factor is not an issue in prevailing literature on the synergistic combination of exploitation and exploration. A Norwegian industrial democracy setting seems favorable for this to be true and seems dependent on several conditions such as participatory leadership, mutual trust between employers and employees, trade unions willing to spend time on business development, a good structure and arrangements for coordinating direct and indirect employee participation in innovation, methods and tools that ensure employee participation in incremental and radical innovation, available time and a sense of need (or even urgency).

11.2 Methodological reflections and limitations

Chapter 7 discussed the appropriateness of using single case studies for building scientific knowledge. I will now critically reflect on my own experiences by referring to the discussions about validity and reliability. I will pay special attention to the bias challenge related to action research.

11.2.1 The quality of a single case study design. The question of validity and reliability

The most commonly used tests to evaluate the quality of empirical research were discussed in Chapter 7, a) construct validity, b) internal validity, c) external validity and d) reliability (Yin 2003). All these four tests were used in my case studies.

Construct validity

Construct validity is about establishing correct operational measures for the concepts being studied (Yin 2003). In all three cases I measured and evaluated results through the proposals for incremental and radical innovations that the interventions produced, including follow-up activities and results. In all three cases I found that the increased innovation activity can be explained by the interventions in question. E.g. in the Titania case, the most important measure was the proposals for incremental and radical innovations that came out of the ED Conference.

Establishing a chain of evidence is one way of improving construct validity (Yin 2003). A chain of evidence, anchored in my research objective, the overall research question and the case specific questions and to the case study conclusions, was established for all cases in the way the case report sections are built up. The discussion following the case study narrative refers to this chain of evidence, and ends up with highlighting what I consider to be the most important findings relative to the research question.

Internal validity

Spurious effects are important threats to validity. To strengthen validity, I used more than one source of information about the same activities. E.g. in the Aker Stord study leaders, shop stewards and employees were asked to give information about the same improvement activities. I found that the way they described the overall picture and explained cause-effect relationships harmonized. In addition to asking different actors from different parts and levels in the organization the same questions, a considerable amount of documents, statistics and reports were collected in my case study protocol and compared to the results from the interviews. I found that the information that was given through the interviews on the overall picture harmonized with other kinds of data sources.

In all three cases I found that the causal and explanatory parts were very straightforward. I was interested in mapping the effects that some specific activities designed to increase participation actually had on employee participation in innovation.

Becoming biased is another threat. There is always the danger that the researchers look for things they want to find and neglect important other information. I found many of the results I expected to find, I got some results that did not support my expectations and I found some unexpected ones. E.g. before I started up at Aker Stord, I believed that it would be easier for staff to come up with improvement projects because of easier access to facilities and office resources. It was not. It was easier for the operators. Staff had problems discriminating between what was a natural part of their job and what were improvement projects. In contrast to my expectations, this made them passive actors in the TKA process.

It is a challenge that I got to know staff and many managers and shop stewards in the three companies very well. I wanted them to succeed with their innovation efforts. It could become difficult to be very critical. However, the company representatives operated in a very professional manner, and gave me the impression that they really wanted the researchers to find areas for improvements. I also felt that some of my most critical comments were well received. However, the bias threat was there all the time, and is a challenge with these kinds of case studies. The researchers involved were aware of it and we reminded each other about it. My experience is that being part of a research team, together with other researchers that are not directly involved, helps a lot. We presented our findings for each other and discussed them critically. I was very often reminded about the danger of becoming biased when I presented results in favor of the objectives of the different interventions. In summary, I consider the likelihood small that my findings are false due bias problems. Most of the data are well documented in company reports and statistics.

External validity

The question is whether the results from the single case studies can be generalized. For example, the Titania ED Conference is the eleventh conference in a row of ED Conferences in other companies, and supports the findings from the other 10 conferences (Tønnessen 2000). However, all the companies participated in ED 2000 and VC 2010, favored ID and are situated in the southwestern part of Norway. It is not certain that the conference would have produced equivalent results in another industrial context.

The Serigstad case builds on experiences from similar efforts to engage employees in process redesign at Aker Stord, and produced equivalent results. However, the tool that was used was not quite the same, the context was different and the scope of work different. Even if the two cases were comparable, external validity is a challenge. It is only two companies and both were part of ED 2000 and VC 2010 and favored ID. An important question is whether a less ID intensive context would have given weaker or other exploitation exploration effects.

Reliability

The tactics used to deal with reliability was to develop a case study protocol and a case study database for each case company. The procedures and data gathering are documented in the protocol. This includes field notes, taped interviews, written interviews, observations, company documents, internal magazines, internal articles, surveys, company statistics and narratives.

Some of the techniques that I used to improve validity also improved the reliability of the information (Yin 2003). E.g. at Aker Stord employees were presented the whole chain of evidence from the initial research questions to the conclusions and approved it with a few comments and corrections. Also the use of multiple sources of evidence/triangulation in all three cases improves both validity and reliability (Yin 2003).

11.2.1.1 Action research and the bias challenge

Action research encounters the same validity and reliability challenges as described for case studies in general (Yin 2003). However, as action researchers engage in both shaping and telling the story, it can be a great challenge to give a valid and unbiased presentation. A critique and a principal threat to validity for action research can be lack of impartiality on the part of the researcher (Coughlan & Coghlan 2002). In the Aker Stord case, the researchers had very little influence on the development process in the company. In the Titania and the Serigstad cases researchers acted as "friendly outsiders" and took part in a co-generative process (Greenwood & Levin 1998). The researchers shared ownership with the insiders to the development processes. I will comment on my experiences with my two action research cases below.

The Titania case

The company needed a good start on their development process. I participated in the planning of the process. I brought some new knowledge and change methodology from outside and into the company. I advised the company on how to follow up the results from the ED Conference, and I also gave advises on how to carry out some of the specific improvement projects. I participated in the documentation of results, and I communicated experiences from the conference to the outside world. I took a pragmatic action researcher role (Greenwood & Levin 1989).

I was the one who advocated the ED Conference, I wanted Titania to succeed, and I also reported the results from the conference together with colleagues from academia. In such a situation you can easily find what you look for, and oversee important facts that you do not look for and do not want to happen. More generally, both the ED Conference and its follow-up include a lot of activities and there are a lot of people involved. The amount of information and data that can be gathered is large, and it can be convenient to pick out the data that support the action researcher's own interest. Among the enormous amount of data that could be gathered, I focused on finding data that seemed most relevant for my research objective, overall research question and case specific questions. It is not easy to avoid the bias problem in this situation. However, I focused on gathering and referring to documentable and objective facts about the change process and its results, thereby avoiding that my own subjective opinions dominated reporting and analysis. It is a fact that very much happened in the company before, during and after the conference. The conference itself documented a high degree of employee participation in the process of developing a considerable number of proposals. The follow-up activities still (2011) involve a large number of employees, and a large number of the proposed ideas have become projects, many of which have been implemented.

There was a documented increase in ideas for both incremental and radical innovations. Many of the same methods, structures and approaches were used to support both incremental and radical innovations. The company also experienced increased productivity and better economic results. The researchers did not study the follow-up processes, except for participating in the WEC/WC meetings, receiving some status information on the phone and by email and by considering official statistics from Brønnøysund Registeret (2009). However, it was very difficult to relate the overall business results directly to the innovation efforts. I was informed about improvements in several areas that were related to the project. But as a researcher I find it very difficult to document the exact relation between the project and the overall business results. The project is composed of a considerable number of activities that are difficult to measure in economic terms. The activities are overlapping with other activities independent of the project, and other internal and external factors, such as changes in market prices and sales, affect the bottom-line results as well. In spite of these shortcomings and the bias challenge, I find that the action research methodology that I used can make me conclude that the intervention at Titania produced the reported exploitation and exploration activity synergistically.

The Serigstad case

Like the Titania case, I took part as an action researcher in a co-generative process. I facility-ated the process. The company representatives adjusted the methodology that I proposed in accordance with their needs, and both management and employees decided the practical content. We all worked on an equal basis, and the setting demanded that no knowledge was considered superior to other knowledge. I used some basic principles of "democratic dialogue" (Gustavsen 1992) as a guideline when I facilitated the process. I also took the "pragmatic" action researcher role, simply meaning in my case, doing what I saw was necessary for me to do as a "friendly outsider" (Greenwood & Levin 1989), and left as much as possible in the hands of the insiders.

Compared to the Titania case, my pragmatic action researcher role was closer to the activities in the Serigstad case. Together with a colleague I became part of the project team in the company for more than a year, and participated weekly in project meetings. I was the external specialist who facilitated the change process together with another action researcher. I brought new change techniques and methods from outside and into the company. I trained management, shop stewards and employees in the practice of running change processes, and on how to use a selection of change techniques. I supported the development projects that were prioritized, commented on ideas and asked critical questions. My colleague and I also participated actively in documentation and analysis, and communicated experiences to the outside world.

On the other hand my colleague and I avoided taking part in the development of practical and detailed solutions. The insiders, with some assistance from external suppliers, were the experts on practical solutions. It was, however, difficult to avoid having some influence on the development of the practical solutions. Indirectly we could not avoid influencing e.g. through asking critical questions.

I wanted the company to succeed, and I wanted that to happen by using my approach and methods. It was difficult not to become biased, and to let subjective and predominantly positive evaluations from my side pull conclusions in a false direction. However, the bias problem was handled by using objective documentable data. Important was the documentable increase in participatory activities, productivity improvements and the work environment analysis performed by an external company.

As in the Titania case, the researchers left as much as possible in the hands of the insiders while avoiding coming into a situation where we did not take our share of the practical tasks that needed to be done, e.g. writing reports from meetings. This balancing between leaving as much as possible to insiders on the one hand and taking the expected share in the work to be done in a partnership on the other hand, can be difficult.

I found it important to leave as much as possible into the hands of the insiders especially for two reasons. First of all, it increases learning, and the action researcher leaves some practical change skills behind when she/he leaves the company. It makes it easier for the company to continue after the researcher has left. Secondly, it increases the research quality. If the action researcher runs most of the change processes, the danger of becoming biased increases and it is difficult to evaluate the performance of the insiders.

The project focused on productivity in the factory, and there is no doubt that the project itself produced both incremental and radical innovations and gave good results. Like in the Titania case, action research methodology gave the researchers detailed knowledge about change processes.

11.2.1.2 Case studies and scientific knowledge

To the research group and the participating companies in the research program ED 2000, the single case study design that was applied in the Aker Stord study met its purpose. It prepared both the researchers and company representatives for the comprehensive action research program in ED 2000 and VC 2010 that started out in 1994, and is still (2011) running in the new national Norwegian research program on innovation – VRI. There is no doubt among skeptics and supporters of single case designs alike that the approach is useful in the initial phase of a research project. You get to know the field and get in-depth information as a basis for developing propositions. All my three case studies contributed to that.

My methodological discussion and experience also show that single case studies can contribute to falsify theories, propose new theories and to support existing theory. One single case can falsify a theory (Popper 1959, Flyvbjerg 2006). All my three single case studies contribute to support existing theory and propose new theory.

Skeptics of single case designs have questioned the subjectivity problem and the bias towards verification (Flyvbjerg 2006). The discussion has shown that this is a problem also for other approaches. Some claim that this may not be a larger problem for case designs than for other research approaches. Some will even claim the contrary. In case studies you are close to the object of the study and can be corrected. In surveys e.g. subjectivity can follow the whole research process, and it is less likely that the objects of the study will "call back" (Ragin 1992). In my three case studies it was not very likely that the researchers could get away with any preconceived notions that the employees disagreed with.

So, in spite of many traps, shortcomings and challenges, I do believe that case studies and action research can contribute to scientific knowledge, preferably, but not necessarily, in combination with other research methods. Case studies and action research can contribute not only by helping researchers to get acquainted with the field and to develop propositions before the "real" research starts, but also as a methodological approach to test and propose new theory.

I find that case studies and action research has many advantages. In my case the most important advantages are the in-depth knowledge, the possibility to study development processes and the dynamics of interactions, and the possibility to use scientific knowledge and scientific feedback as an important ingredient in development processes in a dynamic way. Based on the research I have done and the limitations of my methodological approach, I see the need for considerable further research and development. What I find most important is listed below.

11.3 Further research

My research focuses on the synergistic combining of TQM and BPR, and the overall objective is to contribute to the bulk of knowledge needed to further develop the theory of balancing exploitation and exploration (March 1991). My research results and my theoretical and methodological discussions suggest several directions for further research. Among others, my propositions need to be further researched and verified. I have identified six areas, which also incorporate my propositions.

There is a need 1) for clarifications of important elements of the synergistic approach, 2) to further develop and critically test the synergistic approach, 3) to relate to all innovation areas, 4) to evaluate and further develop approaches to optimize learning, 5) to test ways of better involving middle and first-line managers, and 6) the question of the participatory approach in other industrial settings.

Clarifications of important elements in the synergistic approach to exploitation and exploration

Some of the fundamental elements that constitute the synergistic approach need to be better defined. This can be accomplished along with further empirical research. The borderlines between exploitation and exploration and between incremental and radical innovation need to be better clarified. A change can be considered radical from one point of view and incrementtal from another point of view. My research does not go deeply into these discussions, but I am aware of the difficulties of using these terms. However, I find the terms useful for expressing two fundamentally different development logics. For my discussions I found it appropriate to refer to the wide definitions from March (1991) on exploration and exploitation, the wide innovation definition from Kanter (1988), covering booth incremental and radical change, and Imai's (1986) description of the differences between KAIZEN/ CI/incremental change and (radical) innovation for clarifying and representing characteristics of exploitation and exploration. Also the use of the expression "ongoing" when referring to a synergistic practice needs to be better clarified. E.g. should it mean all the time or only when the optimal situation occurs?

Furthermore, in order to define synergistic maturity and to know how to get there and when you get there, we need to know precisely what to look for. Better clarifications of the content of this term will make it easier to test the idea scientifically. It will make it easier to conclude whether synergistic maturity is an ideal goal that is appropriate to work for, but impossible to reach, or whether it is just hard to reach. It will also make it easier for other researchers to contribute to the further development of the synergistic approach. In my case I have used it as an ideal state where the move from a binary towards a dual synergistic state has been in focus. Full synergistic maturity may never be obtained in practice.

Further develop and test the synergistic approach

Balancing exploitation and exploration is necessary for business success (March 1991). A challenge is the question of time and space. Can a company accomplish both in the same system at the same time and on an ongoing basis, or is this too demanding or even optimal? My theoretical analysis shows that this is an ideal situation and could be a driver for long term business success. My empirical analyses show that this is a challenge, indeed, but promising. The Prefabrication of Pipes case and the action research cases showed that the companies were on their way to develop synergistic capabilities, but were far from reaching any form of synergistic maturity. More research is needed to see how this can be achieved and to identify optimal solutions.

My research challenges March's (1991) theory about the balancing of exploitation and exploration as a trade-off and questions whether this will always be correct and whether a modification of the theory is appropriate. March represents a mostly antithetical approach to the balancing of exploitation and exploration. Taken seriously, this theory can lead companies to develop their exploitation and exploration capabilities separated and in different directions. My approach takes the opposite position and calls for a synergistic approach. My own and others' empirical research show that forms of exploitation capability can improve forms of process exploration capability, which in turn is important for radical innovations and a company's strategic flexibility.

I further find that forms of exploitation and exploration can be pursued synergistically. Some of the same tools, methods, approaches, routines, organizational solutions, training and motivational factors can be used to pursue both development directions simultaneously and in a reciprocally reinforcing way. The Aker Stord case and some of the other empirical studies referred to in my theory discussion show that exploitation capability support exploration capability. However, evidence that exploration capability can improve exploitation capability is lacking. There is a need for considerable theoretical and conceptual development and empirical research to test this position.

All innovation areas

In order to operationalize exploitation and exploration into a practical industrial context, I use TQM and BPR to represent the two. TQM is mostly about improving existing practice. BPR focuses on radical process innovation. However, process is only one of several innovation areas. Even if radical process changes are often combined with radical innovations in areas like technology, product, market, organization and raw materials, my research has not been directed explicitly towards these other innovation areas. Further research, going beyond the way forms of exploitation and exploration were operationalized in this book, is needed.

Learning

Learning seems vital for a synergistic approach. Employees continuously need to develop new skills and knowledge about change processes and techniques, the company's work processes, strategy, customer demands, future possibilities and relevant external developments to be able to contribute to incremental and radical innovations and to see how own contributions can contribute effectively to the whole. I find it necessary to evaluate the approach and the tools that I have recommended for the further development of a synergistic approach from a learning perspective. What kind of further developments are necessary to optimize learning?

My research suggests that March's (1991) theory of fast and slow learners may not always be the best way. Is it true that slow learners of the existing code improve exploration? Is it true that fast learners of the existing code reduce exploration? Can fast learners of the existing code promote both exploitation and exploration synergistically? The Aker Stord case shows that fast learners of the existing code profited from their skills, knowledge, attitudes and motivation when they were challenged by a new code and participated in development processes covering other areas than the immediate shop floor level incremental innovations. Will this be true for other settings than a Norwegian ID context?

The role of management

All my three case studies demonstrate the importance of management focus. The top manager at Aker Stord focused on process exploration at the expense of exploitation. The development process followed that direction. The other two, and also one department manager at Aker Stord, focused on both exploitation and exploration and got some of both.

However, top management commitment is not enough. Engaging middle and first-line managers appears to be a challenge. In my three case companies few middle and first-line managers got important roles in the overall planning of the change processes, and few got specific roles in carrying out and following up the changes. Some openly expressed negative attitudes and felt that their position was threatened by the participatory approach. Some were asked to follow up initiatives from employees that they really did not feel necessary at all. However, the diffusion of the change process to the whole company and to all employees depends a great deal on the engagement of the line of command. In the parts of the organizations where the companies failed to engage middle and first-line managers little happened. Employees need to be motivated, have the ability and the opportunities to participate. Line managers have decisive roles in creating the arenas supportive of these conditions. More research is needed to come up with adequate approaches to participatory leadership (Yukl 2006).

The effect of other ID settings

My research took place within a Norwegian industrial setting, where industrial democracy traditionally has had a strong position. So the setting is ideal for testing out my approach. It is likely that I would have met more resistance in other settings. In spite of that, I believe that my participatory approach is adequate from a work environment and learning perspective, and I also believe that it will increase the success rate of business change processes. More empirical research is needed on the role of varying ID contexts.

11.4 Main findings and conclusion

The dual, synergistic approach to the balancing of exploitation and exploration is in its initial phase of development. The approach needs to be further developed and tested scientifically. My empirical research contributes to the body of knowledge needed to know how and under which circumstances the synergistic approach is the best alternative.

My research shows that an antithetical (Sutcliffe et al. 2000) and a sequential approach to the balancing of forms of exploitation (TQM) and exploration (BPR) work. However, March's (1991) theory of balancing exploitation and exploration mostly as a trade-off may not always be the best way, and should be considered modified. My three case studies show that a more synergistic balancing of the two is promising, but challenging. Characteristics of TQM and BPR can be combined, but it is demanding to focus both in the same system, at the same time and on an ongoing basis.

The case studies showed that forms of exploitation, such as TQM, provide a good basis for the development of forms of exploration. E.g. fast learners of the TQM codes became fast learners of BPR codes. The training and process awareness embedded in TQM prepared employees for BPR. Radical BPR changes profited from being integrated in a continuous change process where radical changes are being further exploited and continuously improved. Consequently, process exploitation capability can support radical process exploration capability.

I find TQM and BPR appropriate concepts to depart from for the building of an integrated approach to process improvement and radical process innovation. In a Norwegian industrial setting two enabling factors seem to be important for this to be true: 1) Leadership focus and a visible leadership direction, and 2) Employee participation and engagement.

In all my three case studies I find leadership focus to be decisive. If the leader promoted, supported and requested a dual, synergistic approach to change. That was the result. If the leader requested either TQM or BPR, that was the result. If a new leader stopped asking for TQM results, TQM activities faded away. I also find that less happen in parts of the organizations where middle and first line managers are reluctant, negative or passive to the actual change process.

A Norwegian industrial democracy context can be favorable for combining TQM and BPR. My case studies show that company-wide employee participation and involvement can support the implementation of TQM and BPR, as well as their integration. Adequate tools for direct and indirect employee participation and a holistic framework for coordinating these two forms of employee participation at different company level contribute to that. A well-functioning indirect channel for employee participation is important for following up ideas from the shop floor level. I also find that direct employee participation in change processes at socio-technical level trains and motivates employees for contributing at strategic/higher company level. However, some basic conditions need to be present if employee participation shall contribute to successful change processes. These are participatory leadership, mutual trust between employers and employees, trade unions willing to spend time on business development, available time and a sense of need.

TQM and BPR literature advise the establishment of specific arrangements such as steering committees, advisory groups, project groups, etc. for concept implementation. My research shows that the functions of these specific arrangements can be performed by existing governing and cooperative bodies such as management groups, works councils, department councils, work environment committees and other arrangements that involve leaders, shop stewards and safety delegates. If arranged for, co-operative arenas ease the introduction and follow-up of the change processes. Additional time-consuming arrangements are not needed.

To support the development of a process centered organization, to build a process oriented culture and to train employees to become process professionals some methodology is needed. Several tools have been tested out through my case studies. E.g., I found the Actor-Process

Diagram that was developed for the research program appropriate for redesigning business – and work-processes. Among others, the Diagram and the way it was applied made people understand that in doing their work they contribute to the performance of a process and not isolated tasks. Smarter use of IT, process-oriented organizational changes and new procedures was a result of the change processes and supported the introduction of new processes. The Actor-Process Diagram proved useful for the efforts of integrating incremental and radical change, among others by integrating tools from TQM and BPR.

Hopefully the discussions of theory, methodological approaches, practical solutions and experiences that have been presented in this book can be of value to the further development and future applications of current change management concepts. Most actual are concepts where the integration of the quality and the process management movement is important, such as Business Process Management (BPM).

References list

Aker Olje og Gass Teknologi (1994)

Det handler om … (It is about…..), PDC, Aurskog

Aker Stord A/S (1994)

Total Kvalitet i Aker Stord (Total Quality at Aker Stord),

Aker Stord (2000), Stord

"Framtidas arbeidsorganisering" (The Future Work Organization), *Aker Stord bulletin* 4/00, Stord

Al-Mashari, M. & Zairi, M. (2000)

"Revisiting BPR: A holistic Review of Practice and Development", *Business Process Management Journal* 6, 1: 10-36

Almond, G.A. & Verba, S. (1965)

The Civic Culture, Little Brown & Co, Boston

Andriopoulos, C. & M.W. Lewis (2009)

"Exploitation-Exploration Tensions and Organizational Ambidexterity: Managing paradoxes of Innovation," *Organization Science* 20, 4: 696-717

Anthony, S. D. (2004)

"Disruption is a moving Target", *Strategy and Innovation* 2, 3: 4-6

Archer, R. (1996)

"BPR Methodology Survey Summary Findings", Paper presented at the European Academic Conference on Business Process Reengineering, Cranfield, United Kingdom

Argyris, C. (1977)

"Double Loop Learning in Organizations", *Harvard Business Review*, September–October: 115 -124

Argyris, C. (1991)

"Teaching smart People how to learn", *Harvard Business Review* May–June: 99-109

Argyris, C. (1992)

On Organizational Learning, Oxford University Press, Cambridge.

Argyris, C., Schön, D.A. (1978)

Organizational Learning, Addison-Wesley, Reading, Massachusetts

Argyris, A. & Schön, D.A. (1996)

Organizational Learning 2, Addison Wesley Publishing Company, Wokingham

Australian Community of Practice (2013)

BMP Collaboration, http://www.bpm-roundtable.com, Quensland University of Technology, Brisbane

Banke, P. & Nordskøv, E.C. (2000)

Job Development in Denmark: The historical Framework for Employee initiated Change Processes, Innoflex project

Bashein, B., Markus M. & Riley, P. (1994)

"Precondition for BPR success and how to prevent Failures", *Information Systems Management,* spring 11, 2: 7-13

Berg, T. (ed.) (1983)

Deltagerdemokratiet (Participatory Democracy), Universitetsforlaget, Oslo

Bergman, B. & Klefsjø, B. (1994)

Quality – from Customer Needs to Customer Satisfaction, Studentlitteratur, Lund

Beveridge, W.I.B. (1951)

The Art of Scientific Investigation, William Heinemann, London

Bjelland, M., Thunes, I. & Gandrud, O. (1997)

Introduksjonshåndbok i prosessfornying (*Introduction Handbook in Process Change*), Aker Stord, Stord

Bjørnevik, P.E., Stevenson, B. & Tønnessen, T. (2005).

Referat fra oppstartskonferanse Titania A/S (Report from the start up Conference at Titania A/S), RF, Stavanger.

Boer, H., Gertsen, F. (2003)

"From Continuous Improvement to Continuous Innovation: A (retro) (per)spective". *International Journal of Technology Management* 26, 8: 805-827

Boer H., Kuhn, J. & Gertsen, F. (2006)

"Continuous Innovation. Managing Dualities through Co-ordination", *CINet Working Paper Series* WP2006-01, Aalborg

Brocke, J.V., Becker, J., Braccini, A. M., Butleris, R. & Hofreiter, B. (2011)

"Current and Future Issues in BPM Research: A European Perspective from the ERCIS meeting 2010", *Communications of the Association for Information Systems* 28: 393-415

Brown, S. & Eisenhardt, K. (1998)

"Product Development: Past Research, present Findings and Future Directions", *Academy of Management Research* 20, 2: 343-378

Bruun, N., Hyden, H., Flodgren, B., Nielsen, R. & Halvorsen, M. (1989)

Den Nordiske Modellen, Liber, Lund

Brønnøysund registeret (2009)

"Titania AS – Regnskapsinformasjon" (Company Accounts), *Brønnøysund Registeret,* Brønnøysund.

Bråten, S. (1972)

"Model Monopoly and Communication: Systems Theoretical Notes on Democratization", *Acta Sociologica* 16, 2: 98-107

Buchanan, D. & Huczunzky, A. (1999)

Organizational Behaviour, Prentice Hall, London

Butler, C. (1994)

"The role of IT facilitating BPR: Observations from the Literature", in: Classon, B., Havryszkiewycs, I., Underwood B., Weber, R. (eds.) *Business Process Reengineering: Information Systems Opportunities and Challenges*, A-54: 60-147, IFIP, Holland

Burns, T., Stalker, G.M. (1961)

The Management of Innovations, Tavistock Publications, London

Buytendorp, H. & Hansen, K. (1997)

Konferanserapport 3 BU 2000 TfS nettverket (Conference Report 3 ED 2000 the TfS Network), Rapport RF-97/046, Stavanger

Børve, H. (1998)

Konferanserapport 6 BU 2000, IfS nettverket (Conference Report 6 ED 2000 TfS Network), Rapport RF-98/199, Stavanger

Cabana, S., Emery, F. & Emery, M. (1995)

"The Search for Effective Strategic Planning is over", *Journal of Quality and Participation,* July/August: 10-19

Cabana S. & Fiero, J. (1995)

"Motorola, Strategic Planning and the Search Conference", *Journal of Quality and Participation,* July/August: 22-31

Campbell, A. (1995)

"Mission, Vision and Strategy Development" in Creiner S. (ed.) *The Financial Time. Handbook of Management*: 139-149, Pitman Publishing, London

Campbell, D.T. (1975)

"Degree of Freedom and Case study", *Comparative political Studies*, 8: 178-193

Cardinal, L.B. (2001)

"Technological Innovation in the Pharmaceutical Industry: The Use of Organizational Control in Managing Research and Development", *Organization Science*, 12, 1: 19-36

Chesbrough, H.W. (2003)

"Thriving in the era of Open Innovation", *MIT Sloan Management Review,* 44, 3: 33-41

Christensen, C.M. (2003)

The Innovator's Solution, Harvard Business School Press, Boston, Massachusetts

CINet (2006)

Continuous Innovation Network, www.continuous-innovation.net

Claussen, T. (2001)

Arbeidslivets klassikere og dagens arbeidsliv (The Work -Life Classics and todays Work Life), Unipub Forlag, Oslo

Coch, L., French, J.R.P. (1953)

"Overcoming Resistance to Change", in Cartwright, D, & Zander A.

(eds) *Group dynamics: Research and Theory,* Tavistock Publications, London

Cole, G.D.H (1920)

Social Theory, Methuen, London

Cole, R. (1993)

"The leadership, Organization and co-determination Program and its Evaluation: a comparative Perspective", in Naschold, F., Cole, R., Gustavsen, B. & Beinum, H.V. *Constructing the new Industrial Society,* Van Gorcum, Assen/Mastricht & Arbetslivscentrum, Stockholm

Cole, R. (1994)

"Reengineering the Corporation: A Review Essay", *Quality Management Journal,* July 1994: 77-85

Cole, R., Bacdayan, P., White, J. (1995)

"Quality, Participation and Competitiveness" in Cole, R. E. (ed.)

The Death and life of the American Quality Movement, Oxford University Press, Oxford

Cole, R. (2001)

"From Continuous Improvement to Continuous Innovation", *Quality Management Journal* 8, 4: 1-22

Cole, R. (2002)

"From Continuous Improvement to Continuous Innovation", *Business Excellence. Make it happen,* Verona: The 7th World Congress for TQM, Proceedings: 3-8

Confederation of Norwegian Businesses and Industry – NHO & Norwegian Confederation of Trade Unions – LO (2006), *Basic Agreement* 2006 – 2009, Oslo

Conti, T. (1997)

"Optimizing Self-assessment", Kanji, G. K. (ed.) *Total Quality Managemen,. Special Issue, Proceedings for the 2nd World Congress for Total Quality Management, The Quality Journey* 8, 2&3 June: 5-15, Carfax Publishers, Abingdon

Continuous Innovation Network (2006)

"Introduction", www.continuous-innovation.net

Cooper, R. & Markus, M. (1995)

"Human engineering", *Sloan Management Review*, Summer: 39-50

Corso, M. (2002)

"From Product Development to continuous Product Innovation: Mapping the Routes of corporate Knowledge", *International Journal of Technology Management* 23, 4: 322-340.

Coughlan, P., Coghlan, C. (2002)

"Action Research for Operation Management", *International Journal of Operations & Production Management* 22, 2: 230-240

Coulsen-Thomas, C. (ed.) (1995)

The Responsive Organization. Reengineering new patterns of Work, Policy Publications, London

Crosby, P.B. (1979)

Quality is free, McGraw-Hill Inc., New York

Csikszentmihalyi, M. (1979)

Creativity: Flow and the Psychology of Discovery and Invention, Harper Collins, New York

Curry, J. & Kenney, M. (1997)

"Beating the Clock: Corporate Response to the rapid Change in the PC Industry", *California Management Review* 42, 1: 8-36

Dahl, R. (1970)

Etter Revolusjonen. Makt og medbestemmelse i et godt samfunn (After the Revolution. Power and Co-determination in a good Society), Dreyer Forlag, Oslo

Dahl, R. (1992)

Økonomisk Demokrati – Den neste utfordringen (Economic Democracy – The next Challenge), Ad Notam Gyldendal, Oslo

Daly, J. (1994)

"For beta or for worse", *Forbes*, ASAP 5, December: 36-40

Davenport, T.H. & Short, J. (1990),

"The New Industrial Engineering: Information Technology and Business Process Redesign", *Sloan Management Review*, Summer 1990, pp. 11-27

Davenport, T.H. (1993)

"Need radical Innovation and Continuous Improvement? Integrated Process Re-engineering and TQM", *Planning Review,* May/June: 6-12

Davenport, T.H. (1993)

Process Innovation. Reengineering Work through Information Technology, Harvard Business School Press, Boston

David, P.A. (1985)

"Clio and Economics of OWERTY", *American Economic Review*, 75: 332-337

Deakins, E. & Makgill, H. (1997)

"What killed BPR? Some Evidence from the Literature", *Business Process Management Journal,* 3, 1: 81-107

Deming, W. E. (1986)

Out of Crisi,. Cambridge University Press, Cambridge, Massachusetts

Duncan, R. (1976)

"The Ambidextrous Organization: Designing dual structures for Innovation", in Kilmann R.H., Pondy P.D. & Slevin D.P. (eds.) (1976), *The Management of Organization Design,* Vol. 1: 167-188, New Holland, New York

Ebeltoft, A. (1991)

Dialogkonferanser (Dialogue Conferences), Arbeidsforskningsinstituttet, Rapport 1/91, Oslo

Edwards, R. (1979)

Contested Terrain. The transformation of the Work Place in the 20th Century, Basic Books Inc., New York

Eidstuen, H. (1997)

Konferanserapport 4. BU 2000 TfS nettverket (Conference Report 4 ED 2000 TfS Network), Rapport RF-97/138, Stavanger

Eidstuen, H. (1997a)

Konferanserapport 5. BU 2000 TfS nettverket (Conference Report 5 ED 2000 TfS Network),. Rapport RF-97/299, Stavanger

Eidstuen, H. (1998)

Konferanserapport 7. BU 2000, IfS nettverket (Conference Report 7 ED 2000 TfS Network),. Rapport RF-98/300, Stavanger

Eisenhardt, K. M. (1989)

"Building Theories from Case Study Research", *Academy of Management Review* 14, 4: 532-550

Emery, M. & Purser, E.R. (1995)

The Search Conference: A Powerful Method for Planning Organizational Change and Community Action, Jossey Bass Publishers, San Francisco

Engelstad, P. (1996)

"The Development Organization as Communicative Instrumentation: Experiences from the Karlstad program", in Toulmin, S., & Gustavsen, *B. Beyond Theory. Changing Organizations through Participation,* John Benjamin's Publishing Company, Amsterdam/Philadelphia

Ernst & Young & Willoch, B.E. (1996)

BPR Norden, Institutet for Process Innovation, Stockholm

European Foundation for Quality Management – EFQM (2003)

The Fundamental Concepts of Excellence, EFQM, Brussels

European Foundation for Quality Management – EFQM (2003a)

EFQM Excellence Model, Large Company, Operational and Business Unit Version, Brussels

European Foundation for Quality Management – EFQM (2006)

Committed to Excellence, Information Brochure, EFQM, Brussels

European Foundation for Quality Management – EFQM (2006a)

The EFQM Excellence Award, Information Brochure for 2006, Brussels

Falkum E., Eldring, L., & Colbjørnsen, T. (1999) (eds)

Medbestemmelse og medvirkning (Co-determination and Participation), FAFO, Oslo

Fang, C., J. Lee & M.A. Schilling (2010)

"Balancing exploration and Explotation through Structural design: The Isolation of Subgroups and Organizational Learning", *Organization Science* 21, 3: 625-642

Feigenbaum, V.A. (1991)

Total Quality Control, McGraw-Hill Inc, New York

Fellesforbundet & TBL (1996)

Verkstedsoverenskomsten 94/96 (The general Agreement for Manufacturing Companies 94/96), Oslo

Fine, C. (1998)

Clock speed, Perseus Books, Reading, Massachusetts

Finne, H. & Nilssen, T. (1998)

"De selvstyrte gruppene ved Siemens i Trondheim" (Self managed Work Groups at Siemens in Trondheim), in Nilssen, T.: *Mot et bedre arbeidsliv (Towards a better Work Life),* Fagbokforlaget, Bergen

Finster, M. P. (2001)

"Creating Continuous Innovation", *Quality Management Journa,* 8, 4: 22-26

Flyvbjerg, B. (2006)

"Five Misunderstandings about Case Study Research", *Qualitative Inquiry,* 12: 219-244

French, J.R.P., Israel, J. & Aas, D. (1960)

"An experiment in Participation in a Norwegian Factory", *Human relations* 13, 1: 3-19

Friberg, O. (1990)

Arbeidsmiljøloven (The Work Environment Act), Tano, Oslo

Florida, R. & Kennedy, M. (1990)

The Breakthrough Illusion, Basic Books, New York

Gandrud (2001).

Pilotering av ny arbeidsorganisering for Aker Stord (Piloting of a new Work Organization for Aker Stord), Aker Maritime, Stord.

Gandrud, O., Haga & Tønnessen, T. (1997)

Ansattes medvirkning i Total Kvalitet. (Employee Participation in Total Quality), Revisjonsrapport Aker Stord KR-97-023, Stord

Gandrud, O., Haga, & Tønnessen, T. (2000)

Prosessforbedring på norsk – Et rammeverk. (Business Process Reengineering the Norwegian Way), Industrinettverket A/S, Stord

Gandrud, O., Haga, & Tønnessen, T. (2004)

Pilotprosjektet "Utviklingsansvarlig i SMB" (The Pilot project "Business Development Facilitator in SME"), Rogalandsforskning, Report RF2004/010, Stavanger

Galloway, D. (1994)

Mapping Work Processes, Quality Press, Milwaukee

Geertz, C. (1995)

After the Fact: Two Countries, four Decades, one Anthropologist, Harvard University Press, Cambridge

Gibson, C.B. & J. Birkinshaw (2004)

"The antecedents, consequences and mediating role of organizational ambidexterity", *Academy of Management Journal* 47, 2:209-226

Goffman, E. (1963)
Behaviour in Public Places: Notes on the social Organization of Gatherings, Free Press, New York

Goldthorpe, J., Lockwood, D., Beckhofer F. & Platt, J. (1968-1969)
The Affluent Worker, Cambridge University Press, Cambridge

Gomory, R. (1989)
"From the Ladder of Science to the Product Development Cycle", *Harvard Business Review* 67, 6: 99-105

Greenwood, D. & Levin, M. (1998)
Introduction to Action Research, Sage Publications, Thousand Oaks, New Dehli/London

Greve, H. (2007)
"Exploration and Exploitation in Product Innovation", *Industrial and Corporate change* 16, 5: 945-975

Grønnhaug, K. & Hansen, K. (eds) (2001)
Medvirkning, læring og konkurranseevne (Participation, Learning and Competitiveness), Fagbokforlaget, Bergen

Gummesson, E. (2000)
Qualitative Methods in Management Research, (2nd.edition), Sage, Thousand Oaks, California

Gustavsen, B. & Hunnius, G. (1981)
New Patterns of Work Reform. The case if Norway, The University Press, Oslo

Gustavsen, B. (1990)
Strategier for utvikling i arbeidslivet (Strategies for Work Place Development), Tano, Lillehammer

Gustavsen, B. (1992)
Dialogue and Development, Arbetslivscentrum, Van Gorcum, Assen Maastricht

Gustavsen, B. (1993)
"Creating Productive Structures: The Role of Research and Development", Naschold, F, Cole R., Gustavsen, B., & Beinum, H., *Constructing the New Industrial Society,* The Swedish Center for Working Life, Van Gorcum Assen, Stockholm/Maastricht

Gustavsen, B., Hofmaier, B., Ekman Philips, M. & Wikman, A. (1996)
Concept driven Development and the Organization of the Process of Change. An Evaluation of the Swedish Working Life Fund, John Benjamins Publishing Company, Amsterdam/Philadelphia

Gustavsen, B., Colbjørnsen, T. & Pålshaugen, Ø. (1998)
Development Coalitions in Working Life, John Benjamin's Publishing Company, Amsterdam/Philadelphia

Gustavsen, B. (2002).
Mellom tekst og virkelighet. Samarbeid om utvikling mellom bedrifter og forskning (Between Texts and Reality. The Cooperation between Enterprises and Research), Norges Forskningsråd, Oslo

Haga, T., Gjerstad, T., Sjursen R., Bjørnevik, P.E., Jøsendal, K. & Tønnessen, T. (2005)

Utviklingskoalisjonen Hordaland/Rogaland. Innovasjon og verdiskaping gjennom regionale partnerskap og lokale nettverk. Sluttrapport 2005 (The Development Coalition Hordaland/Rogaland. Innovation and Value Creation through Regional Partnership and Local Networks. Final report 2005), International Research Institute of Stavanger

Hammond, J. & Morrison, J. (1996)

The Stuff Americans are made of, Macmillan, New York

Hammer, M. & Champy, J. (1993)

Reengineering the Corporation, HarperCollins Publishers, New York

Hammer, M. & Stanton, S.A. (1995)

The Reengineering Revolution. A Handbook, Harper Collins Business, London

Hammer, M. (1996)

Beyond Reengineering, HarperCollingsBusiness, London

Hammer, M. (2010)

"What is Business Process Management?", in Brocke, J,V. & Rosemann M. *Handbook on Business Process Management 1*: 3-16. Springer-Verlag, Berlin Heidelberg

Handfield, R., Gosh, S. & Fawcett, S. (1998)

"Quality-Driven Change and Its Effects on Financial Performance", *Quality Management Journal* 5, 3: 13-30

Hansen, K. (1996)

Referat fra konferansen Forum for Bedriftsutvikling (Report from the Conference Forum for Business Development), Rapport RF-96/367, Stavanger

Hansen, K. & Sun, H. (1997)

Kvalitetsledelse i praksis. Resultater fra en spørreskjemaundersøkelse blant medlemsbedriftene i Norsk Forening for Kvalitet (Quality Management in Practice. Results from a Survey among the Member Companies in the Norwegian Association for Quality Management), RF97/143, Stavanger

Hansen, K. (1997)

"TQM og bedriftsdemokrati" (TQM and Industrial Democracy), in Bredal, D. (ed.) *Kvalitet I Norge (Quality in Norway),* Norden Publishing House Ltd. & Norsk Forening for Kvalitet og ledelse, Oslo

Hansen, K., Sun, H. & Tønnessen, T. (1996)

Internasjonal studie av kvalitetsledelse i praksis. Spørreskjema. (International Study of Quality Management in practice. Questionnaire), Stavanger University College & Rogaland Research, Stavanger

Hardie, N. (1998)

"The Effects of Quality on Business Performance", *Quality Management Journal* 5, 3: 65-83

Harrison, A. (1995)

"Business Processes: Their Nature and Properties", in Burke, G. & Peppard, J. (ed.) *Examining Business Process Re-engineering – current perspectives and Research Directions,* The Cranfield Management Series, London

Hart, C.W. & Bogan, C.E. (1992)

The Baldridge, McGraw-Hill Inc., New York

Hellevik, O. (1994)

Forskningsmetode i sosiologi og statsvitenskap (Research Methodology in Sociology and Political Science), Universitetsforlaget, Oslo

Hendrick K. & Benner, L. (1987)

Investigating Accidents with STEP, Marcel Decker Inc., New York

Henkoff, R. (1989)

"What Motorola learns from Japan", *Fortune* 119, 9: 157-168

Hertog, D.J.F. & Mari, C. (2000)

Management of Change and Human Resources, EUROFER, Brussels

Hertog, D.J.F. & Verbruggen, V. (2000)

Organizational Innovations in the Netherlands, Innoflex project

Hertog, D.J.F, Huzzard, T. & Hague, J. (2004)

"Revisiting QWL. The Problem and Prospects of Convergence in Europe", in *Promoting new Forms of Work Organization and other cooperative Arrangements for Competitiveness. Proceeding of the 1st International Conference*: 7-30, National and Kapodistrian University of Athens, Greece

Hill, S. (1991)

"Why Quality Circles Failed but Total Quality Management Might Succeed", *British Journal of Industrial Relations* 29, 4: 541-568

Holbek, J. (1983)

Foretaksstrategi (Business Strategy), Bedriftsøkonomens Forlag, Oslo

Holter, H. (1964)

"Øket medvirkning – de ansattes muligheter og ønsker" (Increased Participation – Employees' Opportunities and Wishes), in Thorsrud, E. & Emery F (eds.) *Industrielt demokrati (Industrial Democracy):* 134-137, Universitetsforlaget, Oslo

Hovedorganisasjonenes Fellestiltak Bedriftsutvikling – HFB (1993)

Samarbeid om bedriftsutvikling (Business Development Cooperation), NHO/LO, Oslo

Howell W. & Dipboye, R. (1986)

Essentials of Industrial & Organizational Psychology, The Dorsey Press, Chicago

Hussey, D. (1994)

Strategic Management. Theory and Practice, Elsevier, Oxford

Hovedorganisasjonenes Fellestiltak (2005)

Medvirkning til utvikling (Employee Participation in Development), NHO/LO, Oslo

Huzzard, T. (2000)

*The Convergence of QWL and Competitiveness: A current Swedish Literature Review,*Innoflex Project, National Institute of Working Life, Stockholm, Sweden

Leydesdorff, L. & Etzkowitz, H. (1998)

"Emergence of a Triple Helix of University-Industry-Government Relations", *Science and Public Policy* 23: 279-286

Leana, C.R., Locke, E.A. & Schweiger, D.M (1990)

"Fact and Fiction in analyzing Research on Participatory Decision Making: A Critique of Cotton, Vollrath, F., Lengnick-Hall and Jennings", *Academy of Management Review* 15: 137-146

Imai, M. (1986)

Kaizen, McGraw-Hill Publishing Company, New York

Ishikawa, K. (1982)

Guide to Quality Contro,. Asian Productivity Press, Tokyo

Ishikawa, K. (1985)

What is Total Quality Control? Prentice Hall Inc., Englewood Cliffs, New York

Isobe, T. (2000)

Report on Strategies of small Champions, Higashi-Osaka Chambers of Commerce and Industry, Higashi-Osaka, Japan

Isobe, T., Makino, S., M., Montgomery, D.B. & Chian, L.K. (2007)

"Technological Capabilities and Firm Performance: The Case of small Manufacturing Companies in Japan", *Research Paper Series,* Stanford Graduate School of Business, Research Paper No. 1980, Stanford. USA

Juran, J.M. (1989)

Juran on Leadership for Quality, McGraw-Hill, New York

Juran, J.M. (1995)

A History of managing for Quality,. Quality Press, Milwaukee

Jacoby, N. (2005)

"Exploration and Exploitation Strategies. What kind of analytical models?", *Cahiers de la MSE,* 2005.4, Sorbonne Université Paris, France

Kalleberg, R. (1992)

Kontruktiv samfunnsvitenskap (Constructive Social Science), ISO Rapport no. 4, The University of Oslo

Kanter, R. (1988)

"When thousand Flowers bloom", *Research in Organizational Behaviour* 10: 169-211

Karlsen, J. I. & Munkeby, I. (1998)

"Den norske samarbeidsmodellen" (The Norwegian Cooperative Model), in Nilssen, T. (ed.) *Mot et bedre arbeidsliv. (Towards a better Work Life),* Fagbokforlaget, Bergen

Kelada, J.N. (1996)

Integrating Reengineering with Total Quality, ASQC, Wisconsin.

Kelly, J.E. (1982)
Scientific Management, Job Redesign and Work Performance, Academic Press, London

Kemmis, S. & McTaggart, R. (2000)
"Participatory Action Research : Communicative action and Public Sphere", in Denzin, S. & Lincoln Y.S.(eds.) *The Sage Handbook of Qualitative Research,* Sage Publications, Thousand Oaks, London/New Dehli

Knights, D.& Willmott, H. (2000)
"The Reengineering Revolution. An Introduction", in Knights, D. & Willmott, H. (eds.) *The Reengineering Revolution: Critical Studies of Corporate Change,* Sage, London.

Koch, C., Vogelius, P. & Manske, F. (1999)
"Er ledelseskonceptet Business Process Reengineering vejen til et bedre arbejde?" (Is the Management Concept Business Process Reengineering the Way to a better Work Life), in Hvid, H. (ed) *Ressorcer og velfærd I arbejdslivet* (*Resources and Welfare in Work Life),* Frydenlund, Copenhagen

Kobasa, S.C. (1979)
"Stressful Life Events, Personality, and Health: An Inquiry into Hardiness", *Personality and Social Psychology* 37: 1-11

Levinthal, D.A. & March, J.G. (1993)
"The Myopia of Learning", *Strategic Management Journal* 14: 95-112

Levitt, B. & March, J. G. (1988)
"Organizational Learning", *Annual Review of Sociology* 14: 319-340.

Lewin. K. (1946)
"Action Research and Minority Problems", *Journal of social issues* 2, 4: 34-46

Lindøe, P. (1995)
Referat fra konferansen om bedriftsutvikling (Report from the Conference about Business Development), RF report, Stavanger

Lindøe, P. (2001)
Arbeidsmiljø og prosessforbedring (Work Environment and Process Improvement), RF-2991/172, Stavanger

Lindøe, P., Karlsen, J.E. & Lie, T. (2001)
Et nordisk grep på arbeidsmiljøregulering? (A Nordic Model for Work Environment Regulations?), Nordisk Ministerråd, København

Lorange, P., Chakawarthya, B., Roos, J. & van de Ven, A. (1994)
Implementing Strategic Processes. Change, Learning and Co-operation, Blackwell, Oxford

LO & NHO (2005)
Hovedorganisasjonenes Fellestiltak. Strategi (The main Partners' common Project. Strategy), Oslo

LO-NHO (2006)
Hovedavtalen 2003-2006 (The Main Agreement 2003-2005), Oslo.

Lowenthal, J. (1994)
Reengineering the Organization, Quality Press, Milwaukee

Lynn, L. (1982)
How Japan innovates, Westview Press, Boulder, Colorado

Magnusson, K., Kroslid, D. & Bergman, B. (2003)
Six Sigma. The pragmatic Approach, Studentlitteratur, Lund, Sweden

Lund Mannheim, K. (1956)
Essays on the Sociology of Culture, Oxford University Press, Oxford

Mak, W. M. (1999)
"Cultivating a quality Mind Set", *Total Quality Management.* 10: 622– 626

March, J.G. (1991)
"Exploration and Exploitation in Organizational Learning", *Organization Science* 2, 1: 71-87

March, J.G. (1994)
A primer on Decision Making, Free press, New York

March, J.G. (1995)
"The Future, Disposable Organizations and Rigidities of Imagination", *Organization* 2: 427-440.

McDonald, J. (1992)
"Reasons for Failure". *The TQM Magazine*, August: 237-240

Mill, J.S. (1965)
"The Principles of Political Economy with some of their Applications to Social Philosophy", in Robson J.M. (ed.) *The Collected Works of John Stuart Mill*, Volume II, books I-II (originally published 1848), Routledge, London,

Mintzberg, H. (1994)
The Rise and Fall of Strategic Planning, Prentice Hall, New York

Moltu, B., Monteiro, E., Holtan, E. & Sørensen, K. (2000)
BPR is dead. Long live the Process. The Uptake of BPR in Norway, Working paper 2/2000, Senter for Teknologi og Samfunn, SINTEF, Trondheim

Moltu, B. (2000)
"Revolusjonen som spreidde seg. Business Process Reengineering. Om den nyttige dobbelttydinga" (The Revolution that spread. Business Process Reengineering. About the useful double Meaning), in Rolfsen, M. (ed.) *Trendenes Tyranni (The Tyranny of Fashions),* Fagbokforlaget, Bergen

Montgomery, D.B., Isobe, T. & Makino, S. (2007)
"Technological Capabilities and Firm Performance: The Case of Small Manufacturing Firms in Japan", *Stanford University Graduate School of Business Research Paper No. 1980*, December

Naschold, F., Cole, R., Gustavsen, B. & Beinum, H. (1993)
Constructing the New Industrial Society, Van Gorcum, Assen/Maastricht/ Arbetslivscentrum, Stockholm

Neerland, H. (2000)
"Kvalt av kvalitet. Total Kvalitetsledelse" (Choked by Quality. Total Quality Management), in Rolfsen, M. (ed.) *Trendenes Tyranni (The Tyranny of Fashions),* Fagbokforlaget, Bergen

Nietzche, F. (1969)
Ecco Homo, Vintage, New York

Nolan, T.W., Schall, M.W., Berwick, D.M. & Roessner, J. (1996)
Reducing delays and Waiting Times throughout the Health Care System, Institute for Health Care Promotion, Boston

Norsk Hydro (2006)
The Hydro Way, Bergen

Nylehn, B. (1994)
"Den som intet våger, vinner ikke. Om den uoppnåelige medbestemmelsen" (The one who do not dare, do not win. About the unachievable Co-determination), *Sosiologisk Tidsskrift* 1: 3-22

Olberg, D. (ed.) (1995)
Endringer i arbeidslivets organisering (Changes in the Organization of Work Life), FAFO, Oslo

Oshri, I, Pan, S.L. & Newell, S. (2005)
"Trade-offs between Knowledge Exploitation and Exploration Activities", *Knowledge Management Research & Practice* 3: 10-23

Ouchi, W. (1980)
"Market, Bureaucracies and Clans", *Administrative Science Quarterly* 25, 1: 129-141

PA Consulting Group (1991)
Trening for Totalkvalitet (Training for Total Quality), PA Consulting Group, Oslo

Pateman, C. (1970)
Participation and Democratic Theory, Cambridge University Press, Cambridge

Peattie, I. (2001)
"Theorizing Planning: some Comments on Flyvbjerg's Rationality and power", *International Planning Studies* 6, 3: 257-262

Petersen, A.H., Boer, H. & Gertsen, F. (2004)
"Learning in different Modes: The Interaction between incremental and radical Change", *Knowledge and Process Management* 11, 2: 1-11

Pfeffer, J. (1994)
Competitive Advantage through People, Harvard Business School Press, Boston

Politi og Justisdepartementet (2011)
Lov om Aksjeselskaper (The Norwegian Limited Companies act), Oslo

Popper, K. (1959)

The Logic of Scientific Discovery, Basic Books, New York

Pålshaugen, Ø. (1991)

Som sagt, så gjort. (As said, so done), Novus Forlag, Oslo

Pålshaugen, Ø. (1996)

"Globaler Wettbewerd und lokale Entwicklungsorganization", in Denisow, K., Fricke, W. & Stieler-Lorentz, B. (Hrsg.): *Partizipation und Produktivitet. Zu einigen kulturellen Aspekten der Okonomi,* Friedrich EbertStiftung, Bonn.

Pålshaugen, Ø. & Quale, T. (2000)

Forskning og Bedriftsutvikling (Research and Business Development), The Work Research Institute, Oslo

Quale, T.U. (1995)

"Ledelse – fra administrasjon til omstilling" (Management – from Administration to Change), in Olberg, D. (ed.) *Endringer i arbeidslivets organisering. (Changes in the Organization of Work Life),* FAFO rapport 183, Oslo

Ragin, C.C. (1992)

"What is a Case?", Ragin, C.C. & Becker, H.S. (eds.), *What is a Case? Exploring the Foundations of Social Inquiry:* 1-18, Cambridge University Press, Cambridge

Raisch, S., Birkinshaw, J., Probst G. & Tushman, M.L. (2009)

"Organizational Ambidexterity: Balancing Exploitation and Exploration for sustained performance", *Organization Science,* vol. 20, no. 4, July–August

Reason, P. & Bradbury, H. (eds.) (2006)

Handbook of Action Research, Sage Publications, London

Remenyi, D., Williams, B, Money, A. & Schwartz, E. (1999)

Doing Research in Business and Management, Sage, London

Reve, T. (1994)

"Skandinavisk organisasjon og ledelse: Fra konkurranse fordel til konkurranse ulempe" (Scandinavian Organization and Management: From competitive Advantage to a Drawback), *Tidsskrift for samfunnsforskning* 4, 35: 568-582. Universitetsforlaget, Oslo

Riddel, D. (1968)

"Social Self-government: The Background of Theory and Practice in Yugoslav Socialism", *British Journal of Sociology* 19, 1: 47-75

Roberts, L. (1994)

Process Reengineering, ASQC Quality Press, Milwaukee, Wisconsin

Robinson, A. (1991)

Continuous Improvement in Operations, Productivity Press, Cambridge, Massachusetts

Rolfsen, M. (1994)

Japanisme på norsk (Japanisme in Norwegian), Fagbokforlaget, Bergen

Rolfsen, M. (2000)

Trendenes tyranni (The Tyranny of Management Fashions), Fagbokforlaget, Bergen

Roos, G., von Krogh, G. & Roos, J. (1996)
Innføring i strategi (Introduction to Strategy), Fagbokforlaget, Bergen

Rorty, R. (1985)
"Habermas and Lyotard on Postmodernity", in Bernstein, R.J. (ed.) *Cognition and Categorization*: 27-48, Lawrence Earlbaum, Hillsdale, New York

Rosemann, M. & Brocke, J. M. (20109
"The Six Core Elements of Business Process Management", in Brocke, J. V. & Rosemann, M. (eds.) *Handbook on Business Process Management 1*: 107-122, Springer-Verlag, Berlin Heidelberg

Rousseau, J.J. (1968)
The Social Contract, Cranston, M. (transl.), Penguin Books

Ruessmann, T., Preece, I., Peppard, J. (1994)
Tools and Methods in Business Process Redesign, Working paper, Information Systems Research Centre, Cranfield School of Management, Cranfield

Røvik, K.A. (1998)
Moderne organisasjoner (Modern Organizations), Fagbokforlaget, Bergen

Sacquepee, S., Dufau, M. (2001)
The Convergence of Quality of Working Life, Competitiveness and Organizational Change in France, Innoflex Project.

Schein, E.H. (1987)
Process Consultation. It's role in Organization Development, Addison-Wesley Publishing Company, Reading

Schein, E.H. (1992)
Organizational Culture and Leadership, 2d edition Jossey Bass, San Francisco

Schein, E.H. (2002)
"Notes Toward a Better Understanding of Process", *OD Practitioner* 34, 2: 1-6.

Schumpeter, J. (1934)
The Theory of Economic development, Oxford University Press, New York

Scott Morton, M. (ed.) (1991)
The Corporation of the Nineties. Information Technology and Organizational Transformation, Arrow, London

Senge, P. (1990)
The fifth Discipline: The Art and Practice of the Learning Organization, Doubleday/Currency, New York

Sitkin, S.B. & Stockel, D. (1996)
"The Road to Hell: The Dynamics of Distrust in an Era of Quality", in Kramer, R. M. & Tyler, T.R. (eds.) *Trust in Organizations: Frontiers of Theory and Research*, Sage, Thousand Oaks, California

Skjærpe, M. & Ewing, B.M. (2000)

Miljøvernrapport – Serigstad Jernstøperi (Work Environment Report – Serigstad Iron Foundry), Bryne Bedriftshelsetjeneste AS, Bryne

Stacey, R.D. (1992)

Managing the Unknowable: Strategic Boundaries between Order and Chaos in Organizations, Jossey-Bass, San Francisco

Stake, R.E. (1995)

The Art of Case Study Research, Thousand Oaks, California

Stake, R.E. (2000)

"Qualitative Case Studies", in Denzin, N.K. & Lincoln, Y.S. (eds.) *Handbook of Qualitative Research,* Sage, London

Statoil (2006)

We in Statoil, Stavanger

StatoilHydro (2008)

The StatoilHydro Book, Stavanger

Stephen, F.H. (1967)

"Management Structures and Industrial Relations in a Yugoslav shipyard", Unpublished paper, Glasgow University, cited in Pateman, C. (1970) *Participation and Democratic Theory,* Cambridge University Press, Cambridge

Sternberg, R.J. & Koligian, J. (1990)

Competence considered, New Haven, Yale University Press

Sun, H. (1997)

"A survey of Total Quality in Norway", in Bredal, D. (ed.) *Kvalitet i Norge (Quality in Norway),* Norden Publishing House Ltd. & Norsk Forening for Kvalitet og ledelse, Oslo

Sutcliffe, K., Sitkin S.B. & Browning L.D. (1997)

"Perspectives on Process Management: Implications for Research on Twenty-First Century Organizations", in Cooper C.L. & Jackson, S.E. *Creating Tomorrows Organizations,* John Wiley & Sons Ltd., New Jersey

Sutcliffe, K., Sitkin, S.B & Browning, L.D. (2000)

"Tailoring Process Management to situational Requirements; Beyond the Control and Exploration Dichotomy", in Cole, R. & Scott, R. (eds) *The Quality Movement and Organization Theory*, Sage Publications, Thousand Oaks, California

Sørensen, Aa. B. (1992)

"Aktionsforskning i og om arbejdslivet" (Action Research for and about Work Life), *Tidsskrift for Samfunnsforskning* 33: 213-230

Tenner, A.R. & DeToro, I.J. (1997)

Process Redesign. The Implementation Guide for Managers, Addison-Wesley, Reading, Massachussets

The Research Council of Norway (1995)

"Enterprise Development 2000" *Programme Memorandum:* 4, Oslo

The Research Council of Norway (2006)
"Value Creation 2010", *Programme Homepage*: 1. Oslo

Thorsrud, E. & Emery, F. (1964)
Industrielt demokrati (Industrial Democracy), Universitetsforlaget, Oslo

Thorsrud, E. & Emery, F.(1969)
Mot en ny bedriftsorganisasjon (Towards a new Business Organization), Tanum, Oslo

Thorsrud, E. (1982)
Samarbeidsprosjektet LO/NAF og utviklingen av nye former for arbeidsorganisasjon i Norge 1962-82 (The joint Project between the Norwegian Confederation of Trade Unions (LO) and the Confederation of Norwegian Businesses and Industry (NHO) about developing new Forms of Work Organization in Norway in the Period 1962-82), The Work Research Institute (AFI-WRI) Document no 11, Oslo

Trist, E. & Bamforth, K.W. (1951)
"Some Social and Psychological Consequences of the Longwall Method of Coal getting", *Human Relations* 4: 3-38

Tushman, M., Anderson, P. & O'Reilly, C. (1997)
"Technology Cycles, Innovation Streams and ambidextrous Organizations: Organizational Renewal through Innovation Streams and strategic Change", *Managing Strategic Innovation and Change,* Oxford University Press, New York

Tushman, M., Smith W.K., Wood R.C., Westerman, G. & O'Reilly, C. (2006)
Organizational Designs and Innovation Streams, Working paper 8, Harvard Business School, Boston.

Tønnessen, T. (1992)
Ledelsesfilosofien TQM og forbedringsmuligheter i boring. (The Management Philosophy TQM and potentials for improving Offshore Drilling Operations), RF SPS-A-14/85, Stavanger

Tønnessen, T. (1997)
"The International TQM Concept and Scandinavian Traditions of Industrial Democracy – Conflicting or mutually reinforcing Development Strategies?", Kanji, G. (ed.) *Total Quality Management. Special Issue. Proceedings for the 2nd World Congress for Total Quality Management. The Quality Journey* 8, 2&3: 286-290, Carfax Publishers, Abingdon

Tønnessen, T. (1997a)
Revisjonsrapport Aker Stord 1997 (Audit Report Aker Stord 1997), Unpublished, Stavanger

Tønnessen, T. (1999)
Revisjonsrapport Aker Stord 1999 (Audit Report Aker Stord 1999), Unpublished, Stavanger

Tønnessen, T. (2000)
"BU konferansen" (The Enterprise Development Conference), Pålshaugen, Ø. & Quale, T. (eds) *Forskning og Bedriftsutvikling – nye samarbeidsforsøk (Research and Development – new Cooperation Efforts),* The Work Research Institute, Oslo

Tønnessen, T. (2002)

"Innovation through Company Wide Employee Participation", *Business Excellence. Make it happen,* The 7th World Congress for Total Quality Management, Proceedings 1: 473-479

Tønnessen, T. (2005)

"Continuous Innovation through Company Wide Employee Participation", *The TQM Magazine* 17, 2: 195-207

Trist, E. (1963)

Organizational Choice, Tavistock Publications, London

Un, C.A. (2007)

"Managing the Innovators for Exploration and Exploration", *Journal of Technology Management & Innovation* 2, 3: 4-20.

Vakola, M. (2000)

"Critique of existing Business Process Management Methodologies", *Business Process Management Journal* 6, 3: 238-250

Verba, S. (1961)

Small Groups and political Behavior, Princeton University Press, Princeton

Vesey, J.T. (1991)

"The new Competitors: They think in term of Speed and Market", *Academy of Management Executive* 5, 2: 23-33

Volberda, H.W. (1998)

Building the flexible Firm, Oxford University Press, Oxford

Wieviorka, M. (1992)

"Case studies: History or Sociology?", in Ragin, C.C., Becker, H.S. (eds.) *What is a Case? Exploring the fundamental of Social Enquiry:* 159-172, Cambridge University Press, Cambridge

Voss, C., Blackmon, K. (1996)

"Quality Pay Off Comes", *European Quality* 3, 1: 28-30, European Quality Publications Ltd., London

West, M.A. (1998)

Impact of People Management Practices on Business Performance, IPD, London

Weston, F. C. (2001)

"The need for Continuous Innovation and Continuous Improvement", *Quality Management Journal,* 8, 2: 27-30

Weber, M. (1971)

Makt og Byråkrati. (Power and Bureaucracy), Østerberg, D. Translation to Norwegian from Wirtschaft und Gesellschaft 1922, part 3, Chapter 1, Gyldendal, Oslo

Wheatly, M., Szwejczewski, M. & New, C. (1996)

The Making of Britain's Best Factories, Management, Business Intelligence Ltd., London

Willoch, B.E. (1994)

Business Process reengineering – En praktisk innføring og veiledning (Business Process Reengineering – A practical Introduction and Guide), Fagbokforlaget, Bergen

Wood, R. & Bandura, A. (1989)

"Social Cognitive Theory of Organizational Management", *Academy of Management Review,* 14: 247-287

Yalcinkaya, G., Calantone, R. & Griffith, D.A. (2007)

"An Examination of Exploration and Exploitation Capabilities: Implications for Product Innovation and Market Performance", *Journal of International Marketing* 5, 4: 63-93

Yin, R.K. (2003)

Case Study Research. Design and Methods, Sage, Thousand Oaks, California

Yukl, G. (2006)

Leadership in Organizations Persons Prentice Hill, New York

Zairi, M. & Sinclair, D. (1995)

"Business Process re-engineering and Process Management: A Survey of current Practice and future Trends in Integrated Management", *Management Decision* 33: 3-16

Zairi, M., Letza, S.R. & Oakland, J. (1996)

"Does TQM Impact on Bottom Line Results?" *The TQM Magazine* 6, 1: 38-43. University Press, Bradford

Zaltmann, G.R., Duncan J. & Holbek J. (1973)

Innovation and Organizations, John Wiley and Sons, New York

Appendices

Appendix 1

The VC 2010 regional partnership Hordaland & Rogaland and the local networks

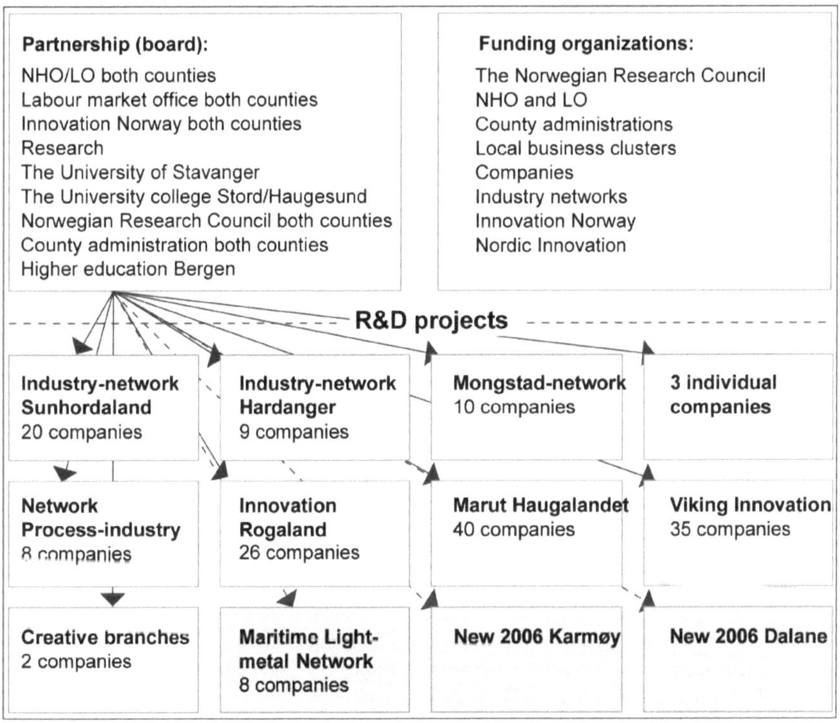

Appendix 2

Employee participation in TKA. Interview guides Aker Stord 1997

Employees

General knowledge about TKA

1. Are you familiar with the improvement process at Aker Stord?
2. Do you know about the 3 large improvement projects at Aker Stord, "Process improvement", "Governing documents" and "Administrative routines"?
3. What do you know about the action plan for "Aker Stord 2000"?
4. What do you know about the large improvement projects in the concern, "Training project managers", "Project start", "The piping process", "Better utilization of facilities in Norway", "Improved support processes", "Strategy for Aker Stord as part of "Large projects in Norway – SPIN""?

Information/training

5. Have you participated in training related to TKA, Department Committees, process improvement, EFQM self-evaluation or customer-client interviews? If no, continue with 9.
6. What do you think about the training you participated in?
7. Where any changes or actions taken after the training? Was the training followed up in any way?
8. Did you get any training in improvement techniques like brainstorming, improvement proposal schemes, cause-effect analysis, brown wall paper method or others? Have these been used afterwards?
9. Have your leader given any information about improvement activities like customer-client interviews, the large improvement projects at Aker Stord, Aker Stord 2000 or the large improvement projects in the concern?
10. Do you get information about the improvement activities in any other ways like bulletin boards, internal magazines, reports from meeting, email or other?
11. Have shop stewards given any information about the improvement process and different activities like customer-client interviews, brainstorming, the large improvement projects in the concern, the large improvement projects at Aker Stord, Aker Stord 2000 or others?
12. How do you get information and take part in TKA training when you work in the large

building projects?

13. Have the information about TKA and the TKA training changed in 1997 compared to 1996?

Idea generation

14. Have you participated in appraisal-interviews? If no, continue on 18.
15. What is prioritized in the appraisal-interviews and how is it followed up?
16. To what extent do you discuss and agree on individual development and improvement plans during the appraisal-interview?
17. What do you think about the appraisal-interview practice?
18. Have you participated in pre-job meetings? If no, continue on 20.
19. What do you think about these kinds of meetings?
20. Have you participated in brainstorming, team meetings, customer-client interviews, improvement project teams or process improvement teams?
21. What do you think about activities like these?
22. Do you know whether improvements proposals from brainstorming, team meetings, department meetings, customer-client interviews or other are being registered in the electronic proposal database?
23. What do you know about the improvement activity in the Department Committee?
24. How do you consider the Department Committee as a body for co-ordinating local improvement activities?
25. Do you propose ideas for improvement and how?
26. Is some of the time in department meetings, team meetings and similar kinds of meetings used for proposing and discussing improvement ideas? In that case, how is this carried out and followed up? What are the experiences with this kind of practice?
27. Do you feel that your own proposals for improvements are handled and followed up in a proper manner?
28. Have you participated in meetings where the goal has been to discuss how the department or gang work with improvements?
29. In that case, what is your experience with that kind of meetings?
30. Should the company increase employee participation in improvement activities?
31. Are you familiar with the suggestion box scheme?
32. What is your opinion about this arrangement?

33. What is your opinion about the system for reporting undesired events? Does reporting have any practical consequences? Does reporting and follow up give improvements?
34. The CEO had monthly meetings with 10 employees from different parts of the company in 1996? This practice has stopped. What is you opinion about that?
35. How is the possibility to participate in improvement activities and propose improvement ideas when you have your daily work in the large building projects?

Implementation/follow up of improvement proposals

36. Have you participated in the planning and implementation of improvement projects?
37. How is this work organized?
38. What is your opinion about the way improvement projects are carried out and followed up?
39. How is this work organized in the large building projects?
40. What is you experience with improvement proposals from employees that have been implemented?
41. Have the implemented proposals resulted in smarter ways to perform work?
42. Have the improvement proposals improved work environment and well being?
43. Has increased employee participation in improvement activities had any effect on shop stewards' work and role?
44. Do you feel that the time spent on employee participation in improvement activities is worthwhile compared to the outcome and the results of these activities?

The importance of the improvement process

45. Have the improvement process changed Aker Stord?
46. Have employee participation in – and attitudes to the improvement process changed compared to 1996?
47. Do you see any changes for you and your colleagues as a consequence of the new process orientation in the organization?
48. Do you think that the improvement process will last?
49. Do you think that the present process improvement activities at Aker Stord will have any consequence for your participation and engagement in the improvement process?
50. The process improvement projects give a flatter organization and fewer leaders than before? Are leaders willing to delegate more? Are employees capable and motivated for increased autonomy?

51. Do you think that the reorganization in the concern will weaken or strengthen Aker Stord's position?
52. What consequences do you think that the reorganization in the concern will have for shop stewards' possibility to have influence on the further development of Aker Stord?
53. Do you think that TKA has had any consequences for Aker Stord's capability to deliver quality to the right time in an efficient way
54. Has this changed in 1997 compared to 1996?
55. Do you have a TKA facilitator in your department or gang?
56. In that case, how is this facilitator role carried out and what is your experiences with this practice?
57. How much time weekly do you think that you use on participating in improvement activities?
58. Do you think that you possibility to influence on the further development of you own work situation has changed as a consequence of the improvement activities?
59. Do you think that shop stewards should participate and be engaged in the improvement activities? Why?
60. Do you think your leaders' willingness to listen to improvement proposals from employees has changed as a consequence of the improvement process?
61. What makes it difficult to further increase employee participation in the improvement activities?

Leaders

General knowledge about TKA

1. Are you familiar with the improvement process at Aker Stord?
2. Do you know about the 3 large improvement projects at Aker Stord, "Process improvement", "Governing documents" and "Administrative routines"?
3. What do you know about the action plan for "Aker Stord 2000"?
4. What do you know about the large improvement projects in the concern, "Training project managers", "Project start", "The piping process", "Better utilization of facilities in Norway", "Improved support processes", "Strategy for Aker Stord as part of "Large projects in Norway – SPIN"?

Information/training

5. Have you participated in training related to TKA, Department Committees, process improvement, EFQM self evaluation or customer-client interviews? If no, continue with 9.
6. What do you think about the training you participated in?
7. Where any changes or actions taken after the training? Was the training followed up in any way?
8. Did you get any training in improvement techniques like brainstorming, improvement proposal schemes, cause-effect analysis, brown wall paper method or others? Have these been used afterwards?
9. Have your leader given any information about improvement activities like customer-client interviews, the large improvement projects at Aker Stord, Aker Stord 2000 or the large improvement projects in the concern?
10. Do you get information about the improvement activities in any other ways like bulletin boards, internal magazines, reports from meeting, email or other?
11. Have shop stewards given any information about the improvement process and different activities like customer-client interviews, brainstorming, the large improvement projects in the concern, the large improvement projects at Aker Stord, Aker Stord 2000 or others?
12. How do you get information and take part in TKA training when you work in the large building projects?
13. Have the information about TKA and the TKA training changed in 1997 compared to 1996?

Idea generation

14. Have you participated in appraisal-interviews? If no, continue on 18.
15. What is prioritized in the appraisal-interviews and how is it followed up?
16. To what extent do you discuss and agree on individual development and improvement plans during the appraisal-interview?
17. What do you think about the appraisal-interview practice?
18. Have you participated in pre-job meetings? If no, continue on 20.
19. What do you think about these kinds of meetings?
20. Have you participated in brainstorming, team meetings, customer-client interviews, improvement project teams or process improvement teams?
21. What do you think about activities like these?

22. Do you know whether improvements proposals from brainstorming, team meetings, department meetings, customer-client interviews or other are being registered in the electronic proposal database?
23. What do you know about the improvement activity in the Department Committee?
24. How do you consider the Department Committee as a body for co-ordinating local improvement activities?
25. Do you propose ideas for improvement and how?
26. Is some of the time in department meetings, team meetings and similar kinds of meetings used for proposing and discussing improvement ideas? In that case, how is this carried out and followed up? What are the experiences with this kind of practice?
27. Do you feel that your own proposals for improvements are handled and followed up in a proper manner?
28. Have you participated in meetings where the goal has been to discuss how the department or gang work with improvements?
29. In that case, what is your experience with that kind of meetings?
30. Should the company increase employee participation in improvement activities?
31. How do you see your role as a leader as regards promoting employee participation in improvement activities
32. What can you as a leader do to promote employee participation in improvement activities?
33. What prevents you from promoting increased employee participation in improvement activities?
34. Should shop stewards have any role in promoting employee participation in improvement activities?
35. In that case, how?
36. Are you familiar with the suggestion box scheme?
37. What is your opinion about this arrangement?
38. What is your opinion about the system for reporting undesired events? Does reporting have any practical consequences? Does reporting and follow up give improvements?
39. The CEO had monthly meetings with 10 employees from different parts of the company in 1996? This practice has stopped. What is your opinion about that?
40. How is the possibility to participate in improvement activities and propose improvement ideas when you have your daily work in the large building projects?

Implementation/follow up of improvement proposals

41. Have you participated in the planning and implementation of improvement projects?
42. How is this work organized?
43. What is your opinion about the way improvement projects are carried out and followed up?
44. How do you follow up improvement proposals from employees in your department or gang?
45. How do you see your role as a leader in promoting employee participation in the implementation of improvement proposals?
46. What can you do to promote employee participation in the implementation of improvement projects?
47. How do you see you role as a leader in promoting improvement projects and groups across department and gang boarders?
48. What prevents you from doing this?
49. What have you done as a leader to promote the initiation of improvement projects and groups across department and gang boarders?
50. How is this work organized in the large building projects?
51. What is you experience with improvement proposals from employees that have been implemented?
52. Have the implemented proposals resulted in smarter ways to perform work?
53. Have the improvement proposals improved work environment and well being?
54. Has increased employee participation in improvement activities had any effect on shop stewards' work and role?
55. Do you feel that the time spent on employee participation in improvement activities is worthwhile compared to the outcome and the results of these activities?

The importance of the improvement process

56. Have the improvement process changed Aker Stord?
57. Have employee participation in – and attitudes to the improvement process changed compared to 1996?
58. Do you see any changes for you and your colleagues as a consequence of the new process orientation in the organization?
59. Do you think that the improvement process will last?

60. Do you think that the present process improvement activities at Aker Stord will have any consequence for your own and employees' participation and engagement in the improvement process?
61. The process improvement projects give a flatter organization and fewer leaders than before? Are leaders willing to delegate more? Are employees capable and motivated for increased autonomy?
62. Do you think that the reorganization in the concern will weaken or strengthen Aker Stord's position?
63. What consequences do you think that the reorganization in the concern will have for shop stewards' possibility to have influence on the further development of Aker Stord?
64. Do you think that TKA has had any consequences for Aker Stord's capability to deliver quality to the right time in an efficient way
65. Has this changed in 1997 compared to 1996?
66. Do you have a TKA facilitator in your department or gang?
67. In that case, how is this facilitator role carried out and what is your experiences with this practice?
68. How much time weekly do you think that you use on participating in improvement activities?
69. Have your engagement in the TKA process had any consequences for your own development as a leader?
70. Have the TKA process had any consequence for the people you are responsible for?

Shop stewards

General knowledge about TKA

1. Are you familiar with the improvement process at Aker Stord?
2. Do you know about the 3 large improvement projects at Aker Stord, "Process improvement", "Governing documents" and "Administrative routines"?
3. Have the trade unions participated in the planning, implementation and follow up of these projects?
4. What do you know about the action plan for "Aker Stord 2000"?
5. Have the trade unions participated in the planning, implementation and follow up of Aker Stord 2000?

6. What do you know about the large improvement projects in the concern, “Training project managers”, “Project start”, “The piping process”, “Better utilization of facilities in Norway”, “Improved support processes”, “Strategy for Aker Stord as part of “Large projects in Norway – SPIN””?

7. Have the trade unions participated in the planning, implementation and follow up of these large improvement projects in the concern?

8. In that case, how?

Information/training

9. Have you participated in training related to TKA, Department Committees, process improvement, EFQM self-evaluation or customer-client interviews? If no, continue with question 13.

10. What do you think about the training you participated in?

11. Where any changes or actions taken after the training? Was the training followed up in any way?

12. Did you get any training in improvement techniques like brainstorming, improvement proposal schemes, cause-effect analysis, brown wall paper method or others? Have these been used afterwards?

13. Have your leader given any information about improvement activities like customer-client interviews, the large improvement projects at Aker Stord, Aker Stord 2000 or the large improvement projects in the concern?

14. Do you get information about the improvement activities in any other ways like bulletin boards, internal magazines, reports from meeting, email or other?

15. Have shop stewards given any information about the improvement process and different activities like customer-client interviews, brainstorming, the large improvement projects in the concern, the large improvement projects at Aker Stord, Aker Stord 2000 or others?

16. Have any information and TKA training been explicitly addressed to the trade unions and shop stewards?

17. Have the trade unions and the shop stewards had any role in the planning of TKA information and training?

18. In that case, how?

19. How do the trade unions consider the TKA information and training and why?

20. Is the TKA information and training given adequate for the trade unions and the shop stewards?

21. If not, what do you miss?
22. What kind of role do you think the trade unions and the shop stewards should have in TKA information and training?
23. Do the trade unions and the shop stewards have that kind of role today?
24. How do you get information and take part in TKA training when you work in the large building projects?
25. Have the information about TKA and the TKA training changed in 1997 compared to 1996?

Idea generation

26. Have you participated in appraisal-interviews? If no, continue with question 30.
27. What is prioritized in the appraisal-interviews and how is it followed up?
28. To what extent do you discuss and agree on individual development and improvement plans during the appraisal-interview?
29. What do you think about the appraisal-interview practice?
30. Have you participated in pre-job meetings? If no, continue with question 32.
31. What do you think about these kinds of meetings?
32. Have you participated in brainstorming, team meetings, customer-client interviews, improvement project teams or process improvement teams?
33. What do you think about activities like these?
34. Do you know whether improvements proposals from brainstorming, team meetings, department meetings, customer-client interviews or other are being registered in the electronic proposal database?
35. What do you know about the improvement activity in the Department Committee?
36. How do you consider the Department Committee as a body for co ordinating local improvement activities?
37. Do you propose ideas for improvement and how?
38. Is some of the time in department meetings, team meetings and similar kinds of meetings used for proposing and discussing improvement ideas? In that case, how is this carried out and followed up? What are the experiences with this kind of practice?
39. Do you feel that your own proposals for improvements are handled and followed up in a proper manner?
40. Have you participated in meetings where the goal has been to discuss how the

department or gang work with improvements?

41. In that case, what is your experience with that kind of meetings?
42. Should the company increase employee participation in improvement activities?
43. What do the union and the shop stewards consider being their role in problem identification and improvement proposals?
44. Have the trade unions and the shop stewards had any specific role?
45. In that case, what?
46. How did the trade unions and the shop stewards get this role? By asking for it or by invitation from management?
47. What is the trade unions opinion about appraisal interviews?
48. Do the trade unions find the appraisal interview scheme satisfactory?
49. If not, how can it be improved?
50. Is there any aspect of the appraisal interview scheme that the trade unions do not like?
51. Has the appraisal interview scheme been a controversial theme for the trade unions?
52. In that case, in what way?
53. What is the trade unions opinion about team meetings?
54. Have the trade unions had any specific role in idea generating meetings?
55. Do you trade union want to have any specific role?
56. In that case, which role?
57. What is the trade unions opinion on idea generating meetings?
58. Have the trade unions had any specific role in customer-client interviews?
59. Do the trade unions want to have any specific role in customer-client interviews?
60. Do the trade unions participate in any other kinds of improvement activities other than those already mentioned?
61. In that case, which?
62. What is the trade unions opinion about these improvement activities?
63. Are you familiar with the suggestion box scheme?
64. What is your opinion about this arrangement?
65. What is your opinion about the system for reporting undesired events? Does reporting

have any practical consequences? Does reporting and follow up give improvements?

66. The CEO had monthly meetings with 10 employees from different parts of the company in 1996? This practice has stopped. What is you opinion about that?

67. How is the possibility to participate in improvement activities and propose improvement ideas when you have your daily work in the large building projects?

Implementation/follow up of improvement proposals

68. Have you participated in the planning and implementation of improvement projects?

69. How is this work organized?

70. What is your opinion about the way improvement projects are carried out and followed up?

71. What kind of role do the trade unions think they should have in the implementation and follow up of improvement projects?

72. Have the trade unions had this role?

73. Have the way improve projects are carried out and followed up been a controversial theme in the trade unions?

74. In that case, in what way?

75. Have the trade union members contacted the unions regarding the implementation and follow up of improvement projects?

76. Have the improvement activities changed the role and work of any of the co-operating bodies that is important to the trade unions like the works council, the work environment committee, the department committee or others?

77. Have these bodies become weaker or gained a more important position?

78. In that case, in what way?

79. Have the improvement process weakened or strengthened the trade unions position in the company?

80. In that case, how?

81. How is improvement work organized in the large building projects?

82. What is you experience with improvement proposals from employees that have been implemented?

83. Have the implemented proposals resulted in smarter ways to perform work?

84. Have the improvement proposals improved work environment and well being?

85. Has increased employee participation in improvement activities had any effect on

shop stewards' work and role?

86. Do you feel that the time spent on employee participation in improvement activities is worthwhile compared to the outcome and the results of these activities?

The importance of the improvement process

87. Have the improvement process changed Aker Stord?

88. Have employee participation in – and attitudes to the improvement process changed compared to 1996?

89. Do you see any changes for you and your colleagues as a consequence of the new process orientation in the organization?

90. Do you think that the improvement process will last?

91. Do you think that the present process improvement activities at Aker Stord will have any consequence for your participation and engagement in the improvement process?

92. The process improvement projects give a flatter organization and fewer leaders than before? Are leaders willing to delegate more? Are employees capable and motivated for increased autonomy?

93. Do you think that the reorganization in the concern will weaken or strengthen Aker Stord's position?

94. What consequences do you think that the reorganization in the concern will have for shop stewards' possibility to have influence on the further development of Aker Stord?

95. Do you think that TKA has had any consequences for Aker Stord's capability to deliver quality to the right time in an efficient way

96. Has this changed in 1997 compared to 1996?

97. Do you have a TKA facilitator in your department or gang?

98. In that case, how is this facilitator role carried out and what is your experiences with this practice?

The improvement process and industrial democracy

99. Do you think that the improvement process has improved or weakened industrial democracy?

100. Have employee participation increased or decreased as a consequence of the improvement process?

101. Have the improvement process motivated or de-motivated employee to take part and engage in the further development of Aker Stord?

102. What stops further employee participation in improvement activities?

103. Do you feel that improvement proposals from employees are treated in a good manner?

104. Have you or other shop stewards been criticized for spending too much time on improvement activities instead of spending the time on negotiations and on improving work conditions?

105. Do you think employees consider it important for shop stewards and trade unions to take an active part in the improvement process?

106. How much of your weekly time do you think you spend on the improvement process?

(My translation)

Appendix 3

Survey. Quality Management in Norway

INTERNATIONAL COMPARATIVE STUDY OF

QUALITY MANAGEMENT PRACTICES

Høgskolen i Stavanger

RF
ROGALAND
RESEARCH

1996

GENERAL INSTRUCTIONS

This questionnaire is a part of a study underway to document the **Quality Management Practices** in different countries, to develop and build a comprehensive, global quality management model.

Most questions require that you choose the alternative that best fits your views on that topic. There are no correct or incorrect answers, and we are interested only in your perceptions. We estimate that it would take you between 25 to 30 minutes to fill this questionnaire. We would like to assure you that the information provided by you will be treated in **strictest confidence.** Your response will be entered in a coded format and in no instance will a company ever be identified as having given a particular response.

Thank you for your cooperation. We believe that, with your assistance, this study can help clarify a number of issues pertaining to quality management practices in an international context. A self-addressed envelope is enclosed for your convenience.

If you like to receive a copy of the summary report please provide your name and address below or attach your business card.

Name: ______________________________

Title: ______________________________

Organization: ______________________________

Address: ______________________________

____

Country: ______________________________

Telephone Number: ______________ Fax Number: ______________

Thank you very much for your cooperation. Please return the filled questionnaire to:

Hongyi Sun
Avd av ØKS
Høgskolen i Stavanger
Pb 2557 Ullandhaug
4004 Stavanger

Telephone: 51831518, Fax: 51831550, e-mail: Hongyi.sun@oks.his.no

Please respond to the following questions by circling one of the numbers [1] to [5] or [X] to the right of the question. The numbers represent the strength or degree of your assessment, agreement, perception or opinion, as the case may be, to the question item.

Scale: [5] Very High [4] High [3] Medium [2] Low [1] Very Low [X] Unable to respond

I. The following questions relate to quality leadership in your company/division.

a. Extent to which the top company executive assumes responsibility for quality performance	5	4	3	2	1	X
b. Acceptance of responsibility for quality by major department heads within the company	5	4	3	2	1	X
c. Degree of participation by top management in the quality improvement process	5	4	3	2	1	X
d. Extent to which the top management has objectives for quality performance	5	4	3	2	1	X
e. Extent to which quality goals are made specific within the company	5	4	3	2	1	X
f. Importance attached to quality by the top management in relation to cost and schedule objectives	5	4	3	2	1	X
g. Amount of review of quality issues in the top management meetings	5	4	3	2	1	X
h. Extent to which the senior management is involved in all activities toward quality excellence	5	4	3	2	1	X
i. Extent to which top managers participate in giving training in TQM	5	4	3	2	1	X

II. The following questions relate to quality information and analysis in your company/division.

a. Availability of data about quality defects such as error rates, defect rates, scrap, rework, returns, etc	5	4	3	2	1	X
b. Extent to which necessary quality data is available on time	5	4	3	2	1	X
c. Extent to which necessary quality data is available to hourly employees	5	4	3	2	1	X
d. Extent to which quality data is available to managers and supervisors	5	4	3	2	1	X
e. Extent to which quality data is used by top management in decision making	5	4	3	2	1	X
f. Extent to which quality data is used by middle management in planning and controlling	5	4	3	2	1	X
g. Extent to which quality data is used by hourly workers in their operations	5	4	3	2	1	X
h. Extent to which quality data and information is available and used in your company/division	5	4	3	2	1	X

III. The following questions relate to strategic planning process of quality management in your company/division.

a. Extent to which quality management is considered in the company/division strategic plan	5	4	3	2	1	X
b. Extent to which quality management is considered in the operational plan (1-2 years)	5	4	3	2	1	X
c. Extent to which customer satisfaction is considered in the company/division strategic plan	5	4	3	2	1	X
d. Extent to which customer satisfaction is considered in the operational plan (1-2 years)	5	4	3	2	1	X
e. Extent to which the top management supports long-term quality improvement process	5	4	3	2	1	X
f. Extent to which quality goals and policy are understood within the company/division	5	4	3	2	1	X
g. Extent to which your company/division integrates quality improvement planning into overall business planning	5	4	3	2	1	X

i. Extent to which policy and strategy are communicated through the whole organisation.................................... 5 4 3 2 1 X

IV. The following questions relate to the support of human resource development in your company/division.

a. Extent to which quality-related training is given to hourly employees throughout the company/division......... 5 4 3 2 1 X

b. Extent to which training in the basic statistical techniques (such as histograms and control charts) is provided in the company/division as a whole .... 5 4 3 2 1 X

c. Extent of training in advanced statistical techniques to those who require it (such as design of experiments etc.) is provided in the company/division... 5 4 3 2 1 X

d. Availability of resources for employee training in the company/division.. 5 4 3 2 1 X

e. Extent to which employee involvement programs are implemented in the company/division............................ 5 4 3 2 1 X

f. Effectiveness of employee involvement programs in the company/division .... 5 4 3 2 1 X

g. Extent to which hourly/nonsupervisory employees participate in quality decisions .... 5 4 3 2 1 X

h. Extent to which employees are held responsible for the output of their process.. 5 4 3 2 1 X

i. Extent to which quality awareness building among employees is ongoing .... 5 4 3 2 1 X

j. Extent to which the company measures employee morale .... 5 4 3 2 1 X

k. Extent to which your company has an environment conducive to full development of your human resource ... 5 4 3 2 1 X

l. Extent of training in improvement techniques in the company............ .... 5 4 3 2 1 X

IV-2: Employee participation and labor union issues related to quality management

a. Extent to which the agreements about improvement activities for the individual employees are made............ 5 4 3 2 1 X

b. Extent to which employees participate in pre-job meetings in order to improve the planning of the job beforehand.. 5 4 3 2 1 X

c. Extent to which employees participate in post-job meetings in order to evaluate the way the job was performance .... 5 4 3 2 1 X

d. Extent to which employees participate in improvement activities inside their own department.......................... 5 4 3 2 1 X

e. Extent to which individual employees participate in the evaluation and improvement of important value adding process (e.g., production, marketing etc.) across department boundaries .... 5 4 3 2 1 X

f. Extent to which your company runs conferences to which all employees are invited to discuss and plan improvement activities .... 5 4 3 2 1 X

g. Extent to which union and their representatives participate in quality improvement process.............................. 5 4 3 2 1 X

h. Extent to which existing bodies like BU, AU or AMU discuss questions concerning quality and quality improvement.. 5 4 3 2 1 X

i. Extent to which existing bodies like BU, or AU or AMU participate in planning and implementing quality improvement .... 5 4 3 2 1 X

j. Extent to which existing bodies like BU, AU or AMU make decisions about quality issues .... 5 4 3 2 1 X

k. Extent to which top management believe the union's participation in quality improvement is important.......... 5 4 3 2 1 X

m. Extent to which union representative participate in giving TQM training .... 5 4 3 2 1 X

V. The following questions relate to quality assurance of products and services in your company/division.

a. Extent to which new product/service design is reviewed before the product/service is produced........ 5 4 3 2 1 X

b. Clarity of product/service specifications .... 5 4 3 2 1 X

c. Clarity of product/service procedures .... 5 4 3 2 1 X

d. Extent to which implementation/producibility is considered in the product/service design process .... 5 4 3 2 1 X

e. Extent to which acceptance sampling is used to accept/reject lots or batches of work........ 5 4 3 2 1 X

f. Amount of preventive equipment maintenance .... 5 4 3 2 1 X

g. Amount of inspection, review, or checking for in-process quality .... 5 4 3 2 1 X

h. Amount of inspection, review, or checking for final quality .... 5 4 3 2 1 X

i. Extent to which process design minimizes the chances of employee errors .... 5 4 3 2 1 X

j. Clarity of work or process instructions given to employees .... 5 4 3 2 1 X

k. Extent to which your company/division uses systematic approaches for assuring quality of goods and services........ 5 4 3 2 1 X

l. Extent to which processes are identified, reviewed and, if necessary, revised .... 5 4 3 2 1 X

VI. The following questions relate to supplier quality of your company/division.

a. Extent to which suppliers are selected based on quality rather than price .... 5 4 3 2 1 X

b. Degree to which your company relies on few dependable suppliers .... 5 4 3 2 1 X

c. Extent to which you educate suppliers .... 5 4 3 2 1 X

d. Extent to which your company provides technical assistance to your suppliers .... 5 4 3 2 1 X

e. Extent to which the supplier is involved in your product development process .... 5 4 3 2 1 X

f. Extent to which the supplier is involved in the improvement of your manufacturing process .... 5 4 3 2 1 X

g. Extent to which you build long term relationships with your suppliers .... 5 4 3 2 1 X

h. Clarity of specifications provided to your suppliers .... 5 4 3 2 1 X

i. Amount of inspection, review, or checking for incoming quality at your plant........ 5 4 3 2 1 X

j. Extent to which your company/division develops relationships with your suppliers that lead to continuous improvements in quality........ 5 4 3 2 1 X

VII. The following questions relate to quality results of your company/division.

a. Extent to which scrap levels have been reduced by quality management........ 5 4 3 2 1 X

b. Extent to which rework levels have been reduced by quality management........ 5 4 3 2 1 X

c. Extent to which warranty costs have been reduced by quality management........ 5 4 3 2 1 X

p. Extent to which your company compares the current process quality levels with those of world leaders........ 5 4 3 2 1 X

q. Overall, extent to which customer satisfaction is considered in your quality management process .... 5 4 3 2 1 X

r. Extent to which your company established joint improvement activities with customers and suppliers............ 5 4 3 2 1 X

IX. The following questions relate to general matters such as public responsibilities, employee training, well-being and morale.

a. Extent to which public health issues are considered as a company/division responsibility .... 5 4 3 2 1 X

b. Extent to which public safety issues are considered as a company/division responsibility .... 5 4 3 2 1 X

c. Extent to which environmental issues are considered as a company/division responsibility .... 5 4 3 2 1 X

d. Extent to which the organization extends its quality leadership to the external community .... 5 4 3 2 1 X

e. Extent to which training in specific work-skills (technical and vocational) is given to hourly employees ...... 5 4 3 2 1 X

f. Extent to which training in specific work-skills (technical and vocational) is given to other employees throughout the company .... 5 4 3 2 1 X

g. Extent to which employees are recognized for superior quality performance .... 5 4 3 2 1 X

h. Extent to which the company measures employee well-being .... 5 4 3 2 1 X

X. Please indicate by choosing the appropriate options in answering the following questions which relate to environmental factors of your organization.

Your company is ___ private _____ public (if public, go to question d.) ____ Other (trust, e.g.)

a. Your competitors are (choose all the types if they apply):
_____ Local _____ Regional _____ Domestic ____International

b. Your customers are (choose all the types if they apply):
_____ Local _____ Regional _____ Domestic _____International

c. Your company/division is:
_____ Wholly Norwegian Owned _____ Joint Venture _____ Wholly Non-Norwegian Owned

d. Your company/division is: _ Unionized _____ Not unionized _____ Partially Unionized

f. The effect of government bureaucratic obstacles red tape and regulations on the operation is:
_____ Very high _____ High _____ Medium _____ Low ____ Very low

h. The level of education of your work force is:
_____ Very high _____ High _____ Medium _____ Low ____ Very low

i. The average age of your work force is: ______________________

j. Do the employees in your company/division have life time employment ? _____ Yes _____ No

k. How long has the quality management initiative been in place in your company/division ?
_____ Less than a year _____ 1 year to 3 years _____ 3 years to 5 years _____ More than 5 years

l. Your company/division has been in business for:
_____ Less than 2 years _____ 2 years to 5 years _____ 6 years to 10 years _____ More than 10 years

m. The position of your company/division vis-à-vis ISO-9000 certification is:

Our organization currently is ISO-9000 certified (Please specify)
____ ISO 9001 certified ____ISO 9002 certified ____ISO 9003 certified
____ We are not certified but plan to go for ISO-9000 certification in the next two years
____ We are not interested in ISO-9000 certification

n. Has your company/division received any quality awards ? ____ Yes ____ No

If Yes give details: ____

o. The most important causes that triggered the quality program were (choose all that apply):
____ Demanding customers ____ Need to reduce cost ____ Increased competition ____ Due to joint ventures
____ Survival of the Company ____ Concern decision ______ Others

p. The ownership of your company is: ____ depedent ____ part of concern ___ part of holding ____ others

XI. Thank you for your assistance in this project. Please give us the following information for statistical purposes.

a. Name of Organization ____________________________

b. Name of Department ____________________________

c. Your position: ____ Top level management ____ Middle level management
____ First level supervisor ____ Professional employee without supervisory responsibilities
____ Other (Please indicate: ____________________________)

d. Please indicate the category which best describe your company's major business (Please check the most appropriate one).

____ Manufacturing ____ Processing ____ Business service ____Other

e. Please indicate the type of business your company is in (e.g. electronics, autoparts, banking, retailing, etc.):
____

f. Please indicate the range of your annual <u>sales (in Norwegian kroner kr):</u>

____ Less than 1 million ____ 1 to 10 million ____ 11 to 25 million ____ 26 to 50 million
____ 51 to 250 million ____ 251 to 500 million ____ 501 to 1000 million ____ More than 1 billion

g. Please indicate the percentage of sales in exports:
____ 0 % ____ Less than 25% ____ 26 to 50% ____ 51 to 75% ____ 76 to 100%

h. Please indicate the range of employees in your organization:

____ Less than 50 ____ 50 to 149 ____ 150 to 499 ____ 500 to 999 ____1000 to 4999 ____ 5000 to 9999
____ 10000 and above

i. Please indicate the level of your education:

____High School ____Undergraduate degree ____Master degree ____Ph.D.

j. You have been working for the current organization for

____ Less than 6 months ____ 6 months to 1 year ____ 1 year to 2 years ____ 2 years to 5 years
____ 5 years to 10 years ____ 10 years and more

MIX
Papier aus verantwortungsvollen Quellen
Paper from responsible sources
FSC® C105338

If you have any concerns about our products,
you can contact us on
ProductSafety@springernature.com

In case Publisher is established outside the EU,
the EU authorized representative is:
Springer Nature Customer Service Center GmbH
Europaplatz 3, 69115 Heidelberg, Germany

Printed by Libri Plureos GmbH
in Hamburg, Germany